ROCKVILLE CAMPUS LIBRARY

PROBLEMS IN COMBINATORICS
AND GRAPH THEORY

WILEY-INTERSCIENCE SERIES IN DISCRETE MATHEMATICS

ADVISORY EDITORS

Ronald L. Graham
AT&T Bell Laboratories, Murray Hill, New Jersey

Jan Karel Lenstra
Mathematisch Centrum, Amsterdam, The Netherlands

Graham, Rothschild, and Spencer
RAMSEY THEORY

Tucker
APPLIED COMBINATORICS

Pless
INTRODUCTION TO THE THEORY OF ERROR-CORRECTING CODES

Nemirovsky and Yudin
PROBLEM COMPLEXITY AND METHOD EFFICIENCY IN OPTIMIZATION
(Translated by E.R. Dawson)

Goulden and Jackson
COMBINATORIAL ENUMERATION

Gondran and Minoux
GRAPHS AND ALGORITHMS
(Translated by S. Vajda)

Fishburn
INTERVAL ORDERS AND INTERVAL GRAPHS: A STUDY OF PARTIALLY ORDERED SETS

Tomescu
PROBLEMS IN COMBINATORICS AND GRAPH THEORY
(Translated by Robert A. Melter)

Palmer
GRAPHICAL EVOLUTION: AN INTRODUCTION TO THE THEORY OF RANDOM GRAPHS

PROBLEMS IN COMBINATORICS AND GRAPH THEORY

IOAN TOMESCU

Faculty of Mathematics
University of Bucharest
Bucharest, Romania

Translated from Romanian by

ROBERT A. MELTER

Department of Mathematics
Long Island University
Southampton, New York

A Wiley-Interscience Publication
JOHN WILEY & SONS
New York · Chichester · Brisbane · Toronto · Singapore

Copyright © 1985 by John Wiley & Sons, Inc.

All rights reserved. Published simultaneously in Canada.

Reproduction or translation of any part of this work
beyond that permitted by Section 107 or 108 of the
1976 United States Copyright Act without the permission
of the copyright owner is unlawful. Requests for
permission or further information should be addressed to
the Permissions Department, John Wiley & Sons, Inc.

Library of Congress Cataloging in Publication Data:

Tomescu, Ioan.
  Problems in combinatorics and graph theory.
  (Wiley-Interscience series in discrete mathematics)
  Translation of: Probleme de combinatorică şi teoria
grafurilor.
  Bibliography: p.
  1. Combinatorial analysis. 2. Graph theory.
I. Title. II. Series.
QA164.T6713   1985   511.6   84-21701
ISBN 0-471-80155-0

Printed in the United States of America

10 9 8 7 6 5 4 3 2 1

Preface

This book is a translation of *Probleme de Combinatorică și Teoria Grafurilor*, which was published in Bucharest, Romania in 1981. In Romania, graph theory is taught in the faculty of mathematics and in particular in what is known as the chair of informatics (=computer science). Students in preparatory schools which specialize in mathematics and physics also receive instruction in this subject. Thus the selection of problems presented includes some which are quite elementary and self-contained and others which were previously accessible only in research journals. The author has used the text to prepare Romanian candidates for participation in International Mathematical Olympiads.

Each problem is accompanied by a complete and detailed solution together with appropriate references to the mathematical literature. This should enable mature students to use the book independently. Teachers of courses in combinatorics and graph theory will also find the text useful as a supplement, since important concepts are developed in the problems themselves. Even in the more elementary problems the reader will learn the important concepts and structures in the field.

Revisions in the original Romanian edition have been made, and about 60 new problems and solutions have been added. The careful translation of this material by Professor Melter will make available to users of English-language texts a unique collection of problems in an expanding field of mathematics which has many scientific applications.

The author and translator are grateful to Ms. Linda Kallansrude for her professional typing and to the staff of John Wiley & Sons for their interest and concern in the development of this book.

IOAN TOMESCU

Bucharest, Romania
February 1985

Contents

GLOSSARY OF TERMS USED ix

PART I. STATEMENT OF PROBLEMS 1

  1. COMBINATORIAL IDENTITIES 3

  2. THE PRINCIPLE OF INCLUSION AND EXCLUSION; INVERSION FORMULAS 10

  3. STIRLING, BELL, FIBONACCI, AND CATALAN NUMBERS 15

  4. PROBLEMS IN COMBINATORIAL SET THEORY 22

  5. PARTITIONS OF INTEGERS 30

  6. TREES 33

  7. PARITY 39

  8. CONNECTEDNESS 41

  9. EXTREMAL PROBLEMS FOR GRAPHS AND NETWORKS 45

  10. COLORING PROBLEMS 52

  11. HAMILTONIAN PROBLEMS 56

  12. PERMUTATIONS 58

  13. THE NUMBER OF CLASSES OF CONFIGURATIONS RELATIVE TO A GROUP OF PERMUTATIONS 62

14. PROBLEMS OF RAMSEY TYPE	65
PART II. SOLUTIONS	69
BIBLIOGRAPHY	335

Glossary of Terms Used

Abel's Identity: See Problem 1.29.

Arborescence with root a: A digraph with the property that for every point $x \neq a$ there is a unique path from a to x.

Articulation point (or cut point): A point x of a connected graph G with the property that the subgraph G_x obtained from G by removing the point x is no longer connected.

Automorphism: An isomorphism of a graph G with itself.

Balanced incomplete block design (BIBD): See Problem 4.37.

Bell's number, denoted B_n: The number of all partitions of a set with n elements; thus $B_n = S(n, 1) + S(n, 2) + \cdots + S(n, n)$ (see Stirling number).

Bicovering of a set X: A family of nonempty subsets of X such that each element of X is contained in exactly two subsets of the family.

Block of a graph G: A maximal 2-connected subgraph of G. Each two blocks of a graph which is not itself 2-connected have in common at most one point, which must be a point of articulation.

Burnside's lemma: See Problem 13.2.

Cardinal number of a finite set X: The number of elements in X; it is denoted $|X|$.

Catalan number, denoted C_n: The number of ways in which parentheses can be inserted in a nonassociative product of n factors. The numerical value of the Catalan numbers is given by the formula $C_n = (1/n)\binom{2n-2}{n-1}$.

Cauchy's formula: See Problem 12.6.

Cauchy's identities: See Problems 3.35 and 12.7.

Cayley's formula: The number of labeled trees with n vertices is equal to n^{n-2}.

Center of a connected graph G: The set of vertices x_0 of minimum eccentricity, i.e., $e(x_0) = \min e(x) = \rho(G)$ where $\rho(G)$ is the radius of G.

Characteristic equation of a linear recurrence relation: See Problem 1.31.

Chromatic index $q(G)$: The minimum number of colors with which the edges of G can be colored if every two edges with a common endpoint are colored with distinct colors.

Chromatic number $\chi(G)$: The minimum number of colors with which the vertices of G can be colored if every two adjacent vertices have distinct colors.

Chromatic polynomial of a graph: See Problem 10.14.

Circuit: A path $D=(x_0, \ldots, x_r)$ in a digraph G such that $x_0=x_r$ and the arcs (x_0, x_1), (x_1, x_2), ..., (x_{r-1}, x_r) are distinct. The circuit is said to be **elementary** if all the vertices of the circuit, with the exception of the first and last (which coincide), are pairwise distinct. The **length** of a circuit is equal to the number of arcs it contains.

Clique: A complete subgraph of a graph.

k-coloring of a graph: Let $G=(X, U)$. A k-coloring of G consists of a partition $X=X_1 \cup \cdots \cup X_k$ of the set of vertices, such that no vertices in the same class are adjacent. It may also be defined as a function $f: U \to \{1, 2, \ldots, k\}$ such that $[i, j] \in U$ implies $f(i) \neq f(j)$.

Combinations of n things taken k at a time, number of, (or simply n take k, or **binomial coefficient**) $\binom{n}{k}=n(n-1)\cdots(n-k+1)/k!$: The number of ways of choosing k objects from a set of n objects. By definition $\binom{n}{k}=1$ if $k=0$ and $n \geqslant k$. Likewise $\binom{n}{k}=0$ if $n<k$. This notation is also utilized when n is rational or $n<0$. Further details are given in the text. (See, e.g., Problem 1.22.)

Numbers of combinations of n things taken k at a time with replacement: The number of increasing words of length k formed from an alphabet A with n letters on which is defined a total order. The words are thus of the form $c_1 c_2 \ldots c_k$, where $c_1 \leqslant c_2 \leqslant \cdots \leqslant c_k$ and $c_i \in A$ for $1 \leqslant i \leqslant k$. The numerical value of n take k with replacement is $n(n+1)\cdots(n+k-1)/k!$. The number of strictly increasing words of length k with letters in A is equal to $\binom{n}{k}$.

Complement of a graph: Let $G=(X, U)$. The complement of G is a graph $\bar{G}=(X, \bar{U})$. It has the same set of vertices X as G. Two vertices are adjacent in $\bar{G}$ if and only if they are not adjacent in G.

(v, k, λ)-configuration: See Problem 4.38.

Covering of a set X: A family of pairwise distinct nonempty subsets, whose union is X. A covering A of X which is made up of k subsets is said to be **irreducible** if the union of every $k-1$ subsets of A is a proper subset of X.

Cut of a network $G=(X, U, c)$ with source a and sink b: The set of arcs $\omega^-(A) = \{(x, y) | x \notin A, y \in A\}$ where $A \subset X$, $a \notin A$, and $b \in A$. The **capacity** of the cut $\omega^-(A)$ is the sum of the capacities of the arcs of $\omega^-(A)$.

(a, b)-cut of a digraph: A set C of arcs with the property that every path from vertex a to vertex b $(a \neq b)$ contains at least one arc of C.

Cycle in a graph: A walk $W=[x_0, x_1, \ldots, x_r]$ with the property that $x_0=x_r$ and all the edges $[x_0, x_1]$, $[x_1, x_2]$, ..., $[x_{r-1}, x_r]$ are pairwise distinct. The cycle is said to be **elementary** if all its vertices (except the first and the last) are distinct. The **length** of a cycle is equal to the number of edges in it.

Degree of a vertex x: (1) In a graph G, the degree of a vertex x, denoted $d(x)$, is the number of edges incident with x. (2) If G is a digraph, then the **indegree** $d^-(x)$ of a vertex x is the number of arcs which terminate at x, i.e., of the form (y, x); its **outdegree** $d^+(x)$ is the number of arcs of the form

(x, z), that is, which originate at x. It follows that $d(x)=d^+(x)+d^-(x)$.

Diameter of a connected graph G: The maximum distance between a pair of vertices of G. The diameter is denoted $d(G)$.

Distance between vertices x and y: Let x and y be vertices of a connected graph G. The distance $d(x, y)$ is the length of a shortest walk in G from x to y.

Dixon's formula: See Problem 1.8.

Eccentricity of a vertex x of a connected graph: $e(x)=\max_y d(x, y)$, where $d(x, y)$ is the distance between x and y.

Erdös–Ko–Rado theorem: See Problem 4.22.

Eulerian cycle (circuit): A cycle (circuit) in a graph G which passes through all the edges (arcs) of G.

Euler's formula for a planar graph G: If G is connected and contains n vertices and m edges, then every planar representation of G has $m-n+2$ faces.

Euler's function $\varphi(n)$: The number of positive integers smaller than n and prime to n. If the decomposition of n into prime factors contains distinct prime factors $p_1, \ldots, p_q$, then

$$\varphi(n)=n\left(1-\frac{1}{p_1}\right)\left(1-\frac{1}{p_2}\right)\cdots\left(1-\frac{1}{p_q}\right).$$

Euler's identities: See Problems 1.37 (or 2.9) and 5.6.

Eulerian number: See Problem 12.22.

Euler's Pentagonal Theorem: The recurrence relation for the number $P(n)$ of partitions of an integer n. See Problem 5.8.

Face of a planar representation of a planar graph G: A connected component of the topological space obtained by removing from the plane the edges and vertices of the planar representation of G. The boundary of each face is a closed Jordan curve, consisting of the edges of an elementary cycle of G. The infinite face is the unique unbounded connected component obtained in this way.

Ferrers diagram: Let $n=n_1+n_2+\cdots+n_k$ be a partition of an integer n. A Ferrers diagram is a table with n cells. In the first line there are n_1 cells, in the second line n_2 cells, etc. The cells are arranged beneath each other and are aligned on the left. A Ferrers diagram is **symmetric** if there are the same number of cells in line i and column i for every $i\geqslant 1$. This symmetry is with respect to the principal diagonal of the diagram.

Fibonacci numbers: Numbers defined by $F_0=F_1=1$ and $F_{n+2}=F_{n+1}+F_n$ for each $n\geqslant 0$.

Filter basis: A system S of nonempty subsets of a set X with the property that for every $A, B \in S$ there exists $C \in S$ such that $C \subset A \cap B$.

Flow in a network: See Problem 9.18.

Ford–Fulkerson algorithm: See Problem 9.20.

Ford–Fulkerson theorem: In every network the maximum value of the exit flow is equal to the minimum capacity of a cut (see Problem 9.19).

Gauss's number $\begin{bmatrix}n\\k\end{bmatrix}_q$: The number of subspaces of dimension k of an n-dimensional vector space over a field with q elements, where q is a power of a prime. An expression for the value of Gauss's number is given in Problem 3.33.

Generating function: Let (a_n) be a sequence of numbers. Its generating function is the sum of the series $\sum_{n=0}^{\infty} a_n x^n$. The expression $\sum_{n=0}^{\infty} a_n x^n / n!$ is called the exponential generating function of the sequence (a_n). These series are considered as formal series to which algebraic operations can be applied, without consideration of their convergence. In general the generating functions in actual use are defined by means of series which are convergent for all real numbers or for an interval of real numbers of positive length, but the convergence of the series will not be established in this book. The series representations which we use depend on the expansions of e^x and $\ln(1+x)$, on Newton's generalized binomial formula, and on the sum of an infinite geometrical progression.

Girth of a graph G: The length, denoted $g(G)$, of the shortest elementary cycle in the graph G.

Graph: (1) a graph G is an ordered pair of sets (X, U), where X is a finite set called the set of vertices or nodes, and U contains unordered pairs of distinct elements of X called edges. If an edge is denoted $[x, y]$, then x, y are called its endpoints, the vertices x and y are said to be adjacent in the graph G, and the vertices x and y are by definition incident with the edge $[x, y]$. (2) A **digraph** (directed graph) G is an ordered pair of sets (X, U), where X is called the set of vertices or nodes, and U contains ordered pairs of distinct elements of X, called arcs. If an arc is denoted $u=(x, y)$, then x is called its **initial vertex** and y its **terminal vertex**; the arc is said to be **directed** from x to y. One also says that the vertices x and y are **adjacent** in G and **incident** with the arc (x, y). A **spanning graph** of a graph $G=(X, U)$ is a graph $G_1=(X, V)$ where $V \subset U$. It is thus a graph G_1 obtained from G by suppressing certain edges (arcs). A **subgraph** of a graph G is a graph $H=(Y, V)$ where $Y \subset X$; the edges (arcs) of V are those edges (arcs) in U which have both endpoints in the set of vertices Y. A subgraph H of G **induced**, or **generated**, by the set of vertices Y is obtained from G by suppressing all the vertices of $X \setminus Y$ and all the arcs incident with them.

Graph, bipartite A graph $G=(X, U)$ for which there exists a partition of X in the form $X=A \cup B$, $A \cap B=\emptyset$, such that each edge u of the graph has one endpoint in A and the other in B. A bipartite graph is said to be complete if it contains all edges of the form $[a, b]$ where $a \in A$ and $b \in B$. If $|A|=p$ and $|B|=q$, the complete bipartite graph is denoted $K_{p,q}$.

Graph, k-chromatic: A graph G with chromatic number $\chi(G)=k$.

Graph, k-colorable: A graph G with chromatic number $\chi(G) \leq k$. It is thus a graph whose vertices can be colored with k colors so that each two adjacent vertices have different colors.

Graph, complete, on n vertices: A graph, denoted K_n, in which every two vertices are adjacent. It has $\binom{n}{2}$ edges. The complete graph on a denumerably infinite set of vertices is denoted K_∞. A digraph is complete if each two distinct vertices x and y are adjacent with respect to either the arc (x, y), the arc (y, x), or both. In the complete digraph, denoted K_n^*, each two distinct vertices x and y are joined by both the arcs (x, y) and (y, x). It has $n(n-1)$ arcs.

Graph, connected: A graph G with the property that every two vertices are the endpoints of a walk in G. If G is not connected, then it has at least two connected components (maximal connected subgraphs, which are pairwise disjoint with respect to vertices). A connected graph with at least $k+1$ vertices is **k-connected** if the graph obtained by suppressing every set Y of vertices of cardinality $|Y| \leqslant k-1$ is connected.

Graph, strongly connected: A digraph G with the property that for every two vertices x and y there is a path $D_1 = (x, \ldots, y)$ and a path $D_2 = (y, \ldots, x)$ in G.

Graph, Hamiltonian: A graph which contains a Hamiltonian cycle, or a directed graph which has a Hamiltonian circuit.

Graph, multipartite: A graph $G = (X, U)$ whose vertex set can be partitioned as $X = A_1 \cup \cdots \cup A_k$, so that each edge has its endpoints in two distinct sets of the partition. A multipartite graph is complete if each pair of vertices located in different partition sets is adjacent.

Graph, planar: A graph G whose vertices can be represented as points in the plane; the edges become arcs of a Jordan curve which join points corresponding to adjacent vertices. Two such arcs have in common at most one endpoint.

Graph, k-regular: A graph in which each vertex x has degree $d(x) = k$, or a digraph with the property that $d^-(x) = d^+(x) = k$ for every vertex x.

Graphs, isomorphic: The graphs $G = (X, U)$ and $H = (Y, V)$ are isomorphic if there exists a bijection $f : X \to Y$ such that $[x, y] \in U$ if and only if $[f(x), f(y)] \in V$.

Hamiltonian cycle (circuit): An elementary cycle (circuit) which contains all the vertices of the graph.

Independence number $\alpha(G)$ of a graph G: The maximum number of vertices in an independent set of G.

Independent (internally stable) set: A subset of vertices which induces a subgraph consisting only of isolated vertices.

Inversion of a permutation $p \in S_n$: A pair $\{p(i), p(j)\}$ with the property that $1 \leqslant i < j \leqslant n$ and $p(i) > p(j)$.

König's theorem: See Problem 9.23.

Li-Jen-Shu formula: See Problem 1.5(h).

Lucas numbers L_n: Defined by $L_0=2$, $L_1=1$, and $L_{n+2}=L_{n+1}+L_n$ for every $n\geqslant 0$.

Matching of a graph G: A set of edges such that no two have a common end point. The maximum number of edges in a matching of G is denoted $v(G)$.

Matroid: For a definition in terms of independent sets see Problem 6.30.

Moebius function: See Problem 2.20.

Moebius inversion formula: See Problem 2.22.

Multigraph: If in the definition of a graph $G=(X, U)$ the set of edges is replaced by a multiset, then a multigraph is obtained. A multigraph can contain many edges having the same endpoints.

Multinomial formula: See Problem 1.16.

Multinomial number: A number of the form $\binom{n}{n_1,\ldots,n_p}=n!/n_1!\cdots n_p!$ where $n_1,\ldots,n_p\geqslant 0$ and $n_1+\cdots+n_p=n$. For $p=2$ this reduces to the binomial coefficients

$$\binom{n}{n_1, n_2}=\binom{n}{n_1}=\binom{n}{n_2}.$$

If $n_1+\cdots+n_p\neq n$, the multinomial number is by definition equal to zero.

Multiset X, or *collection*, of type $1^{k_1} 2^{k_2} \cdots n^{k_n}$: A set X together with a partition of itself of type $1^{k_1} 2^{k_2} \cdots n^{k_n}$, that is, containing k_j classes with j elements for $j=1, 2, \ldots, n$; the elements belonging to a class with p elements are identified for $2\leqslant p\leqslant n$.

Network: See Problem 9.18.

Newton's generalized binomial formula: $(x+a)^\alpha = a^\alpha + \alpha a^{\alpha-1}x + \{\alpha(\alpha-1)/2!\} \times a^{\alpha-2}x^2 + \cdots + \binom{\alpha}{k}a^{\alpha-k}x^k + \cdots$, where $a>0$. This series is convergent for every real number α and every real number x with $|x|<a$. If α is a positive integer one obtains Newton's binomial formula.

Nörlund's formula: See Problem 3.2(b).

Orbit of a permutation group: If $G\subset S_n$ is a group of permutations of the set $X=\{1, \ldots, n\}$ and $x, y \in X$, then x is equivalent to y with respect to the group G if there exists a permutation $f \in G$ such that $y=f(x)$. The equivalence classes for this equivalence relation are called the orbits of the group G.

Partition: (1) A partition of a set X is a representation of X in the form $X=A_1\cup A_2\cup\cdots\cup A_k$ where the nonvoid sets $A_1, \ldots, A_k$ are pairwise disjoint; these sets are called the classes of the partition. The partition does not depend on the order of writing the classes nor on the order of the elements in each class. A partition has **type** $1^{k_1} 2^{k_2} \cdots n^{k_n}$ if it contains k_j classes with j elements. (2) A partition of an integer n is a representation of n in the form $n=n_1+n_2+\cdots+n_k$; the integers $n_1, n_2, \ldots, n_k$ are called the **parts** of the partition and satisfy the inequalities $n_1 \geq n_2 \geq \cdots \geq n_k \geq 1$. The number of partitions of n into k parts is denoted $P(n, k)$; the total number of partitions of n is denoted $P(n)$.

Path: Let $G = (X, U)$ be a digraph. A path is a sequence of vertices $D = (x_0, x_1, \ldots, x_r)$ such that $(x_0, x_1), (x_1, x_2), \ldots, (x_{r-1}, x_r) \in U$, i.e., are arcs of the graph. The vertices x_0 and x_r are called **endpoints** of the path D. The **length** of a path is equal to the number of arcs it contains. The path D is said to be **elementary** if its vertices $x_0, x_1, \ldots, x_r$ are pairwise distinct.

Permanent of a matrix A: Let $A = (a_{ij})_{i,j=1,\ldots,n}$. The permanent of A is denoted $\text{per}(A)$ and is defined by $\text{per}(A) = \sum_{p \in S_n} a_{1p(1)} a_{2p(2)} \cdots a_{np(n)}$.

Permutation of a set $X = \{1, \ldots, n\}$: A bijection $p: X \to X$; it can be written either as $p(1)\, p(2) \cdots p(n)$ or in the form

$$\begin{pmatrix} 1 & 2 & \cdots & n \\ p(1) & p(2) & \cdots & p(n) \end{pmatrix}$$

A permutation is of the type $1^{k_1} 2^{k_2} \cdots n^{k_n}$ if it contains k_j cycles with j elements, when $k_1 + 2k_2 + \cdots + nk_n = n$. Every permutation has a unique representation (if we ignore the order of the factors) as a product of cycles which do not have any common elements; the product is composition of functions.

Permutation, circular, or **cycle** with r elements: A permutation in S_n with a unique cycle of length r, the remainder of the $n - r$ cycles being of length one. A cycle is thus a permutation of type $1^{n-r} r^1$.

Permutations, conjugate: Two permutations s, t in S_n are conjugate if there exists $g \in S_n$ such that $s = gtg^{-1}$, or equivalently if s and t have the same cycle structure.

Petersen graph: See Figure 8.3.

Pólya's theorem: See Problem 13.6.

Polynomial, cycle index of a permutation group $G \subset S_n$: A polynomial in n variables

$$\frac{1}{|G|} \sum_{g \in G} x_1^{\lambda_1(g)} \cdots x_n^{\lambda_n(g)},$$

where $\lambda_i(g)$ is the number of cycles of length i of g for $1 \leq i \leq n$.

Principle of inclusion and exclusion: See Problem 2.2.

Projective plane, finite: A symmetric BIBD with parameters $(v, v, k, k, 1)$ where $v \geq 4$. (See Problem 4.50.)

Prüfer code associated with a tree: See Problem 6.15.

Radius of a connected graph: The smallest eccentricity of its vertices.

Ramsey number $R(p, q)$ with two parameters: The smallest integer t with the property that each graph with t vertices contains either a complete subgraph with p vertices or an independent set with q vertices.

Schur number $S(k)$: The largest number r with the property that the set $\{1,\ldots,r\}$ can be partitioned into k possibly empty subsets, with the property that none of them contains numbers x, y, z such that $x+y=z$.

Sperner's theorem: See Problem 4.21.

Star: A complete bipartite graph of the form $K_{1,p}$.

Steiner triple system: If X is a set with $v \geqslant 3$ elements where $v \equiv 1$ or $v \equiv 3 \pmod 6$, a Steiner system of order v is a family of three-element sets of X, called triples, such that each two-element subset of X is contained in a unique triple.

Stirling number of the first kind, $s(n, k)$: The coefficient of x^k in the expansion

$$x(x-1)\cdots(x-n+1) = \sum_{k=0}^{n} s(n,k) x^k.$$

Stirling number of the second kind, $S(n, k)$: The number of partitions of a set with n elements into k classes.

Support or transversal set of the edges of a graph: A set S of vertices with the property that every edge has at least one endpoint in S. The smallest cardinal number of a support for a graph G is denoted $\tau(G)$.

Surjections, number of: The number of surjections $f: X \to Y$ where $|X|=m$ and $|Y|=n$ is denoted $s_{m,n}$.

Symmetric difference: Let A and B be sets. Their symmetric difference is defined as $A \triangle B = (A \setminus B) \cup (B \setminus A)$.

System of distinct representatives (SDR) of a family of sets: A family of subsets $M(S) = \{S_1, S_2, \ldots, S_m\}$ of a set S has an SDR if there is an injective function $f: \{S_1, \ldots, S_m\} \to S$ such that $f(S_i) \in S_i$ for every i, $1 \leqslant i \leqslant m$.

Tournament: A complete, antisymmetric, directed graph. Thus between each two vertices $x \neq y$, there is one and only one arc (x, y) or (y, x).

Transposition: A permutation $g \in S_n$ which has $n-2$ fixed points and is therefore of type $1^{n-2} 2^1$.

Tree: A connected graph without cycles.

Tree, spanning, of a connected graph G: A spanning graph of G which is a tree.

Triangulation: A planar representation of a planar graph in which each face is a triangle (cycle with three vertices).

Triangulation of an elementary cycle with n vertices: The graph which consists of the cycle and the $n-3$ diagonals which do not intersect in the interior of the cycle.

Turán's theorem and **Turán's number** $M(n, k)$: See Problem 9.9.

Vandermonde's formula: See Problem 3.2(a).

Van der Waerden's number $W(k, t)$: The smallest natural number n with the property that if the set $\{1, \ldots, n\}$ is partitioned into k classes, then there exists a class of the partition which contains an arithmetic progression with $t+1$ terms.

Vertex, isolated: A vertex of degree zero in a graph.

Vertex, terminal: A vertex of degree one in a graph.

Vizing's theorem: The chromatic index of every graph G is equal to D or $D+1$, where D is the maximum degree of the vertices of G. (See Problem 10.19.)

Walk: (1) In a graph $G=(X, U)$ a walk is a sequence of vertices $W = [x_0, x_1, \ldots, x_r]$ with the property that each two successive vertices are adjacent, that is $[x_0, x_1], [x_1, x_2], \ldots, [x_{r-1}, x_r] \in U$. The vertices x_0 and x_r are called the **endpoints** of the walk, and r is the **length** of the walk. If the vertices $x_0, x_1, \ldots, x_r$ are pairwise distinct, then W is said to be **elementary**. (2) If G is a directed graph, a walk $W = [u_1, u_2, \ldots, u_p]$ is a sequence of arcs, with the property that for each i, each two successive arcs u_i and u_{i+1} have one common endpoint, $1 \leqslant i \leqslant p-1$. The endpoint of u_1 which is not common to u_2 and the endpoint of u_p which is not common to u_{p-1} are called the **endpoints** of the walk.

Walk (path), Hamiltonian: An elementary walk (path) which contains all the vertices of the graph.

Part I

STATEMENTS OF PROBLEMS

1

Combinatorial Identities

1.1 Show that the following identities hold for every natural number n:

(a) $\sum_{k=0}^{[n/2]} \left\{ \binom{n}{k} - \binom{n}{k-1} \right\}^2 = \frac{1}{n+1} \binom{2n}{n}$,

where $[x]$ is the greatest integer $\leq x$;

(b) $\sum_{k=0}^{n} \binom{n+k}{n} \frac{1}{2^k} = 2^n$.

1.2 Prove the equalities listed below:

(a) $\sum_{p=k-3}^{n-3} \binom{p}{k-3} \binom{n-p-1}{2} = \binom{n}{k}$;

(b) $\sum_{p=0}^{n} (-1)^p \binom{2n-p}{p} = \begin{cases} 1 & \text{for } n \equiv 0 \pmod{3}, \\ 0 & \text{for } n \equiv 1 \pmod{3}, \\ -1 & \text{for } n \equiv 2 \pmod{3}; \end{cases}$

(c) $\sum_{p=1}^{n} p \binom{n}{p}^2 = n \binom{2n-1}{n-1}$.

1.3 Let $S_k(n) = 1^k + 2^k + \cdots + n^k$, where k is a non-negative integer. Show that

$$1 + \sum_{k=0}^{r-1} \binom{r}{k} S_k(n) = (n+1)^r.$$

1.4 Prove that for natural numbers m and n, there exists a natural number p such that the identity $(\sqrt{m} + \sqrt{m-1})^n = \sqrt{p} + \sqrt{p-1}$ holds.

1.5 Prove the following combinatorial identities:

(a) $\sum_{k=0}^{m} \binom{p}{k} \binom{q}{m-k} = \binom{p+q}{m}$;

(b) $\sum_{k=m}^{n} \binom{k}{m} \binom{n}{k} = \binom{n}{m} 2^{n-m}$;

(c) $\sum_{k=0}^{m} (-1)^k \binom{n}{k} = (-1)^m \binom{n-1}{m}$;

3

(d) $\sum_{k=0}^{m} \binom{m}{k}\binom{n+k}{m} = \sum_{k=0}^{m} \binom{m}{k}\binom{n}{k} 2^k;$

(e) $\sum_{k=0}^{m} \binom{m}{k}\binom{n+k}{m} = (-1)^m \sum_{k=0}^{m} \binom{m}{k}\binom{n+k}{k}(-2)^k;$

(f) $\sum_{k=0}^{p} \binom{p}{k}\binom{q}{k}\binom{n+k}{p+q} = \binom{n}{p}\binom{n}{q};$

(g) $\sum_{k=0}^{p} \binom{p}{k}\binom{q}{k}\binom{n+p+q-k}{p+q} = \binom{n+p}{p}\binom{n+q}{q};$

(h) $\sum_{k=0}^{p} \binom{p}{k}^2 \binom{n+2p-k}{2p} = \binom{n+p}{p}^2$ (Li-Jen-Shu formula).

1.6 Prove the identity

$$\sum_{k}(-1)^k \binom{2l}{l+k}\binom{2m}{m+k}\binom{2n}{n+k} = \frac{(l+m+n)!(2l)!(2m)!(2n)!}{(l+m)!(m+n)!(n+l)!l!m!n!}$$

for non-negative integers l, m, n, where the summation is taken over all integer values of k.

1.7 Show that

$$\sum_{i=0}^{p} \binom{p}{i}\binom{q}{i} a^{p-i} b^i = \sum_{i=0}^{p} \binom{p}{i}\binom{q+i}{i}(a-b)^{p-i} b^i.$$

1.8 Prove Dixon's formula:

$$\sum_{k=0}^{2n}(-1)^k \binom{2n}{k}^3 = (-1)^n \frac{(3n)!}{(n!)^3}.$$

1.9 Given the expansion

$$(1+x+x^2)^n = a_0 + a_1 x + a_2 x^2 + \cdots + a_k x^k + \cdots + a_{2n} x^{2n},$$

show that:

(a) $a_n = 1 + \frac{n(n-1)}{(1!)^2} + \frac{n(n-1)(n-2)(n-3)}{(2!)^2} + \cdots = \sum_{k \geq 0} \frac{(2k)!}{(k!)^2}\binom{n}{2k};$

(b) $a_0 a_1 - a_1 a_2 + a_2 a_3 - \cdots - a_{2n-1} a_{2n} = 0;$

(c) $a_0^2 - a_1^2 + a_2^2 - \cdots + (-1)^{n-1} a_{n-1}^2 = \frac{1}{2}[a_n + (-1)^{n-1} a_n^2];$

(d) $a_p - \binom{n}{1} a_{p-1} + \binom{n}{2} a_{p-2} - \cdots + (-1)^p \binom{n}{p} a_0$

$= \begin{cases} 0 & \text{if } p \text{ is not a multiple of 3,} \\ (-1)^k \binom{n}{k} & \text{if } p = 3k; \end{cases}$

(e) $a_0 + a_2 + a_4 + \cdots = \frac{1}{2}(3^n + 1).$

and
$$a_1+a_3+a_5+\cdots=\tfrac{1}{2}(3^n-1);$$
(f) $a_0+a_3+a_6+a_9+\cdots=a_1+a_4+a_7+a_{10}+\cdots$
$$=a_2+a_5+a_8+a_{11}+\cdots=3^{n-1}.$$
(g) With respect to the summations
$$a_0+a_4+a_8+\ldots,\ a_1+a_5+a_9+\ldots,$$
$$a_2+a_6+a_{10}+\ldots,\ a_3+a_7+a_{11}+\ldots$$
show that three are equal and that the fourth differs from their value by one.

(h) Verify the inequalities
$$1=a_0<a_1<\cdots<a_n \quad \text{and} \quad a_n>a_{n+1}>\cdots>a_{2n}=1$$
for every $n \geqslant 2$.

1.10 Suppose that
$$(1+x+x^2+\cdots+x^m)^n=a_0+a_1x+a_2x^2+\cdots+a_{mn}x^{mn},$$
and set
$$S_i=a_i+a_{i+(m+2)}+a_{i+2(m+2)}+\cdots$$
for every i, $0 \leqslant i \leqslant m+1$. Show that:
$$S_i=\frac{(m+1)^n+(-1)^{n-1}}{m+2}+(-1)^n \quad \text{for} \quad n+i\equiv 0 \pmod{m+2}$$
and in the opposite case,
$$S_i=\frac{(m+1)^n+(-1)^{n-1}}{m+2}.$$

1.11 What is the coefficient of x^k in the expansion of
$$(1+x+x^2+\cdots+x^{n-1})^2?$$

1.12 Given positive integers n, and r show that there also exist r unique integers
$$a_1>a_2>\cdots>a_r\geqslant 0$$
such that
$$n=\binom{a_1}{r}+\binom{a_2}{r-1}+\cdots+\binom{a_r}{1}.$$

1.13 Show that
$$(x+y)^n=\sum_{k=1}^{n}\binom{2n-k-1}{n-1}(x^k+y^k)\left(\frac{xy}{x+y}\right)^{n-k}.$$

1.14 Prove the identity
$$\sum_{k=0}^{n-1}\binom{n-1}{k}n^{n-1-k}(k+1)!=n^n.$$

1.15 Show that the number of arrangements of a set of n objects in p boxes such that the jth box contains n_j objects, for $j=1,\ldots,p$ is equal to the multinomial number
$$\binom{n}{n_1,n_2,\ldots,n_p}=\frac{n!}{n_1!n_2!\cdots n_p!},$$
where $n_i\geq 0$ and $n_1+n_2+\cdots+n_p=n$.

1.16 Prove the multinomial formula
$$(a_1+a_2+\cdots+a_p)^n=\sum_{\substack{n_1,\ldots,n_p\geq 0 \\ n_1+\cdots+n_p=n}}\binom{n}{n_1,n_2,\ldots,n_p}a_1^{n_1}a_2^{n_2}\cdots a_p^{n_p},$$
where $a_1,\ldots,a_p$ are elements of a commutative ring.

1.17 Justify the identity
$$\sum_{k\geq 1}\sum_{(s_1,\ldots,s_k)}\binom{h}{s_1,\ldots,s_k}=\binom{m-1}{h-1},$$
where the second summation is taken over all choices of the numbers $s_1,\ldots,s_k\geq 0$ which satisfy the relations
$$s_1+s_2+\cdots+s_k=h;$$
$$s_1+2s_2+\cdots+ks_k=m.$$

1.18 A function $f:\{1,\ldots,n\}\to\{1,\ldots,r\}$ is said to be increasing if $f(i)\leq f(j)$ for every i,j, $1\leq i<j\leq n$. Show that the number of increasing functions defined on the set $\{1,\ldots,n\}$ with values in the set $\{1,\ldots,r\}$ is equal to
$$\frac{[r]^n}{n!}=\frac{r(r+1)\cdots(r+n-1)}{n!}.$$
This number is also called the number of combinations with replacement of r things taken k at a time.

1.19 In how many ways can a natural number m be written as the sum of n non-negative integers
$$m=u_1+u_2+\cdots+u_n,$$
where two sums are considered to be different even if they differ only in the order of their terms? What is the result if $u_i>0$ for all i?

1.20 Determine the number of monomials in the expansion of the polynomial
$$(x_1+x_2+\cdots+x_p)^n.$$

Combinatorial Identities

1.21 Let $P(x)$ be a polynomial of degree n such that $P(x)=2^x$ for every $x=1, 2, \ldots, n+1$. Determine $P(n+2)$.

1.22 Verify the identity

$$\sum_{\substack{c_1+c_2+\cdots+c_k=n \\ c_i \geq 0}} c_1 c_2 \cdots c_k = \frac{n(n^2-1^2)\cdots(n^2-(k-1)^2)}{(2k-1)!}.$$

1.23 Prove the identity

$$(1+x+x^2+\cdots)^n = \sum_{r \geq 0} \binom{n+r-1}{r} x^r.$$

1.24 Show that

$$\sum (-1)^{j_1+j_2+\cdots+j_n+1} \frac{(j_1+j_2+\cdots+j_n-1)!}{j_1! j_2! \cdots j_n!} = \frac{1}{n}$$

for every positive integer n, where the sum is taken over all partitions of n of the form $j_1 + 2j_2 + \cdots + nj_n = n$ and $j_i \geq 0$ for $1 \leq i \leq n$.

1.25 Show that for $h \geq 2$ the following relations hold:

(a) $\displaystyle \max_{(n_1,\ldots,n_k)} \sum_{i=1}^{k} \binom{n_i}{h} = \binom{n-k+1}{h};$

(b) $\displaystyle \min_{(n_1,\ldots,n_k)} \sum_{i=1}^{k} \binom{n_i}{h} = (k-r)\binom{t}{h} + r\binom{t+1}{h}.$

The maximum [minimum] is taken over all representations of n of the form

$$n = n_1 + \cdots + n_k \quad \text{and} \quad n_1, \ldots, n_k \geq 1, \quad t = \left\lfloor \frac{n}{k} \right\rfloor,$$

and r is the remainder when n is divided by k.

1.26 Evaluate:

(a) $\displaystyle \max_{1 \leq k \leq n} \max_{n_1+\cdots+n_k=n} n_1 n_2 \cdots n_k;$

(b) $\displaystyle \max_{1 \leq k \leq n} \max_{n_1+\cdots+n_k=n} \binom{n_1}{2}\binom{n_2}{2}\cdots\binom{n_k}{2}.$

1.27 If $n_1 \geq 3h$, $n_2 \geq 3h$, $h \geq 1$, where n_1, n_2, h are integers, show that for all integers x, y such that $0 \leq x \leq h$, $0 \leq y \leq h$, the following inequality holds:

$$\binom{n_1-x}{n_1-h}\binom{n_2-y}{n_2-h} \geq \binom{2h}{x+y}.$$

1.28 For every $p \leq k$ justify the identity

$$\sum_{\alpha_1+2\alpha_2+\cdots+n\alpha_n=p} \frac{k!}{\alpha_1!\cdots\alpha_n!\{k-(\alpha_1+\cdots+\alpha_n)\}!} \binom{n}{1}^{\alpha_1}\cdots\binom{n}{n}^{\alpha_n} = \binom{nk}{p}.$$

1.29 Prove Abel's identities:

(a) $\sum_{k=0}^{n} \binom{n}{k} x(x+k)^{k-1}(y+n-k)^{n-k} = (x+y+n)^n$;

(b) $\sum_{k=0}^{n} \binom{n}{k} (x+k)^{k-1}(y+n-k)^{n-k-1} = \left(\frac{1}{x}+\frac{1}{y}\right)(x+y+n)^{n-1}$;

(c) $\sum_{k=1}^{n-1} \binom{n}{k} k^{k-1}(n-k)^{n-k-1} = 2(n-1)n^{n-2}$.

1.30 Show that

$$n^{n-1} + \sum_{k=1}^{n-1} \binom{n}{k} k^{k-1}(n-k)^{n-k} = n^n.$$

1.31 Given a recurrence relation of the form

$$f(n+2) = af(n+1) + bf(n),$$

where a, b are real numbers with $b \neq 0$ and $n = 0, 1, 2, \ldots$, the quadratic equation

$$r^2 = ar + b$$

is called the characteristic equation of the given recurrence relation. Show that:

(a) If the characteristic equation has two distinct roots r_1 and r_2, then the general solution of the recurrence relation has the form

$$f(n) = C_1 r_1^n + C_2 r_2^n,$$

where the constants C_1 and C_2 are determined from the initial conditions by solving the system of equations

$$C_1 + C_2 = f(0),$$
$$C_1 r_1 + C_2 r_2 = f(1).$$

(b) If the characteristic equation has a double root equal to r_1, the general solution of the recurrence relation has the form

$$f(n) = r_1^n (C_1 + C_2 n),$$

where $C_1 = f(0)$ and $C_2 = [f(1) - r_1 f(0)]/r_1$.

1.32 A pupil has $\$n$. Every day he buys exactly one of the following products: a bun which costs $\$1$, an ice cream which costs $\$2$, or a pastry which costs $\$2$, until he has no more money. In how many ways can he use up these $\$n$?

1.33 Let $U(n)$ be the number of ways in which one can cover a 3-by-n rectangle $ABCD$ with dominoes (rectangles with sides 1 and 2). Show that $U(n) = 0$ if n is odd and that for n even U is given by the formula

$$U(2m) = \frac{1}{2\sqrt{3}} \{(\sqrt{3}+1)(2+\sqrt{3})^m + (\sqrt{3}-1)(2-\sqrt{3})^m\}.$$

Combinatorial Identities

1.34 How many words of length n can be formed with letters of the alphabet $A = \{a, b, c, d\}$ so that the letters a and b are not adjacent?

1.35 Let
$$a_n = \sum_{k=0}^{[n/2]} \binom{n-k}{k} z^k.$$
Show that
$$a_n = \frac{1}{\sqrt{1+4z}} \left\{ \left(\frac{1+\sqrt{1+4z}}{2}\right)^{n+1} - \left(\frac{1-\sqrt{1+4z}}{2}\right)^{n+1} \right\}$$
for $z \neq -\frac{1}{4}$, and $a_n = (n+1)/2^n$ for $z = -\frac{1}{4}$.

1.36 Consider the polynomial
$$f(x) = a_0 x^n + a_1 x^{n-1} + \cdots + a_n.$$
Let
$$S_n(x_1, \ldots, x_n) = f(x_1 + \cdots + x_n) - \sum f(x_1 + \cdots + x_{n-1}) \\ + \sum f(x_1 + \cdots + x_{n-2}) - \cdots + (-1)^n f(0),$$
where the first summation is taken over the $\binom{n}{n-1}$ sums of $n-1$ variables x_j, the second summation is taken over the $\binom{n}{n-2}$ sums of $n-2$ variables x_j, etc. Show that $S_n(x_1, \ldots, x_n) = a_0 n! x_1 x_2 \cdots x_n$.

1.37 Let $p(x_1, \ldots, x_n)$ be a polynomial in n variables of degree m. Denote by $z^k p$ the polynomial obtained by replacing k of the variables $x_1, \ldots, x_n$ in p with 0 in all possible ways and then summing the $\binom{n}{k}$ polynomials thus obtained. Show that
$$p - z^1 p + z^2 p - \cdots = \begin{cases} 0 & \text{if } m < n; \\ c \cdot x_1 \cdots x_n & \text{if } m = n, \end{cases}$$
where c is the coefficient of the monomial $x_1 \cdots x_n$ in the expansion of the polynomial p. By taking $p(x_1, \ldots, x_n) = (x_1 + \cdots + x_n)^k$ and setting $x_1 = \cdots = x_n = 1$, deduce Euler's identity:
$$\sum_{i=0}^{n} (-1)^i \binom{n}{i} i^k = \begin{cases} 0 & \text{for } 0 \leq k < n; \\ (-1)^n n! & \text{for } k = n. \end{cases}$$

1.38 Show that the following identity holds for all positive integers n, p with $n \geq p$:
$$p! = n^p - \binom{p}{1}(n-1)^p + \binom{p}{2}(n-2)^p - \cdots + (-1)^p \binom{p}{p}(n-p)^p.$$

1.39 Show that
$$\frac{\binom{n}{0}}{x} - \frac{\binom{n}{1}}{x+1} + \cdots + (-1)^n \frac{\binom{n}{n}}{x+n} = \frac{n!}{x(x+1)\cdots(x+n)}.$$

2

The Principle of Inclusion and Exclusion; Inversion Formulas

2.1 In a Romanian high-school class there are 40 students. Among them 14 like mathematics, 16 like physics, and 11 like chemistry. It is also known that 7 like mathematics and physics, 8 like physics and chemistry and 5 like mathematics and chemistry. All three subjects are favored by 4 students. How many students like neither mathematics, nor physics, nor chemistry?

2.2 Justify the following formula, known as the principle of inclusion and exclusion:

$$\left|\bigcup_{i=1}^{q} A_i\right| = \sum_{i=1}^{q} |A_i| - \sum_{1 \leq i < j \leq q} |A_i \cap A_j| + \cdots + (-1)^{q+1} \left|\bigcap_{i=1}^{q} A_i\right|.$$

2.3 If $A_1, A_2, \ldots, A_q \subset X$, prove that the number of elements in X which belong to p of the sets A_i is equal to

$$\sum_{k=p}^{q} (-1)^{k-p} \binom{k}{p} \sum_{\substack{K \subset Q \\ |K| = k}} \left|\bigcap_{i \in K} A_i\right|$$

(the sieve formula of C. Jordan).

2.4 Let n be a positive integer and $\varphi(n)$ the value of Euler's function, i.e., the number of positive integers less than or equal to and prime to n. If $n = p_1^{i_1} p_2^{i_2} \cdots p_q^{i_q}$ is the decomposition of n into q distinct prime factors, show that

$$\varphi(n) = n \left(1 - \frac{1}{p_1}\right)\left(1 - \frac{1}{p_2}\right) \cdots \left(1 - \frac{1}{p_q}\right).$$

2.5 Let p be a permutation of a set $X = \{1, \ldots, n\}$. A fixed point of p is a number i such that $p(i) = i$ $(1 \leq i \leq n)$. Show that the number $D(n)$ of permutations of X without fixed points is given by

$$D(n) = n! \left(1 - \frac{1}{1!} + \frac{1}{2!} - \frac{1}{3!} + \cdots + \frac{(-1)^n}{n!}\right).$$

How many permutations of a set of n objects have p fixed points?

2.6 Let $X = \{1, 2, \ldots, n\}$, and let $D(n)$ be the number of permutations of the set X without fixed points. If $E(n)$ represents the number of even permutations of X without fixed points, show that

$$E(n) = \tfrac{1}{2}\{D(n) + (-1)^{n-1}(n-1)\}.$$

2.7 Show that $\sum_{d|n} \varphi(d) = n$, where φ is Euler's function.

2.8 Show that the number of square matrices of order 3 with non-negative integer elements for which every row sum and every column sum is equal to r is given by

$$\binom{r+2}{2}^2 - 3\binom{r+3}{4}.$$

2.9 Verify that

$$\sum_{n=0}^{\infty} D(n) \frac{t^n}{n!} = \frac{e^{-t}}{1-t},$$

$$D(n+1) = (n+1)D(n) + (-1)^{n+1},$$

$$D(n+1) = n\{D(n) + D(n-1)\}.$$

2.10 Show that the number $s_{n,m}$ of surjective functions $f : X \to Y$ with $|X| = n$ and $|Y| = m$ is given by the expression

$$s_{n,m} = m^n - \binom{m}{1}(m-1)^n + \binom{m}{2}(m-2)^n + \cdots + (-1)^{m-1}m.$$

Deduce from this that if $E(n, m)$ denotes the right side of this expression, then $E(n, n) = n!$ and $E(n, m) = 0$ for $n < m$ (Euler's identities).

2.11 Denote by $s_{n,m,r}$ the number of functions $f : X \to Y$ which have the property that $f(X) \supset Z$, where $|X| = n$, $|Y| = m$, and $Z \subset Y$, $|Z| = r$. Verify the formula

$$s_{n,m,r} = m^n - \binom{r}{1}(m-1)^n + \binom{r}{2}(m-2)^n - \cdots + (-1)^r(m-r)^n.$$

2.12 Let A be an alphabet formed of n pairs of identical letters a_1, a_1; a_2, a_2; $\cdots$; a_n, a_n. Different pairs contain different letters. Form all the words which use all $2n$ letters of the alphabet A so that no adjoining letters are identical. Show that the number of words formed in this way is equal to

$$\frac{1}{2^n}\left\{(2n)! - \binom{n}{1}2(2n-1)! + \binom{n}{2}2^2(2n-2)! - \cdots + (-1)^n 2^n n!\right\}.$$

2.13 Let a_n be the number of digraphs with n vertices which are labeled with numbers from the set $\{1, \ldots, n\}$ and which do not contain a circuit. Show that the numbers a_n satisfy the recurrence relation

$$a_n = \sum_{k=1}^{n} (-1)^{k-1} \binom{n}{k} 2^{k(n-k)} a_{n-k}$$

if, by definition, $a_0 = 1$.

2.14 A set X is said to be a collection of objects of type $1^{\lambda_1} 2^{\lambda_2} \cdots n^{\lambda_n}$ if there exists a partition of the set X which contains λ_j classes with j elements, for $j = 1, \ldots, n$. Objects which belong to the same class of the partition are identified. An arrangement of the objects in cells is a function $f : X \to A$, where A is the set of cells. If $f(x) = a_i$, we shall say that the object $x \in A$ is arranged in cell a_i. By definition, two arrangements are equivalent if one can be obtained from the other by a permutation of the objects in the same classes of the partition of X. Classes of this equivalence relation are called arrangement schemes of objects in cells.

Denote by $A_\varnothing(1^{\lambda_1} 2^{\lambda_2} \cdots n^{\lambda_n}; 1^m)$ the number of arrangement schemes of a collection of objects of type $1^{\lambda_1} 2^{\lambda_2} \cdots n^{\lambda_n}$ in m distinct cells, and by $A(1^{\lambda_1} 2^{\lambda_2} \cdots n^{\lambda_n}; 1^m)$ the number of arrangement schemes which leave no cell empty. Show that

$$A_\varnothing(1^{\lambda_1} 2^{\lambda_2} \cdots n^{\lambda_n}; 1^m) = \binom{m}{1}^{\lambda_1} \binom{m+1}{2}^{\lambda_2} \cdots \binom{m+n-1}{n}^{\lambda_n},$$

$$A(1^{\lambda_1} 2^{\lambda_2} \cdots n^{\lambda_n}; 1^m) = \sum_{k=1}^{m} (-1)^{m-k} \binom{m}{k} \binom{k}{1}^{\lambda_1} \binom{k+1}{2}^{\lambda_2} \cdots \binom{k+n-1}{n}^{\lambda_n}.$$

2.15 Find the number of possible ways of writing a natural number p as a product of m factors different from one, for which two products are also considered to be different if the order of the factors is different; the decomposition of the number p in prime factors contains λ_1 factors of exponent 1, λ_2 factors of exponent 2, ..., λ_n factors of exponent n.

2.16 Let

$$M(p, q) = (2^q - 1)^p - \binom{q}{1}(2^{q-1} - 1)^p + \binom{q}{2}(2^{q-2} - 1)^p \cdots + (-1)^{q-1} q.$$

Show that $M(p, q) = M(q, p)$.

2.17 Prove the inverse binomial formula: If the numbers $a_0, a_1, \ldots, a_n$ and $b_0, b_1, \ldots, b_n$ satisfy the relations $a_k = \sum_{i=0}^{k} \binom{k}{i} b_i$ for $k = 0, 1, \ldots, n$, then the numbers $b_0, b_1, \ldots, b_n$ are given by the relations

$$b_k = \sum_{i=0}^{k} \binom{k}{i} (-1)^{k-i} a_i.$$

2.18 Count in two different ways the number of representations of m as a sum of n integers, $m = u_1 + u_2 + \cdots + u_n$, where $u_i \geq 2$ for $1 \leq i \leq n$. Two sums are

The Principle of Inclusion and Exclusion; Inversion Formulas

also considered to be distinct if they differ only in the order of their terms. Use this to obtain the identity

$$\sum_{i=0}^{n-1} (-1)^i \binom{n}{i}\binom{m-i-1}{n-i-1} = \binom{m-n-1}{n-1}$$

for any $m \geqslant n+1 \geqslant 2$.

2.19 Let $V = \{x_1, \ldots, x_n\}$ be a finite set on which is defined a partial order $\leqslant$ which, by definition, satisfies the following three properties:

(a) $x_i \leqslant x_i$ for every $i = 1, \ldots, n$ (reflexivity);

(b) $x_i \leqslant x_j$ and $x_j \leqslant x_i$ imply $x_i = x_j$ (and thus $i = j$) for $1 \leqslant i, j \leqslant n$ (antisymmetry);

(c) $x_i \leqslant x_j$ and $x_j \leqslant x_k$ imply $x_i \leqslant x_k$ for $1 \leqslant i, j, k \leqslant n$ (transitivity).

A square matrix of order n whose elements are real numbers $(a_{ij})_{i,j=1,\ldots,n}$ will be said to be compatible with the partial order defined on V, or simply compatible, if $a_{ij} \neq 0$ implies $x_i \leqslant x_j$ for every $i, j = 1, \ldots, n$. Show that the sum and the product of two compatible matrices is a compatible matrix, and show that if a compatible matrix is nonsingular, then its inverse is compatible.

2.20 If V is the partially ordered set of the preceding problem, show that there exists a function μ defined on $V \times V$ with the following properties:

(a) $\mu(x, y) = 0$ if x is not less than or equal to y;

(b) $\mu(x, x) = 1$ for every $x \in V$;

(c) $\sum_{x \leqslant y \leqslant z} \mu(x, y) = 0$ for every $x < z$ if $x, z \in V$.

The function μ is called the Moebius function of the set V.

2.21 Evaluate the function $\mu(x, y)$ introduced in the preceding problem if V is:

(a) the family of all subsets of a finite set S with respect to the partial order relation of nonstrict inclusion, denoted $X \subset Y$;

(b) the set of integers $1, 2, \ldots, n$ where $x \leqslant y$ is defined by $x \mid y$ (i.e., x is a divisor of y);

(c) an arborescence, where the relation of partial order between vertices is defined by $x \leqslant y$ if the unique path which joins the root of the arborescence with the vertex y passes through x.

2.22 Let $f(x)$ be a real-valued function defined on a set V, and $\mu(x, y)$ the Moebius function on V. Let

$$g(x) = \sum_{z \leqslant x} f(z).$$

Show that

$$f(x) = \sum_{z \leqslant x} \mu(z, x) g(z).$$

(This is the Moebius inversion formula).

2.23 If p and q are two partitions of a set X with n elements, then define a partial order relation by $p \leqslant q$ if the partition q is a refinement of the partition p. That is, every class of q is contained in some class of p. Denote by 0 the partition with a single class consisting of X, and by 1 the partition with n classes each formed by a singleton of X; it follows that $0 \leqslant p \leqslant 1$ for each partition p of X. Show that

$$\mu(0, 1) = (-1)^{n-1}(n-1)!.$$

2.24 If the partially ordered set V contains a first element 0 and a last element 1 such that $0 \leqslant x \leqslant 1$ for each $x \in V$, denote by p_k the number of walks of the form

$$a_0 = 0 < a_1 < \cdots < a_k = 1.$$

Show that

$$\mu(0, 1) = -p_1 + p_2 - p_3 + \cdots.$$

2.25 For $n \geqslant 2$ subsets $A_1, \ldots, A_n \subset X$ let

$$S_k = \sum |A_{i_1} \cap A_{i_2} \cap \cdots \cap A_{i_k}|$$

for $1 \leqslant k \leqslant n$, where $\sum$ denotes summation over all integers $1 \leqslant i_1 < i_2 < \cdots < i_k \leqslant n$ and $S_k = 0$ if $k \geqslant n+1$. Prove that for every $p \geqslant 1$ the following inequalities hold:

$$\sum_{k=1}^{2p} (-1)^{k-1} S_k \leqslant \left| \bigcup_{i=1}^{n} A_i \right| \leqslant \sum_{k=1}^{2p-1} (-1)^{k-1} S_k.$$

These inequalities are known as the Bonferroni inequalities.

3

Stirling, Bell, Fibonacci, and Catalan Numbers

3.1 For every real number x and every natural number n let
$$[x]_n = x(x-1)\cdots(x-n+1),$$
$$[x]^n = x(x+1)\cdots(x+n-1),$$
where, by definition, $[x]_0 = [x]^0 = 1$. The Stirling number of the first kind, $s(n, k)$, is defined as the coefficient of x^k in the expansion of $[x]_n$, that is,
$$[x]_n = \sum_{k=0}^{n} s(n, k) x^k.$$
Show that
$$[x]^n = \sum_{k=0}^{n} |s(n, k)| x^k.$$

3.2 Show that the following equalities hold:

(a) $[x+y]_n = \sum\limits_{k=0}^{n} \binom{n}{k} [x]_k [y]_{n-k}$ and

(b) $[x+y]^n = \sum\limits_{k=0}^{n} \binom{n}{k} [x]^k [y]^{n-k}$,

where $[x]_0 = [x]^0 = 1$.

3.3 Prove the following identities:

(a) $x^n = \sum\limits_{k=1}^{n} S(n, k) [x]_k$ and

(b) $[x]^n = \sum\limits_{k=1}^{n} \dfrac{n!}{k!} \binom{n-1}{k-1} [x]_k$,

where $S(n, k)$ are Stirling numbers of the second kind.

3.4 Show that the Stirling numbers of the second kind can be expressed as a function of the number of surjective functions by the relation

$$S(n, m) = \frac{1}{m!} s_{n,m}.$$

Show that the Stirling numbers also satisfy the recurrence relation

$$S(n+1, m) = S(n, m-1) + mS(n, m),$$

where $S(n, 1) = S(n, n) = 1$.

3.5 Justify the following recurrence relations for the Stirling numbers of the second kind, $S(n, m)$, and for the Bell numbers B_n:

(a) $S(n+1, m) = \sum_{k=m-1}^{n} \binom{n}{k} S(k, m-1);$

(b) $B_{n+1} = \sum_{k=0}^{n} \binom{n}{k} B_k,$ where $B_0 = 1.$

3.6 Show that the number of partitions of an n-element set of type $1^{k_1} 2^{k_2} \cdots n^{k_n}$ (i.e., which contain k_j classes with j elements, $j = 1, 2, \ldots, n$) is equal to

$$\text{Part}(1^{k_1} 2^{k_2} \cdots n^{k_n}; n) = \frac{n!}{(1!)^{k_1} k_1! \, (2!)^{k_2} k_2! \, \cdots (n!)^{k_n} k_n!}.$$

Further, the number of permutations $p \in S_n$ of type $1^{k_1} 2^{k_2} \cdots n^{k_n}$ which contain k_j cycles with j elements for $j = 1, 2, \ldots, n$ is equal to

$$\text{Perm}(1^{k_1} 2^{k_2} \cdots n^{k_n}; n) = \frac{n!}{1^{k_1} k_1! \, 2^{k_2} k_2! \, \cdots n^{k_n} k_n!},$$

where $k_1 + 2k_2 + \cdots + nk_n = n$ and $k_i \geq 0$ for $i = 1, \ldots, n$.

3.7 Establish the following recurrence relations for the Stirling numbers of the first and second kinds:

$$\binom{i+j}{j} s(n, i+j) = \sum_{k=0}^{n} \binom{n}{k} s(k, i) s(n-k, j),$$

$$\binom{i+j}{j} S(n, i+j) = \sum_{k=0}^{n} \binom{n}{k} S(k, i) S(n-k, j).$$

3.8 Show that

$$\sum_{k=0}^{n} s(n, k) S(k, m) = \sum_{k=0}^{n} S(n, k) s(k, m) = \delta_{n,m},$$

where $\delta_{n,m}$ is the Kronecker symbol.

3.9 Let $M(n) = \max\{k \mid S(n, k) \text{ is maximum}; 1 \leq k \leq n\}$. Show that the sequence of Stirling numbers of the second kind is unimodal for every natural number n, that is, they satisfy one of the following formulae:

(1) $1 = S(n, 1) < S(n, 2) < \cdots < S(n, M(n)) > S(n, M(n)+1) > \cdots > S(n, n) = 1$;

(2) $1 = S(n, 1) < S(n, 2) < \cdots < S(n, M(n)-1) = S(n, M(n)) > \cdots > S(n, n) = 1$,

and $M(n+1) = M(n)$ or $M(n+1) = M(n)+1$.

3.10 Let $a = (a_0, a_1, a_2, \ldots)$ be an infinite sequence of real numbers. The generalized Stirling numbers are defined as follows:

(1) Stirling numbers of the first kind $s_a(n, k)$ by the identity

$$(x|a)_n = (x-a_0)(x-a_1)\cdots(x-a_{n-1}) = \sum_{k=0}^{n} s_a(n, k) x^k \quad \text{and} \quad (x|a)_0 = 1;$$

(2) Stirling numbers of the second kind by the identity

$$x^n = \sum_{k=0}^{n} S_a(n, k)(x-a_0)(x-a_1)\cdots(x-a_{k-1}) = \sum_{k=0}^{n} S_a(n, k)(x|a)_k.$$

Show that:

(a) $s_a(n, k) = s_a(n-1, k-1) - a_{n-1} s_a(n-1, k)$;

(b) $S_a(n, k) = S_a(n-1, k-1) + a_k S_a(n-1, k)$;

(c) $s_a(n, k) = \sum_{r=k}^{n} s_a(n+1, r+1) a_n^{r-k}$;

(d) $S_a(n, k) = \sum_{r=k}^{n} S_a(r-1, k-1) a_k^{n-r}$;

(e) $s_a(n, k) = \sum_{r=k}^{n} (-1)^{n-r} s_a(r-1, k-1) \prod_{j=r}^{n-1} a_j$;

(f) $\sum_{k=0}^{n} s_a(n, k) S_a(k, m) = \sum_{k=0}^{n} S_a(n, k) s_a(k, m) = \delta_{n,m}$

(Kronecker symbol).

3.11 Show that the generating function for the Stirling numbers of the second kind associated with the sequence $(a_0, a_1, a_2, \ldots)$ can be expressed as

$$\sum_{n=k}^{\infty} S_a(n, k) t^n = \frac{t^k}{(1-a_0 t)(1-a_1 t)\cdots(1-a_k t)}.$$

3.12 Let $S_i(n, k)$ denote the number of partitions of a set X with n elements into k classes, each one of which contains at least i elements. Show that:

(a) $S_i(n, k) = k S_i(n-1, k) + \binom{n-1}{i-1} S_i(n-i, k-1)$;

(b) $S_i(n, k) = \dfrac{1}{k!} \sum\limits_{(j_1,\ldots,j_k)} \dfrac{n!}{j_1! \cdots j_k!}$,

where the sum is taken over all integral solutions of the equation $j_1 + \cdots + j_k = n$ which satisfy $j_s \geq i$ for $s = 1, \ldots, k$.

3.13 Show that the Stirling numbers of the second kind satisfy the following relations:

(a) $S(n, 2) = 2^{n-1} - 1$;

(b) $S(n, n-1) = \dbinom{n}{2}$;

(c) $S(n, 1) - 1!S(n, 2) + 2!S(n, 3) - 3!S(n, 4) + \cdots + (-1)^{n-1}(n-1)! = 0$ for $n \geq 2$.

3.14 A bicovering with k classes of a set X with n elements is a family of k nonempty subsets of X such that each element $x \in X$ is contained in exactly two subsets of the family. If $c(n, k)$ denotes the number of bicoverings with k classes of X, show that

$$c(n, 3) = \tfrac{1}{2}(3^{n-1} - 1).$$

3.15 A partial partition of a set X is a partition of a subset:

$$Y \subset X, \quad Y \neq \varnothing.$$

Show that the number of partial partitions of a set X with n elements is equal to $B_{n+1} - 1$.

3.16 Show that the exponential generating function for the Bell numbers is given by

$$\sum_{n=0}^{\infty} \dfrac{B_n}{n!} t^n = \exp(\exp(t) - 1).$$

3.17 Show that the Bell numbers B_n satisfy

$$B_n = \dfrac{1}{e} \sum_{k=1}^{\infty} \dfrac{k^n}{k!}.$$

Also show that the difference between the number of partitions with an even number of classes and the number of partitions with an odd number of classes of a set with n elements is equal to

$$e \sum_{k=1}^{\infty} \dfrac{(-1)^k k^n}{k!}.$$

3.18 Let $f(n, k)$ denote the number of subsets of the set $X = \{1, \ldots, n\}$ which contain k elements, no two of which are consecutive integers. Show that

$$f(n, k) = \dbinom{n-k+1}{k}.$$

Stirling, Bell, Fibonacci, and Catalan Numbers

If $F_{n+1} = \sum_{k \geq 0} f(n, k)$, then $F_0 = F_1 = 1$. Show that $F_n = F_{n-1} + F_{n-2}$ for every $n \geq 2$. The numbers F_n are called Fibonacci numbers.

3.19 Suppose that $f^*(n, k)$ denotes the number of k-element subsets of $X = \{1, \ldots, n\}$ which contain neither two consecutive integers nor 1 and n simultaneously. Show that

$$f^*(n, k) = \frac{n}{n-k} \binom{n-k}{k}.$$

If $L_n = \sum_{k \geq 0} f^*(n, k)$ for $n \geq 1$, then $L_1 = 1$, $L_2 = 3$, and $L_{n+1} = L_n + L_{n-1}$ for every $n \geq 2$. The numbers L_n are called Lucas numbers.

3.20 Show that the Fibonacci numbers satisfy the identity

$$F_{n+1} F_{n-1} - F_n^2 = (-1)^{n+1}.$$

3.21 In how many ways u_n can one mount a staircase with n steps if every movement involves one or two steps? Show that the generating function is

$$\sum_{n=0}^{\infty} u_n x^n = \frac{1}{1 - x - x^2},$$

where, by definition, $u_0 = 1$.

3.22 Show that every natural number $n \geq 1$ can be written as a sum of pairwise distinct Fibonacci numbers which are not consecutive numbers F_p and F_{p+1} of the Fibonacci sequence.

3.23 Show that the generating function of the Catalan numbers C_n satisfies the equation

$$f(x) = C_1 x + C_2 x^2 + \cdots + C_n x^n + \cdots = \frac{1 - \sqrt{1-4x}}{2}.$$

Use this fact to obtain an expression for the number C_n.

3.24 Show that the number of sequences $(x_1, x_2, \ldots, x_{2n-2})$ such that $x_i \in \{-1, 1\}$ for $i = 1, 2, \ldots, 2n-2$ and which satisfy

(1) $x_1 + x_2 + \cdots + x_k \geq 0$ for every $1 \leq k \leq 2n-2$, and
(2) $x_1 + x_2 + \cdots + x_{2n-2} = 0$

is equal to $(1/n)\binom{2n-2}{n-1}$.

3.25 A triangulation of a convex polygon $A_1 A_2 \cdots A_{n+1}$ with $n+1$ vertices is a set formed of $n-2$ diagonals which do not intersect in the interior of the polygon but only at vertices, and which divide the surface of the polygon into $n-1$ triangles. Show that the number of triangulations of a convex polygon with $n+1$ vertices is equal to

$$C_n = \frac{1}{n} \binom{2n-2}{n-1}.$$

3.26 Show that the number of increasing functions
$$f:\{1,\ldots,n\}\to\{1,\ldots,n\}$$
which satisfy the condition $f(x)\leqslant x$ for every $1\leqslant x\leqslant n$ is equal to the Catalan number
$$C_{n+1}=\frac{1}{n+1}\binom{2n}{n}.$$

3.27 Let $A_1A_2\cdots A_n$ be a convex polygon. In how many ways can this polygon be triangulated with $n-3$ diagonals which do not intersect in the interior of the polygon, so that each triangle has one or two sides in common with the convex polygon?

3.28 Show that the number of sequences $(a_1, a_2, \ldots, a_{k+1})$ formed of nonnegative integers with the properties
$$a_1=0 \quad \text{and} \quad |a_i - a_{i+1}|=1 \quad \text{for } 1\leqslant i\leqslant k$$
is equal to $\binom{k}{[k/2]}$.

3.29 Let $g_0(n+1)$ be the number of sequences $(a_1, a_2, \ldots, a_{n+1})$ of nonnegative integers such that $a_1=0$ and
$$|a_i - a_{i+1}|\leqslant 1 \quad \text{for } i=1, 2, \ldots, n.$$
Show that
$$g_0(n+1)=c(n,n)+c(n,n+1),$$
where
$$(1+x+x^2)^m = \sum_{k\geqslant 0} c(m,k)x^k.$$

3.30 Show that the number of sequences $(a_1, a_2, \ldots, a_{2n+1})$ of nonnegative integers with the property that $a_1=a_{2n+1}=0$ and $|a_i - a_{i+1}|=1$ for $i=1, \ldots, 2n$ is equal to the Catalan number
$$\frac{1}{n+1}\binom{2n}{n}.$$

3.31 Show that the number of sequences
$$(x_1, \ldots, x_r) \quad \text{with} \quad 1\leqslant x_i\leqslant n$$
which contain at most $i-1$ terms smaller than or equal to i for $i=1, \ldots, n$ is equal to $(n-r)n^{r-1}$ for every $1\leqslant r\leqslant n$.

3.32 Let S_n be the number of functions $f:\{1,\ldots,n\}\to\{1,\ldots,n\}$ with the property that if f takes on the value i, then f takes on the value j for $1\leqslant j\leqslant i$. Show that
$$S_n = \sum_{k=0}^{\infty} \frac{k^n}{2^{k+1}},$$

Stirling, Bell, Fibonacci, and Catalan Numbers

and deduce that the exponential generating function of the numbers S_n is equal to

$$\sum_{n=0}^{\infty} \frac{S_n}{n!} x^n = \frac{1}{2-e^x}, \quad \text{where} \quad S_0 = 1.$$

3.33 Let $\begin{bmatrix} n \\ k \end{bmatrix}_q$ be the number of subspaces of dimension k of an n-dimensional vector space V over a finite field F with q elements, where q is a power of a prime. This number is called the Gauss coefficient. Show that

$$\begin{bmatrix} n \\ k \end{bmatrix}_q = \frac{(q^n-1)(q^{n-1}-1)\cdots(q^{n-k+1}-1)}{(q^k-1)(q^{k-1}-1)\cdots(q-1)}.$$

3.34 Demonstrate the following properties of the Gauss coefficients:

(a) $\lim_{q \to 1} \begin{bmatrix} n \\ k \end{bmatrix}_q = \binom{n}{k}$;

(b) $\begin{bmatrix} n \\ k \end{bmatrix}_q = \begin{bmatrix} n \\ n-k \end{bmatrix}_q$;

(c) $\begin{bmatrix} n \\ k \end{bmatrix}_q = \begin{bmatrix} n-1 \\ k-1 \end{bmatrix}_q + q^k \begin{bmatrix} n-1 \\ k \end{bmatrix}_q$.

3.35 Let q be a power of a prime number. Show that Cauchy's identity holds:

$$y^n = 1 + \sum_{k=0}^{n-1} \begin{bmatrix} n \\ k \end{bmatrix}_q (y-1)(y-q)\cdots(y-q^{n-k-1}).$$

3.36 For $n \geq 2$ let $f(n, k)$ denote the number of sequences of k integers $1 \leq a_1 < a_2 < \cdots < a_k = n$ which satisfy

$$a_2 - a_1 \equiv a_3 - a_2 \equiv \cdots \equiv a_{k-1} - a_{k-2} \equiv 1 \pmod{2}$$

and

$$a_k - a_{k-1} \equiv 0 \pmod{2}.$$

Show that:

(a) $f(n, k) = \left(\begin{bmatrix} \frac{n+k-3}{2} \\ k-1 \end{bmatrix} \right)$;

(b) $\sum_{k \geq 1} f(n, k) = F_{n-1}$.

4

Problems in Combinatorial Set Theory

4.1 Let X be a collection of n objects ($n \geq 1$) which are not necessarily pairwise distinct. If $n \geq a^2 + 1$, where a is a non-negative integer, show that one or more of the following two statements is valid:

(1) At least $a+1$ objects are identical.
(2) At least $a+1$ objects are pairwise distinct.

4.2 In how many ways can one arrange k rooks on a chessboard with m rows and n columns so that no rook can attack another?

4.3 Let A be a set formed from 19 pairwise distinct integers which belong to the arithmetic progression $1, 4, 7, \ldots, 100$. Show that there are two distinct integers in A whose sum is equal to 104.

4.4 Let $k \geq 1$ be a natural number. Determine the smallest natural number n with the following property: For every choice of n integers there exist at least two whose sum or difference is divisible by $2k+1$.

4.5 Let $A = (A_i)_{1 \leq i \leq n}$, $B = (B_i)_{1 \leq i \leq n}$, $C = (C_i)_{1 \leq i \leq n}$ be three partitions of a finite set M. If for each i, j, k the following inequality is satisfied:

$$|A_i \cap B_j| + |A_i \cap C_k| + |B_j \cap C_k| \geq n,$$

show that $|M| \geq n^3/3$, with equality holding if $n \equiv 0 \pmod{3}$.

4.6 A mapping $f: X \to X$ is said to be idempotent if $f(f(x)) = f(x)$ for every $x \in X$. If $|X| = n$ prove that:

(a) the number $i(n)$ of idempotent mappings $f: X \to X$ is equal to

$$i(n) = \sum_{k=1}^{n} \binom{n}{k} k^{n-k};$$

(b) $\qquad 1 + \sum_{n=1}^{\infty} i(n) \dfrac{x^n}{n!} = \exp(xe^x).$

4.7 Let P be a partially ordered set. A subset S of P is called a chain if every two elements of S are comparable with respect to the order relation. If S is an antichain, every two elements of S are noncomparable with respect to the order.

For a natural number m, show that if P does not contain a chain of cardinality $m+1$, then P can be represented as a union of m antichains.

4.8 A chain of length n in the family of partitions of an n-element set X is a sequence of pairwise distinct partitions which satisfy

$$P_1 < P_2 < \cdots < P_n.$$

The partition P_1 has a single class formed of X, while P_n has n classes which each contain a single element of X.

Show that the number of chains of length n in the family of partitions of an n-element set X is equal to

$$\frac{(n-1)!\,n!}{2^{n-1}}.$$

4.9 Let $F = \{E_1, \ldots, E_s\}$ be a family of r-element subsets of a set X. If the intersection of each $r+1$ subsets of F is nonempty, show that the intersection of all the subsets of F is also nonempty.

4.10 Let $S = (X_i)_{1 \leq i \leq r}$ be a family of pairwise distinct subsets of X with the property that $X_i \cap X_j \neq \emptyset$ for every $i, j = 1, \ldots, r$. If the set X has n elements, show that max $r = 2^{n-1}$.

4.11 Let X be a nonempty set and let F be a family of m distinct subsets of X where $m \geq 2$. Show that the collection of subsets of the form $A \triangle B$ (symmetric difference of A and B) where $A, B \in F$ contains at least m pairwise distinct sets.

4.12 A covering of a set S is a family of nonempty pairwise distinct subsets of S whose union is equal to S. Show that the number $A(n)$ of coverings of an n-element set is given by the formula

$$A(n) = \sum_{j=0}^{n} (-1)^j \binom{n}{j} 2^{2^{n-j}-1}.$$

4.13 A covering A of a set S by k nonempty subsets is said to be irreducible if the union of every $k-1$ subsets of A is a proper subset of S. If $I(n, k)$ denotes the number of irreducible coverings by k subsets of an n-element set, show that

$$I(n, k) = \sum_{i=k}^{n} \binom{n}{i} (2^k - k - 1)^{n-i} S(i, k),$$

where $S(i, k)$ is the Stirling number of the second kind. In particular verify that

$$I(n, n-1) = \tfrac{1}{2} n(2^n - n - 1)$$

and

$$I(n, 2) = S(n+1, 3).$$

4.14 Let $A_1, \ldots, A_n$ be a collection of n pairwise distinct sets, and

$A_{i_1}, \ldots, A_{i_r}$, a subfamily of maximal cardinality with the property that it does not contain the union of the sets, that is,

$$A_i \cup A_j \neq A_k$$

for each three pairwise distinct indices $i, j, k \in \{i_1, \ldots, i_r\}$. Let $f(n) = \min r$, where the minimum is taken over all families of n pairwise distinct sets. Show that

$$\sqrt{2n} - 1 \leq f(n) \leq 2\sqrt{n} + 1.$$

4.15 Let $A_1, A_2, \ldots, A_n$ be finite sets such that

$$|A_1| = |A_2| = \cdots = |A_n|$$

and let $\bigcup_{i=1}^n A_i = S$. Suppose that for fixed k ($1 \leq k \leq n$) the union of every k sets of this family is equal to S, and the union of at most $k-1$ sets of the family is a proper subset of S. Show that $|S| \geq \binom{n}{k-1}$. When equality holds, it follows that $|A_i| = \binom{n-1}{k-1}$ for every $i = 1, \ldots, n$.

4.16 Let $(X_i)_{1 \leq i \leq k}$ be a family of k-element subsets of a set X. Show that $\min |\bigcup_{i=1}^k X_i|$ is equal to the smallest integer m such that $k \leq \binom{m}{h}$.

4.17 Show that

$$\sum_{A_1, \ldots, A_k} |A_1 \cup A_2 \cup \cdots \cup A_k| = n(2^k - 1) 2^{k(n-1)},$$

where the sum is taken over all choices of subsets $A_1, \ldots, A_k$ of an n-element set X.

4.18 Show that

$$\sum |A_1 \cup \cdots \cup A_k| = (2^k - 1) \sum |A_1 \cap \cdots \cap A_k|,$$

where the sum is taken over all choices of subsets $A_1, \ldots, A_k$ of an n-element set X.

4.19 A collection S of nonempty distinct subsets of an n-element set X is called a filter basis if for every $A, B \in S$ there is a set $C \in S$ such that $C \subset A \cap B$. Show that the number of filter bases of X is equal to

$$\sum_{k=0}^{n-1} \binom{n}{k} 2^{2^k - 1}.$$

4.20 Let $(A_i)_{1 \leq i \leq m}$ and $(B_i)_{1 \leq i \leq m}$ be two families of sets with the property that $|A_1| = \cdots = |A_m| = p$, $|B_1| = \cdots = |B_m| = q$ and $A_i \cap B_j = \varnothing$ if and only if $i = j$. Show that

$$m \leq \binom{p+q}{p}.$$

4.21 Let X be an n-element set, and $G = \{A_1, \ldots, A_p\}$ a family of subsets of X which are noncomparable with respect to inclusion; that is, $A_i \not\subset A_j$ for every $i,j=1,\ldots,p$ with $i\neq j$. Show that

$$\max p = \binom{n}{[n/2]}.$$

This result is called Sperner's theorem.

4.22 Let X be an n-element set, and $F=\{A_1,\ldots,A_p\}$ a family of subsets of X which satisfy the following conditions:

(1) $|A_i|=r\leq n/2$ for every $i=1,\ldots,p$;
(2) $A_i\cap A_j\neq\emptyset$ for every $i,j=1,\ldots,p$.

Show that $\max p = \binom{n-1}{r-1}$.

This result is known as the Erdös–Ko–Rado theorem.

4.23 Let X be a finite set, and $E_1,\ldots,E_m$ a family of subsets of X with the property that the intersection of two distinct sets E_i and E_j never has cardinality exactly equal to one. Further assume that $|E_i|\geq 2$ for $i=1,\ldots,m$.

Show that under these circumstances one can color the elements of X with two colors so that no subset E_i has all its elements colored with the same color.

4.24 Let $F=\{E_1,\ldots,E_n\}$ be a family of r-element subsets of a set X, where $n\leq 2^{r-1}$. Show that it is possible to color the elements of X with two colors so that no subset in the family F has its elements colored with the same color.

4.25 Let M be a set with $n\geq 5$ elements and F, a family of pairwise distinct three-element subsets of M. If F contains at least $n+1$ subsets, show that there are at least two distinct subsets which have exactly one element in common.

4.26 Consider two collections (multisets) of integers $\{a_1,\ldots,a_n\}\neq\{b_1,\ldots,b_n\}$ such that an integer can appear several times in each collection. Assume that the collections

$$\{a_i+a_j\,|\,1\leq i<j\leq n\} \quad \text{and} \quad \{b_i+b_j\,|\,1\leq i<j\leq n\}$$

are equal. Show that n is a power of 2.

4.27 Let X be an n-element set $(n\geq 1)$. Suppose that $F=\{A_1,\ldots,A_m\}$ is a family of subsets of X with the property

$$|A_i\cap A_j|=1 \quad \text{for every} \quad 1\leq i,j\leq m \text{ and } i\neq j.$$

Show that $m\leq n$.

4.28 One is given n distinct points in the plane. Show that there exist fewer than $n\sqrt{n}$ pairs of these points which have their distance equal to 1.

4.29 One is given n points in space, no four of which are coplanar. Consider the set of $\binom{n}{3}$ planes determined by each triple of points with the property that no two planes are parallel. Determine the number of lines of intersection of these planes.

4.30 How many triangles can be formed from the n vertices of a convex polygon if no side of a triangle can be a side of the polygon?

4.31 Consider a convex polygon with n vertices. There are $n(n-3)/2$ diagonals of the polygon with the property that no two are parallel and no three are concurrent other than at vertices of the polygon. Show that the number of points of intersection located outside of the polygon is equal to

$$\frac{n(n-3)(n-4)(n-5)}{12}.$$

4.32 Suppose there are n points on a circle with the property that no three of the $n(n-1)/2$ chords they determine are concurrent in the interior of the circle. Show that in this case these chords delimit

$$\binom{n}{4}+\binom{n}{2}+1$$

regions in the interior of the circle.

4.33 A set of simple closed curves is drawn in the plane. They do not intersect themselves, but each two curves intersect in at least two points. Let n_p be the number of points in which exactly p of the curves intersect.

Show that the number of closed regions of the plane which are bounded by arcs of these curves and which do not contain such an arc in their interior is equal to

$$1+n_2+2n_3+\cdots+(p-1)n_p+\cdots.$$

4.34 Let S be a set, and denote by $M(S)=(S_1, S_2, \ldots, S_m)$ an ordered family of its subset. A system of distinct representatives (SDR) for $M(S)$ is an m-tuple $(a_1, a_2, \ldots, a_m)$ such that $a_i \in S_i$ for $1 \leq i \leq m$ and $a_i \neq a_j$ if $i \neq j$ for $1 \leq i, j \leq m$.

Show that $M(S)$ has an SDR if and only if $|S_{i_1} \cup S_{i_2} \cup \cdots \cup S_{i_k}| \geq k$ for all choices of pairwise distinct numbers

$$\{i_1, \ldots, i_k\} \subset \{1, \ldots, m\}, \quad \text{where } 1 \leq k \leq m.$$

4.35 Let X be an n-element set, and F a family of h-element subsets of X. Denote by $M(n, k, h)$ the minimal number of h-element sets in F which has the property that each k-element subset of X contains at least one set of the family F ($n \geq k \geq h \geq 1$). Show that:

(a) $M(n, k, h) \geq \dfrac{n}{n-h} M(n-1, k, h)$;

(b) $M(n, k, h) \leq M(n-1, k-1, h-1) + M(n-1, k, h)$;

(c) $\dbinom{n}{h} \Big/ \dbinom{k}{h} \leq M(n, k, h) \leq \dbinom{n-k+h}{h}$.

4.36 Show that

(a) $M(n, k, 2) = \binom{n}{2} - \frac{k-2}{k-1} \cdot \frac{n^2 - r^2}{2} - \binom{r}{2}$

for every $n \geq k \geq 2$, where $r \equiv n \pmod{k-1}$ and $0 \leq r \leq k-2$;

(b) $M(n, n-h, k) = h+1$ for every $n \geq k(h+1)$ and $k \geq 1$.

4.37 A balanced incomplete block design (BIBD) is a set B formed of v objects (also called varieties) together with a family F of b subsets of B (called blocks), such that:

(1) each block contains exactly k objects;
(2) each object belongs to exactly r blocks;
(3) each pair of distinct objects is contained in exactly λ blocks.

Show that the parameters (b, v, r, k, λ) of a BIBD satisfy the following relations:

$$bk = vr \quad \text{and} \quad r(k-1) = \lambda(v-1).$$

4.38 Let X be a set with v elements, and let $X_1, X_2, \ldots, X_v$ be a family of v subsets of X. This family is called a (v, k, λ)-configuration if it satisfies the following conditions:

(1) $|X_i| = k$ for $i = 1, 2, \ldots, v$;
(2) $|X_i \cap X_j| = \lambda$ for every $i \neq j$;
(3) $0 < \lambda < k < v-1$.

The incidence matrix $A = (a_{ij})_{1 \leq i, j \leq v}$ of this configuration is a square matrix defined as follows: $a_{ij} = 1$ if the element i of X belongs to the set X_j, and $a_{ij} = 0$ otherwise. Show that

$$A^T A = (k - \lambda)I + \lambda J,$$

if and only if A is the incidence matrix of a (v, k, λ)-configuration where A^T is the transpose of A, and J is a square matrix of order v all of whose elements are 1. The matrix I is the identity matrix of order v.

4.39 Show that every (v, k, λ)-configuration is a BIBD with parameters (v, v, k, k, λ).

4.40 Let X be a v-element set, $v \geq 3$. A Steiner triple system of order v is a family of 3-element subsets of X called triples, such that each two element subset of X is contained in a single triple.

Show that a necessary condition for the existence of a Steiner triple system of order v is that $v \equiv 1$ or $3 \pmod{6}$.

4.41 A BIBD is said to be symmetric if $v = b$ (and hence $k = r$). Show that if a symmetric BIBD has parameters (v, v, k, k, λ) where v is even, then $k - \lambda$ is a perfect square.

4.42 Let X be an n-element set, and Y a k-element subset of X. Show that the maximal number of pairwise distinct subsets of X which are noncomparable with respect to inclusion, and which contain exactly r elements of Y is equal to

$$\binom{k}{r}\binom{n-k}{[(n-k)/2]}.$$

4.43 Consider the functions $f: X \to X$ such that $f(f(x)) = a$ for every $x \in X$, where a is a fixed element of X. If $|X| = n \geq 2$, prove that the set of all such functions has cardinality

$$\sum_{p=1}^{n-1} \binom{n-1}{p} p^{n-p-1}.$$

4.44 Consider the r-element subsets of the set $\{1, \ldots, n\}$. Select the minimum element of each subset. Show that the arithmetical mean of the numbers obtained in this way is equal to $(n+1)/(r+1)$.

4.45 Let $M = \{1, 2, \ldots, 2048\}$. Show that for any subset $X \subset M, |X| = 15$, there are two disjoint subsets $A, B \subset X$ such that

$$\sum_{i \in A} i = \sum_{j \in B} j.$$

Does this property hold for 12-element subsets of M?

4.46 Let $x = (x_1, \ldots, x_n)$ and $y = (y_1, \ldots, y_n)$ be two vectors. It is said that x covers y if $x = y$ or $x_i = y_i$ for $n-1$ values of i. Let F denote the set of p^n vectors $(y_1, \ldots, y_n)$ where $1 \leq y_i \leq p$ for $i = 1, \ldots, n$. A set H of vectors $h_1, h_2, \ldots$ is called a covering set if every vector y in F is covered by at least one vector h_i in H. Let $\sigma(n, p)$ be the minimum number of vectors which such a covering set H can contain.

Prove that $\sigma(2, p) = p$ and $\sigma(n, p) \geq p^n / \{n(p-1) + 1\}$.

4.47 Given a set of $n+1$ positive integers, none of which exceeds $2n$, show that at least one member of the set must divide another member of the set.

4.48 Let X be a finite set containing at least four elements, and let $A_1, \ldots, A_{100}$ be subsets of X which are not necessarily distinct, and are such that $|A_i| > \frac{3}{4}|X|$ for any $i = 1, \ldots, 100$. Show that there exists $Y \subset X, |Y| \leq 4$, with $Y \cap A_i \neq \emptyset$ for every $i = 1, \ldots, 100$.

4.49 The digital plane D is the set of all points (digital points) in the Euclidean plane which have integral coordinates. For any two points $P_1(x_1, y_1)$ and $P_2(x_2, y_2)$ from D the city-block distance is defined by

$$d_4(P_1, P_2) = |x_2 - x_1| + |y_2 - y_1|,$$

which yields a metric for D. For any $F \subset D$ a subset $B \subset F$ is said to be a metric basis for F if for any $x, y \in F, x \neq y$, there exists $b \in B$ such that $d_4(x, b) \neq d_4(y, b)$.

Prove that:

(a) D has no finite metric basis;

(b) for any natural number $n \geq 3$, there exists $E_n \subset D$ such that the minimum number of elements in a metric basis for E_n is equal to n.

4.50 A finite projective plane is a symmetric BIBD Δ with parameters (v, v, k, k, λ) where $v \geq 4$ and $\lambda = 1$. It is traditional in this context to substitute the terms point for object and line for block. From Problem 4.37 one can deduce that

$$v = k^2 - k + 1.$$

The number $n = k - 1$ is called the order of Δ. Thus for a finite projective plane of order n it can be seen that

$$b = v = n^2 + n + 1, \qquad r = k = n + 1.$$

It is convenient to say that a set of points is collinear if it is contained in some line.

Show that a set system $\Delta = (V, E)$ where E is a family of subsets of V is a finite projective plane if and only if the following three conditions hold:

(1) Every pair of points is contained in exactly one line.

(2) Every pair of lines intersects in exactly one point.

(3) There exists a 4-subset of V no 3-subset of which is collinear.

5

Partitions of Integers

5.1 In how many ways can three numbers be selected from the set $\{1, 2, \ldots, 3n\}$ so that their sum is divisible by 3?

5.2 Show that the number $P(n, m)$ of partitions of an integer n into m parts satisfies the recurrence relation

$$P(n+k, k) = P(n, 1) + P(n, 2) + \cdots + P(n, k),$$

with $P(n, 1) = P(n, n) = 1$.

5.3 Show that the number of partitions of an integer n into pairwise distinct parts is equal to the number of partitions of n into odd parts.

5.4 Verify that the number of partitions of a positive integer n into m pairwise distinct parts is equal to

$$P\left(n - \binom{m}{2}, m\right).$$

5.5 For a positive integer n consider partitions of n such that every integer between 1 and n can be uniquely represented as a partial sum of the partition. For which numbers n is

$$n = 1 + 1 + \cdots + 1$$

the unique partition with this property?

5.6 Prove Euler's identity:

$$(1-x)(1-x^2)(1-x^3)\cdots(1-x^n)\cdots = 1 + \sum_{n=1}^{\infty} \psi(n) x^n$$

$$= 1 - x - x^2 + x^5 + x^7 - x^{12} - x^{15} + x^{22} + \cdots,$$

where $\psi(n) = (-1)^k$ if $n = (3k^2 \pm k)/2$ and $\psi(n) = 0$ if n cannot be represented in the form $n = (3k^2 \pm k)/2$ (k is an integer).

5.7 Justify the following expressions for generating functions:
  (a) the generating function for the number $P(n)$ of all partitions of an integer n is

Partitions of Integers

$$\sum_{n=0}^{\infty} P(n)x^n = \frac{1}{(1-x)(1-x^2)(1-x^3)\cdots},$$

where $P(0)=1$;

(b) the generating function of the number $P(n, m)$ of partitions of an integer n into m parts is

$$\sum_{n=0}^{\infty} P(n, m)x^n = \frac{x^m}{(1-x)(1-x^2)\cdots(1-x^m)};$$

(c) the generating function for the number of partitions of n into odd parts is

$$\frac{1}{(1-x)(1-x^3)(1-x^5)\cdots};$$

(d) the generating function for the number of partitions of n into pairwise distinct parts is $(1+x)(1+x^2)(1+x^3)\cdots$;

(e) the generating function for the number of partitions of n into pairwise distinct odd parts is $(1+x)(1+x^3)(1+x^5)\cdots$.

5.8 Prove Euler's Pentagonal Theorem:

$$P(n) = P(n-1) + P(n-2) - P(n-5) - P(n-7) + \cdots$$
$$= \sum_{k \geq 1} (-1)^{k-1} \left\{ P\left(n - \frac{3k^2-k}{2}\right) + P\left(n - \frac{3k^2+k}{2}\right) \right\},$$

for every $n \geq 3$.

5.9 Show that the number of partitions of n such that no integer appears more than twice as a part is equal to the number of partitions of n into parts which are not divisible by 3. For example, for $n=6$, these two sets of partitions are, respectively $\{6, 5+1, 4+2, 4+1+1, 3+3, 3+2+1, 2+2+1+1\}$ and $\{5+1, 4+2, 4+1+1, 2+2+2, 2+2+1+1, 2+1+1+1+1, 1+1+1+1+1+1\}$.

5.10 Let $P(n)$ and $Q(n)$ be the number of partitions of n and the number of partitions of n into odd parts, respectively. Show that the following recurrence relations hold:

(a) $Q(n) = \sum_{i \geq 0} (-1)^i Q(i) Q(2n-i)$, where $Q(0)=1$;

(b) $P(n) = \sum_{i \geq 0} P(i) Q(n-2i)$, where $P(0) = Q(0) = 1$.

5.11 Show that $P(n, m) = P(n-m)$ for $m \geq n/2$.

5.12 Consider the number of noncongruent triangles with pairwise distinct integral sides and perimeter equal to $2n$. Show that this number is equal to the number $Q(n, 3)$ of partitions of n into three pairwise distinct parts. Also show that $Q(n, 3) = [(n^2 - 6n + 12)/12]$.

5.13 Justify the identity

$$(1+x)(1+x^3)(1+x^5)\cdots = \sum_{k=0}^{\infty} \frac{x^{k^2}}{(1-x^2)(1-x^4)\cdots(1-x^{2k})}$$

by counting symmetric Ferrers diagrams in two different ways. Use the same method to prove Euler's identity:

$$(1+xy)(1+x^3y)(1+x^5y)\cdots = \sum_{k=0}^{\infty} \frac{x^{k^2}y^k}{(1-x^2)(1-x^4)\cdots(1-x^{2k})}.$$

5.14 Denote by $B(n)$ the number of the partitions of n into parts which are powers of two. For example $B(6)=6$, and the corresponding partitions are the following:

$$1+1+1+1+1+1 = 2+1+1+1+1 = 2+2+2$$
$$= 4+2 = 4+1+1.$$

Prove that:
  (a) $B(2n+1) = B(2n)$;
  (b) $B(2n) = B(2n-1) + B(n)$;
  (c) $B(n)$ is even for any $n \geq 2$.

5.15 Show that $P(n) \geq 2^{[\sqrt{n}]}$ for every $n \geq 2$.

6
Trees

6.1 Let $A=(X, U)$ be a tree and $A_1=(X_1, U_1), \ldots, A_p=(X_p, U_p)$ a set of subtrees of A. If $B=\bigcap_{i=1}^{p} X_i \neq \emptyset$, show that B is the set of vertices of a subtree of A.

6.2 Let $G_1, \ldots, G_k$ be a collection of subtrees of a tree G with the property that each two subtrees have at least one vertex in common. Show that the entire collection has at least one vertex in common.

6.3 Let $d_1, \ldots, d_n$ be integers such that

$$0 < d_1 \leq \cdots \leq d_n.$$

Show that there exists a tree with n vertices of degrees $d_1, \ldots, d_n$, if and only if

$$d_1 + \cdots + d_n = 2n - 2.$$

6.4 Let A_1, A_2 be two spanning trees of a connected graph G. Show that there exists a sequence of trees

$$A_1 = B_1, B_2, \ldots, B_r = A_2$$

such that B_{i+1} is obtained from B_i by suppressing an edge u and adjoining another edge v between two nonadjacent vertices of B_i, for $i=1, \ldots, r-1$.

6.5 For a connected graph G let $d(x, y)$ denote the distance between vertices x and y, that is, the number of edges contained in the shortest walk which joins x and y. Further define the eccentricity of a vertex x by $e(x) = \max_y d(x, y)$. The center of a graph G consists of those vertices x_0 with the property that $e(x_0) = \min_x e(x) = \rho(G)$. [$\rho(G)$ is called the radius of G.] The diameter of a connected graph G, denoted $d(G)$, is defined by $d(G) = \max_x e(x)$.

(a) Show that the center of a tree consists of a single vertex or two adjacent vertices.

(b) If G is a tree show that $e(x)$ is a convex function in the sense that if y, z are adjacent to x, then

$$2e(x) \leq e(y) + e(z).$$

(c) Show that for every connected graph G,

$$d(G) \leq 2\rho(G).$$

6.6 Show that every tree with n vertices and with diameter greater than or equal to $2k-3$ contains at least $n-k$ walks of length equal to k.

6.7 Suppose that G is a tree with vertex set X. For $x \in X$ let

$$s(x) = \sum_{y \in X} d(x, y).$$

(a) Show that the function $s(x)$ is strictly convex in the sense that if y and z are two vertices adjacent to x, then

$$2s(x) < s(y) + s(z).$$

(b) Prove that the function $s(x)$ attains its minimum for a single vertex or two adjacent vertices of the tree G.

6.8 Determine the trees G on n vertices for which $\sum_{x,y} d(x, y)$ is minimal (maximal).

6.9 Let $x_1, \ldots, x_r$ be terminal vertices of a tree A, and set $d_{ij} = d(x_i, x_j)$. Show that:

(a) for every three indices i, j, k the following relations hold:

$$d_{ij} + d_{jk} - d_{ik} \geq 0$$

and

$$d_{ij} + d_{jk} - d_{ik} \equiv 0 \pmod{2}.$$

(b) for every four indices i, j, k, l, two of the numbers $d_{ij} + d_{kl}$, $d_{ik} + d_{jl}$, $d_{il} + d_{jk}$ are equal, and the third is less than or equal to the two equal numbers.

6.10 Let A and B be two trees whose terminal vertices are labeled with numbers from the set $\{1, 2, \ldots, r\}$. If the distances between these terminal vertices are the same for A and B, that is,

$$d_A(i, j) = d_B(i, j)$$

for every $1 \leq i \leq j \leq r$, show that the trees A and B are isomorphic.

6.11 Let $G = (X, U)$ be a tree and $f : X \to X$ a function with the property that if $[x, y] \in U$ then $f(x) = f(y)$ or $[f(x), f(y)] \in U$. Show that f has a fixed point or a fixed edge.

6.12 Let A be a tree with vertex set X such that $|X| = 2n+1$. An automorphism of A is a bijection $f : X \to X$ which preserves the adjacency of vertices, that is, $[x, y]$ is an edge of the tree A if and only if $[f(x), f(y)]$ is an edge of A. Show that f has at least one fixed point.

6.13 Let $A_1 = (X, U_1)$ and $A_2 = (X, U_2)$ be two trees which have the same vertex set A. Suppose that for every vertex $x \in X$ the subgraph obtained from A_1 by suppressing the vertex x and the edges incident with x is isomorphic to the subgraph obtained from A_2 by the same operation. Show that the trees A_1 and A_2 have the same diameter.

6.14 Let G be a tree with vertex set $X = \{x_1, \ldots, x_n\}$, and set

$$D = (d_{ij})_{i,j=1,\ldots,n},$$

where $d_{ij} = d(x_i, x_j)$ is the distance between x_i and x_j in G. Show that

$$\det D = (-1)^{n-1}(n-1)2^{n-2}.$$

6.15 Let A be a tree with vertices $x_1, \ldots, x_n$. Suppress the terminal vertex (of degree 1) which has the smallest index, together with the edge incident with it, and let A_{n-1} be the tree thus obtained and a_1 the index of the vertex adjacent to the suppressed vertex. Repeat this procedure for the tree A_{n-1}, and determine the index a_2 of the vertex adjacent to the terminal vertex of minimal index of A_{n-1} and so forth, until one comes to a tree consisting of two adjacent vertices. One thus obtains a sequence $(a_1, a_2, \ldots, a_{n-2})$ of $n-2$ numbers $1 \leq a_i \leq n$ for $1 \leq i \leq n-2$, associated with the tree A. (It is called the Prüfer code of A.)

Show that:
  (a) the correspondence thus defined is a bijection between the set of trees A with n vertices $x_1, \ldots, x_n$ and the set of n^{n-2} sequences $(a_1, \ldots, a_{n-2})$ which can be formed with numbers from the set $\{1, \ldots, n\}$;
  (b) there are n^{n-2} trees on n vertices $x_1, \ldots, x_n$ (this result is known as Cayley's formula);
  (c) the number of trees whose vertices $x_1, \ldots, x_n$ have degrees $d_1, \ldots, d_n$ which satisfy $d_i \geq 1$ and $d_1 + \cdots + d_n = 2n - 2$ is equal to

$$\binom{n-2}{d_1 - 1, \ldots, d_n - 1}.$$

6.16 Let t_n denote the number of trees with n labeled vertices. Show that

$$t_n = \sum_{k=1}^{n-1} k \binom{n-2}{k-1} t_k t_{n-k},$$

and obtain Cayley's formula from this identity by using Abel's identity.

6.17 Find the number of trees with n labeled vertices and exactly p terminal vertices.

6.18 Consider the ladder graph of Figure 6.1 with $2n$ vertices.
  (a) In how many ways can n of its edges be chosen so that no two have endpoints in common?
  (b) Show that this graph has

$$\frac{1}{2\sqrt{3}} \{(2+\sqrt{3})^n - (2-\sqrt{3})^n\}$$

spanning trees.

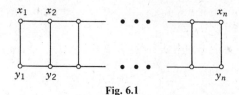

Fig. 6.1

6.19 The distance matrix of a connected graph with p vertices is a square matrix $D=(d_{ij})$ with d_{ij} equal to the distance between vertices i and j. Show that a square matrix D is a distance matrix if and only if it has the following properties:

(1) $d_{ij} \geqslant 0$ for every $1 \leqslant i, j \leqslant p$ and d_{ij} is an integer;
(2) $d_{ij} = 0$ if and only if $i = j$;
(3) D is symmetric;
(4) $d_{ij} \leqslant d_{ik} + d_{kj}$ for every i, j, k and
(5) If $d_{ij} > 1$, there is an index $k \neq i, j$ such that $d_{ij} = d_{ik} + d_{kj}$.

6.20 Prove that the following properties are equivalent for a graph G:

(1) G is a tree;
(2) G is connected, and the deletion of any edge of G results in a graph G_1 which is not connected;
(3) G has no cycle, and if x and y are any two nonadjacent vertices of G, then the graph G_1 obtained from G by inserting the edge $[x, y]$ contains cycles.

6.21 Prove that the number of arborescences having n labeled vertices is equal to n^{n-1}.

6.22 Show that for $n \geqslant 3$ there are n^{n-3} different trees with n unlabeled vertices and $n-1$ edges labeled $1, 2, \ldots, n-1$.

6.23 Let G denote a graph with $n \geqslant 2$ labeled vertices denoted $1, 2, \ldots, n$ and m edges. Label the edges of G with the numbers $1, 2, \ldots, m$, and give each edge an arbitrary direction. The incidence matrix of G is the n-by-m matrix $A = (a_{ij})$, where $1 \leqslant i \leqslant n$ and $1 \leqslant j \leqslant m$, in which a_{ij} equals $+1$ or -1 if the edge j is directed away from or towards the vertex i, and zero otherwise.

Prove that if the graph G has n vertices and is connected, then the rank of its incidence matrix A is equal to $n-1$.

6.24 Show that if B is any nonsingular square submatrix of A, then the determinant of B is $+1$ or -1.

6.25 The reduced incidence matrix A_r of a connected graph G with n vertices is the matrix obtained from the incidence matrix A by deleting some row, say the nth. Prove that a square submatrix B of order $n-1$ of A_r is non-

Trees

singular if and only if the edges corresponding to the columns of B determine a spanning tree of G.

6.26 Prove the Matrix Tree Theorem: If A_r is a reduced incidence matrix of the graph G, then the number of spanning trees of G equals the determinant of $A_r A_r^T$, where A^T denotes the transpose of A.

6.27 Let G be a graph with n vertices, and let $C=(c_{ij})$ for $1 \leq i, j \leq n$ be a matrix defined as follows: c_{ii} is equal to the number of vertices adjacent to i in G; $c_{ij} = -1$ if $i \neq j$ and vertices i and j are adjacent in G; $c_{ij} = 0$ if $i \neq j$ and vertices i and j are not adjacent in G.

Show that $A_r A_r^T$ is the matrix obtained from the matrix C by deleting a row (say the nth) and the column with the same index. Use this property to obtain another proof of Cayley's formula, since the number t_n of trees with n labeled vertices is equal to the number of spanning trees of the complete graph K_n.

6.28 Let R_n be the tree with $2n$ vertices as illustrated in Figure 6.2. Show that the number I_n of independent sets of vertices of this graph is equal to

$$I_n = \frac{3+2\sqrt{3}}{6}(1+\sqrt{3})^n + \frac{3-2\sqrt{3}}{6}(1-\sqrt{3})^n.$$

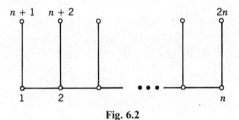

Fig. 6.2

6.29 Let T be a tree having vertex set $\{1,\ldots,n\}$ and edge set denoted by $E(T)$. If $A_1,\ldots, A_n \subset X$, prove that

$$|A_1 \cup \cdots \cup A_n| \leq \sum_{i=1}^{n} |A_i| - \sum_{[i,j] \in E(T)} |A_i \cap A_j|,$$

where the last sum contains $n-1$ terms.

6.30 A matroid M is a pair $(E, \mathscr{I})$, where E is a nonempty finite set, and $\mathscr{I}$ is a nonempty collection of subsets of E, called independent sets, satisfying the following properties:

(1) any subset of an independent set is independent;
(2) if I and J are independent sets containing k and $k+1$ elements respectively, then there is an element e contained in J but not in I, such that $I \cup \{e\}$ is independent.

A base is defined to be any maximal independent set, and the rank function ρ

is an integer-valued function defined on the set of subsets of E by
$$\rho(S) = \max_{I \in \mathscr{I}} |I \cap S|.$$
For a connected graph $G=(X, U)$ with n vertices let E be the set of edges of G and take as independent sets the sets of edges $I \subset U$ which are such that the spanning graph of G with edge set I does not contain a cycle.

Show that $M=(E, \mathscr{I})$ is a matroid [called the circuit matroid of G, and denoted by $M(G)$] whose bases are spanning trees of G. Prove that if $S \subset U$, then its rank $\rho(S) = n - p$, where p is the number of components in the spanning graph (X, S) of G.

7

Parity

7.1 Let U be the set of edges of K_n, the complete graph on n vertices. Let $f: U \to \{-1, 1\}$. An edge u with $f(u) = 1$ will be said to be positive, and one with $f(u) = -1$ will be said to be negative. A triangle (elementary cycle with three vertices) is positive if it contains an even number of negative edges; otherwise it is negative.

If $|f^{-1}(-1)| = p$, that is, there exist p negative edges in U, show that the number $n(f)$ of negative triangles satisfies the relation

$$n(f) \equiv np \pmod{2}.$$

7.2 Let G be a planar graph all of whose faces are triangular, and suppose that the vertices of G are colored with three colors. Show that the number of faces whose vertices are colored with all three colors is even.

7.3 An Eulerian circuit of a digraph G is a circuit which contains every arc of G. Show that a graph G which does not contain isolated vertices has an Eulerian circuit if and only if it is connected and for every vertex x the indegree is equal to the outdegree, that is,

$$d^-(x) = d^+(x).$$

7.4 If the digraph G has at least one vertex x whose outdegree $d^+(x) \geq 3$, then the number of Eulerian circuits of G is even. (Two Eulerian circuits are considered to be identical if they induce the same circular permutation of the arcs of G.)

7.5 If a graph G is such that the degree of each of its vertices is even, then the edges of G can be directed so that in the resulting directed graph each vertex x satisfies

$$d^-(x) = d^+(x).$$

7.6 Show that a graph G has an Eulerian cycle if and only if it is connected and each vertex has even degree. Prove that if G is connected and has $2k$ vertices of odd degree, then it is the union of k walks which are disjoint with respect to edges and which do not contain the same edge twice.

7.7 If $G = (X, U)$ is a directed graph, the complementary graph $\bar{G} = (X, \bar{U})$

is defined as follows: The arc $(x, y) \in \bar{U}$ if and only if $(x, y) \notin U$ for every $x, y \in X$ with $x \neq y$.

Let $h(G)$ denote the number of Hamiltonian paths of the graph G. Show that

$$h(G) \equiv h(\bar{G}) \pmod{2}.$$

This property remains true in the case of a nondirected graph with $n \geq 4$ vertices.

7.8 Show that each tournament contains an odd number of Hamiltonian paths.

7.9 Suppose that the graph G has all of its vertices of odd degree. Show that each edge of G belongs to an even number of Hamiltonian cycles.

7.10 Let $G = (X, U)$ be a connected graph with m edges and n vertices. Show that the number of spanning graphs of G such that every vertex has even degree is equal to 2^{m-n+1}.

7.11 The set X of vertices of any graph can be partitioned into two classes X_1 and X_2 (one of which may be empty) so that the subgraphs with vertex set X_1 (X_2) have all their vertices of even degree. Show that this property remains true if the degrees of the vertices of the subgraph generated by X_1 are even and the degrees of the vertices of the subgraph generated by X_2 are odd.

7.12 Let C be a collection of pairwise distinct subsets of a nonempty finite set X with $n \geq 2$ elements. Show that the only collections C with the property that every proper subset of X intersects an even number of sets from C are $P(X)$ and $P(X) \setminus \{\varnothing\}$, where $P(X)$ is the family of all subsets of X.

8
Connectedness

8.1 Let $d_1 \leq d_2 \leq \cdots \leq d_n$ be the degrees of the vertices of a graph G, and suppose that $d_k \geq k$ for every $k \leq n - d_n - 1$. Show that G is connected.

8.2 Let G be a connected graph with n vertices and $1 \leq k \leq n$. Show that G contains a connected subgraph with k vertices.

8.3 Let G be a graph with n vertices, m edges, and p connected components. Show that

$$p + m \geq n.$$

8.4 Prove that in a connected graph G every two elementary walks of maximal length have at least one vertex in common. If G is a tree, show that all walks of maximal length of G have at least one vertex in common.

8.5 A graph G is said to be bipartite if there exists a partition of its set of vertices,

$$X = A \cup B,$$

such that each edge of the graph has one endpoint in A and the other in B. Show that a graph is bipartite if and only if each elementary cycle in G has an even number of vertices.

8.6 Does there exist a graph with 10 vertices whose vertices have the degree sequence

$$1, 1, 1, 3, 3, 3, 4, 6, 7, 9?$$

8.7 Let $d_1, \ldots, d_n$ be integers such that

$$0 \leq d_1 \leq \cdots \leq d_n.$$

Show that these numbers are the degrees of the vertices of a multigraph with n vertices if and only if

(1) $d_1 + \cdots + d_n$ is even and
(2) $d_n \leq d_1 + \cdots + d_{n-1}.$

8.8 Which numbers can be the number of vertices of a regular graph of degree k?

8.9 Consider a graph G with n vertices which does not contain a complete subgraph with three vertices. Suppose further that for every two nonadjacent vertices x and y there are exactly two vertices which are adjacent to both x and y.

Show that there is an integer $p \geq 0$ such that $n = 1 + \binom{p+1}{2}$. Also show that the graph G is regular of degree p.

8.10 Given natural numbers $r \geq 2$ and $g \geq 3$, show that there exists a graph G which is regular of degree r and with girth $g(G) = g$.

8.11 Let G be a regular graph of degree r with n vertices and $g(G) = g$. Show that

$$n \geq 1 + r + r(r-1) + \cdots + r(r-1)^{(g-3)/2} \quad \text{for } g \text{ odd}$$

and

$$n \geq 2\{1 + (r-1) + \cdots + (r-1)^{g/2-1}\} \quad \text{for } g \text{ even.}$$

8.12 Determine the regular graphs G of degree 3, with minimal number of vertices, such that the smallest length of an elementary cycle is:
   (a) $g(G) = 4$;
   (b) $g(G) = 5$.

8.13 Every connected graph G contains at least one vertex x which has the property that the subgraph G_x obtained from G by suppressing the vertex x, and the edges incident with x, is connected. Does this remain true if instead of connectedness one considers strong connectedness?

8.14 A directed graph G is strongly connected if and only if for every subset A of vertices, $A \neq \varnothing$, there exists at least one arc of G of the form (x, y) where $x \in A$ and $y \notin A$. Show that this statement remains true if instead of the arc (x, y) one takes (y, x) where $y \notin A$ and $x \in A$.

8.15 Show that if a tournament G contains a circuit, then G contains a circuit with three vertices.

8.16 For a tournament with n vertices, $x_1, \ldots, x_n$, let r_i denote the number of arcs which enter x_i, and let s_i denote the number of arcs which leave x_i. Show that:

   (a) $\sum_{i=1}^{n} r_i = \sum_{i=1}^{n} s_i = \binom{n}{2}$;

   (b) $\sum_{i=1}^{n} r_i^2 = \sum_{i=1}^{n} s_i^2$.

8.17 Show that every tournament G contains a vertex x such that every other vertex can be reached from x by a path with at most two arcs.

8.18 Every digraph G contains a set S of pairwise nonadjacent vertices such that every vertex $x \notin S$ can be reached by leaving from a vertex $y \in S$ and traversing a path of length at most equal to 2.

8.19 A tournament T is said to be transitive if, whenever (u, v) and (v, w) are arcs of T, then (u, w) is also an arc of T. Show that an increasing sequence $S: s_1 \leqslant s_2 \leqslant \cdots \leqslant s_n$ of $n \geqslant 1$ non-negative integers is the sequence of outdegrees of a transitive tournament with n vertices if and only if S is the sequence $0, 1, \ldots, n-1$.

8.20 Show that the number $C(n)$ of connected graphs with n labeled vertices satisfies the recurrence relation

$$C(n) = 2^{\binom{n}{2}} - \frac{1}{n} \sum_{k=1}^{n-1} k \binom{n}{k} 2^{\binom{n-k}{2}} C(k)$$

for $n \geqslant 2$ and $C(1) = 1$.

8.21 Show that almost all graphs with n vertices have diameter equal to 2 for $n \to \infty$. This means that if $d_2(n)$ denotes the number of graphs with n vertices and with diameter equal to 2, then

$$\lim_{n \to \infty} \frac{d_2(n)}{2^{\binom{n}{2}}} = 1.$$

8.22 Define a binary relation $\sim$ on the set U of edges of a graph in the following way: Let $u_i \sim u_j$ if $i = j$ or if the edges u_i and u_j are found on the same elementary cycle. Show that $\sim$ is an equivalence relation on U.

8.23 An articulation point of a connected graph G is a vertex x such that the subgraph G_x obtained from G by suppressing the vertex x and the edges incident with x is not connected. A connected graph G which does not contain an articulation point is said to be 2-connected.

Show that the following properties are equivalent for a graph G with $n \geqslant 3$ vertices:

   (1) G is 2-connected;
   (2)   every two vertices of G belong to an elementary cycle;
   (3) G does not have isolated vertices, and each two edges of G lie on some elementary cycle.

8.24 Let G be a 2-connected graph. If G contains two elementary cycles of maximal length, show that these cycles have at least two vertices in common.

8.25 Consider a graph G and two of its vertices x and y. Let $G - x - y$ denote the subgraph of G obtained by suppressing the vertices x and y. Suppose that G_1 and G_2 are two graphs with the same vertex set X and that $|X| \geqslant 4$.

If the graph $G_1 - x - y$ is isomorphic to $G_2 - x - y$ for all choices of vertices $x, y \in X$, show that the graphs G_1 and G_2 are identical.

8.26 Consider a chessboard with n rows and n columns (n odd). Can a knight make a tour of the board by passing once and only once through each of the n^2 squares on the board and returning to the point of departure? (Euler).

8.27 Consider a graph G_n consisting of a zigzag line of n hexagons (as illustrated in Figure 8.1 for $n=5$). This graph has $p=4n+2$ vertices and $q=5n+1$ edges and represents the molecular graph of a cata-condensed benzenoid polycyclic hydrocarbon (a catafusene) with the molecular formula $C_{4n+2}H_{2n+4}$. For $n=1,\ldots,5$ these catafusenes are called benzene, naphthalene, phenanthrene, chrysene and picene.

A perfect matching of G_n is a matching which contains $p/2=2n+1$ edges. Denote by $K(n)$ the number of perfect matchings of G_n (in chemistry this represents the number of Kekulé structures of the catafusene).

Prove that for any $n \geq 1$

$$K(n)=F_{n+1}.$$

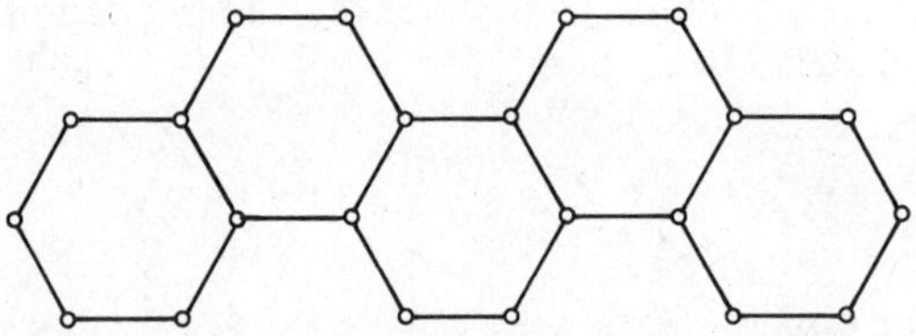

Fig. 8.1

8.28 Let G and G' be connected graphs. A set S of vertices of G is said to be isometrically embeddable in G' if there is a set S' of vertices of G' and a bijection $f:S \to S'$ which preserves distances, that is,

$$d_G(x, y)=d_{G'}(f(x), f(y))$$

for any x, y in S.

Prove that:

(a) if every subset of vertices S of a connected graph G with $|S| \leq 4$ is isometrically embeddable in a tree, then G itself is a tree;

(b) if G is a connected bipartite graph, then any set S of its points, $|S| \leq 3$, is isometrically embeddable in a tree;

(c) if every three points of a connected graph G are isometrically embeddable in a bipartite graph, then G is bipartite.

8.29 The cities $C_1,\ldots,C_N$ are served by n international airlines $A_1,\ldots,A_n$. There is a direct line (without stops) between any two of these localities, and all airlines provide service in both directions. If $N \geq 2^n+1$, prove that at least one of the airlines can offer a round trip with an odd number of landings. Does this property hold for $N=2^n$?

9

Extremal Problems for Graphs and Networks

9.1 In a graph G it will be said that an edge u covers a vertex x if x is one of the endpoints of u. A set of edges forms a matching if no two have a common endpoint. We denote by $v(G)$ the maximum number of edges in a matching, and by $\rho(G)$ the minimum number of edges of G which cover all the vertices of G.
  Show that if G has n nonisolated vertices, then
$$v(G) + \rho(G) = n.$$

9.2 Show that:

(a) If G is regular of degree k and has n vertices, then the number of triangles in G and $\bar{G}$ is equal to
$$\binom{n}{3} - \frac{nk}{2}(n-k-1).$$

(b) If a graph G has n vertices, then G and the complementary graph $\bar{G}$ together contain at least
$$\frac{n(n-1)(n-5)}{24}$$
triangles.

9.3 Show that a graph G with n vertices and m edges contains at least $(4m/3n)(m - n^2/4)$ triangles.

9.4 Show that a tournament with n vertices contains at most $\frac{1}{4}\binom{n+1}{3}$ circuits with three vertices. Prove that this limit is attainable for n odd.

9.5 Show that a tournament with n vertices contains at least one Hamiltonian path and at most $n!/2^{n/2}$ Hamiltonian paths.

9.6 Show that every graph with n vertices and $m > (n/4)(1 + \sqrt{4n-3})$ edges contains at least one elementary cycle with four vertices.

9.7 Show that if a graph with n vertices does not contain a complete subgraph with k vertices ($k \geq 2$) then it contains at least $m = \{n/(k-1)\}$ vertices of degree less than or equal to $p=[(k-2)n/(k-1)]$, where $\{x\}$ is the least integer $\geqslant x$.

9.8 In a set M containing 1001 people, each subset of 11 people contains at least two individuals who know each other. Show that there exist at least 101 people each of whom knows at least 100 persons in the set M.

9.9 Let G be a graph with n vertices and without a complete subgraph with k vertices. Show that the maximum number of edges in G is equal to

$$M(n,k) = \frac{k-2}{k-1} \cdot \frac{n^2 - r^2}{2} + \binom{r}{2}$$

if $n=(k-1)t+r$ and $0 \leqslant r \leqslant k-2$.

The graph G for which this maximum number of edges is attained is unique up to isomorphism. G is made up of $k-1$ classes of vertices. There are r classes which contain $t+1$ vertices; the remaining classes each contain t vertices. Each vertex x is adjacent to all the vertices which do not belong to the class which contains x. This result is called Turán's theorem.

9.10 Suppose that a set M contains $3n$ points in the plane and that the maximal distance between the points is equal to 1. Show that at most $3n^2$ of the distances between the points are larger than

$$\frac{1}{\sqrt{2}}.$$

9.11 Given $2n$ points in the plane with no three collinear, show that the maximum number of line segments which can be constructed with endpoints in this set of points and so that no triangles are formed is equal to n^2.

9.12 Find the maximum number of maximal complete subgraphs (with respect to inclusion) in a graph with n vertices.

9.13 One wants to construct a telephone network connecting points in n cities. Let $c(u)$ be the cost of constructing the line for an edge $u=[x_i, x_j]$ in the complete graph G thus defined. It is desired to minimize the total cost of constructing the network. One must therefore find a spanning tree A of G such that the sum of the costs associated with the edges of A is minimal.

Show that the following algorithm produces a minimal spanning tree of G:

(1) select the edge of G of minimal cost;
(2) among the unchosen edges, select an edge which does not form a cycle with the edges already chosen and which has minimal cost.

Repeat step (2) of the algorithm until a set of edges of cardinality $n-1$ has been chosen.

9.14 Suppose that all the $\binom{n}{2}$ edges of the graph of the preceding problem have different costs. Show that in this case the spanning tree of minimal cost is unique.

9.15 Denote by E the set of vectors $x=(x_1, x_2, \ldots, x_n) \in \mathscr{R}^n$ such that $x_i \geq 0$ for $i=1,\ldots,n$ and $x_1 + \cdots + x_n = 1$. Show that if $G=(X, U)$ is a graph with n vertices then the following equality holds:

$$\max_{x \in E} \sum_{[i,j] \in U} x_i x_j = \frac{1}{2}\left(1 - \frac{1}{k}\right),$$

where k is the maximum number of vertices of a complete subgraph of G.

9.16 Let G be a strongly connected graph. Associate with each arc u in G a non-negative number $c(u) \geq 0$. If a, b are two distinct vertices of G an (a, b)-cut is a set C of arcs with the property that every path from a to b contains at least one arc from C.
Show that

$$\max_D \min_{u \in D} c(u) = \min_C \max_{u \in C} c(u),$$

where D runs through the set of all paths $D=(a,\ldots,b)$ and C includes the set of all (a, b)-cuts of G.

9.17 For a digraph $G=(X, U)$ let a, b be two distinct vertices of G, and c a function $c: U \to \mathscr{R}$ such that $c(u) \geq 0$ for every arc $u \in U$. Suppose that g is a function $g: X \to \mathscr{R}$ which satisfies the following two conditions:

(1) $g(a) = 0$;
(2) $g(y) - g(x) \leq c(x, y)$ for every arc $(x, y) \in U$.

The value of the function c for a path D in the graph G is defined to be the sum of the values of c for all the arcs of G, that is,

$$c(D) = \sum_{u \in A(D)} c(u),$$

where $A(D)$ represents the set of arcs of the path D.
Show that

$$\min_D c(D) = \max_g g(b),$$

where $D=(a,\ldots,b)$ runs through the set of paths from a to b (we assume that there exists at least one path from a to b).

9.18 A digraph $G=(X, U)$ is said to be a network if it satisfies the following conditions:

(1) There is a unique vertex $a \in X$ which no arc enters, i.e., $\omega^-(a) = \varnothing$, where $\omega^-(a)$ denotes the set of arcs which enter the vertex a.
(2) There is a unique vertex $b \in X$ which no arc leaves, that is, $\omega^+(b) = \varnothing$. Here $\omega^+(b)$ denotes the set of arcs which leave the vertex b.
(3) G is connected, and there is a path from a to b in G.
(4) There is a function $c: U \to \mathscr{R}$ such that $c(u) \geq 0$ for each arc $u \in U$.

Vertex a is called the source, vertex b the sink, and $c(u)$ the capacity of the arc u.

A function $f: U \to \mathcal{R}$ such that $f(u) \geq 0$ for each arc u is called a flow in the network G with capacity function c [denoted $G=(X, U, c)$] if the following two conditions are satisfied:

(C) *Condition of conservation of flow:* For every vertex $x \neq a, b$ the sum of the flows of the arcs which enter x is equal to the sum of the flows of the arcs which leave x, that is,
$$\sum_{u \in \omega^-(x)} f(u) = \sum_{u \in \omega^+(x)} f(u) \quad \text{for every} \quad x \in X \setminus \{a, b\}.$$

(B) *Condition of boundedness of flow:* The inequality $f(u) \leq c(u)$ holds for every arc $u \in U$.

For every set of vertices $A \subset X$ define a cut
$$\omega^-(A) = \{(x, y) \mid x \notin A, y \in A, (x, y) \in U\};$$
a cut is thus the set of arcs which enter the set A of vertices. Further let
$$\omega^+(A) = \{(x, y) \mid x \in A, y \notin A, (x, y) \in U\}.$$
$\omega^+(a)$ is thus the set of arcs which leave the set A of vertices. The capacity of the cut $\omega^-(A)$ is defined by
$$c(\omega^-(A)) = \sum_{u \in \omega^-(A)} c(u).$$

Show that:

(a) $$\sum_{u \in \omega^+(a)} f(u) = \sum_{u \in \omega^-(b)} f(u).$$

Henceforth the common value of these two sums will be denoted f_b.

(b) For each set of vertices $A \subset X$ such that $a \notin A$ and $b \in A$, the flow f_b at the exit of the network satisfies
$$f_b = \sum_{u \in \omega^-(A)} f(u) - \sum_{u \in \omega^+(A)} f(u) \leq c(\omega^-(A)).$$

9.19 Prove the Ford–Fulkerson theorem: For every network $G=(X, U, c)$ with source a and sink b, the maximal value of the exit flow is equal to the minimal capacity of a cut, that is,
$$\max_f f_b = \min_{A \mid a \notin A, b \in A} c(\omega^-(A)).$$

9.20 Consider the following algorithm for obtaining a maximal flow at the exit b of a network $G=(X, U, c)$. Assume that the capacity function $c(u) \geq 0$ takes on only integer values:

(1) Define the initial flow as having zero component on each arc of the network, i.e., $f(u)=0$ for each $u \in U$.

(2) Determine the unsaturated walks from a to b on which the flow can be augmented by the following labeling procedure:
   (a) mark the entry a with $[+]$;
   (b) after marking a vertex x, proceed to mark
      (i) with $[+x]$ each unmarked vertex y with the property that the arc $u=(x, y)$ is unsaturated, that is, $f(u)<c(u)$;
      (ii) with $[-x]$ each unmarked vertex y with the property that the arc $u=(y, x)$ has a nonzero flow, that is, $f(u)>0$.

If by this marking procedure the exit b is labeled, then the flow f_b obtained at the current step is not maximal. Now consider a walk v formed of labeled vertices (with the sign $+$ or $-$ respectively) which joins a and b; it can easily be found by following the labels of its vertices from b to a. Denote by v^+ the set of arcs (x, y) where the marking of y has the sign $+$; the arcs are thus directed from a to b. Denote by v^- the set of arcs (x, y) where the marking of y has the sign $-$, these arcs are directed from b to a.

Find the value of

$$\varepsilon = \min \left(\min_{u \in v^+} \{c(u) - f(u)\}, \min_{u \in v^-} f(u) \right).$$

From the method of labeling it follows that $\varepsilon > 0$.

Increase by ε the flow on each arc $u \in v^+$ and decrease by ε the flow on each arc $u \in v^-$. At the exit one therefore obtains a flow equal to $f_b + \varepsilon$. Now repeat step (2) with the new flow.

If by using this labeling procedure the exit b cannot be marked, then prove that the flow obtained has a maximal value f_b at the exit; the set of arcs which join the marked and unmarked vertices constitutes a cut of minimum capacity. Show that this occurs after a finite number of steps.

9.21 The algorithm of the preceding problem does not have a finite number of steps and does not lead to the maximum flow at the exit if the capacity function for arcs $c: U \to \mathcal{R}$ has irrational values. To see this, let

$$a_n = r^n,$$

where $r = (\sqrt{5} - 1)/2$ is an irrational number less than 1 which satisfies the recurrence relation

$$a_{n+2} = a_n - a_{n+1}$$

for every $n \geq 0$. Consider the network with 10 vertices illustrated in Figure 9.1. It contains the arcs $A_1 = (x_1, y_1)$ of capacity a_0, $A_2 = (x_2, y_2)$ of capacity a_1, $A_3 = (x_3, y_3)$ of capacity a_2, and $A_4 = (x_4, y_4)$ of capacity a_2, together with the arcs (y_i, y_j), (x_i, y_j), (y_i, x_j) for $i, j = 1, \ldots, 4$ and $i \neq j$, and the arcs (a, x_i) and (y_i, b) for $i = 1, \ldots, 4$. In order not to overcomplicate the figure, two arcs in the opposite senses (x, y) and (y, x) have been represented by a single nondirected edge $[x, y]$.

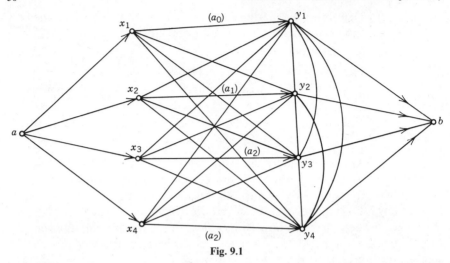

Fig. 9.1

All the arcs of the network other than A_1, A_2, A_3 and A_4 have a capacity equal to

$$c = \sum_{n=0}^{\infty} a_n = \frac{1}{1-r} = \frac{3+\sqrt{5}}{2} > 2.$$

Apply the algorithm of the preceding problem, using the order of the walks $[a, x_1, y_1, b]$, $[a, x_2, y_2, x_3, y_3, b]$, $[a, x_2, y_2, y_1, x_1, y_3, x_3, y_4, b]$, and so on—that is, so that the flow f_b always increases; it will be equal to $a_0 + a_1 + a_2 + a_3 + \cdots$, and strictly less than the maximum flow $\max_f f_b = 4c$.

9.22 For a graph G, denote by $\nu(G)$ the maximum number of edges in a matching (a set of edges which has no endpoints pairwise in common); let $\tau(G)$ be the minimum number of vertices of a support S of G (a set of vertices such that every edge has at least one endpoint in S). If G is bipartite, show that $\nu(G) = \tau(G)$ by applying the Ford–Fulkerson theorem to a network constructed in a suitable way from the graph G.

9.23 Let $A = (a_{ij})_{i=1,\ldots,n,\, j=1,\ldots,m}$ be a binary matrix with n rows and m columns. Show that the maximum number of elements equal to 1 which are found in different rows and columns of the matrix is equal to the minimum number of rows and columns which together contain all elements equal to 1 in the matrix. This result is called König's theorem.

9.24 For a graph G with n vertices define the following two operations:

(α) delete an edge between two adjacent vertices;

(β) insert an edge between two nonadjacent vertices.

Let $\delta_2(G)$ denote the minimum number of operations (α) and/or (β) needed to transform G into the union of two disjoint cliques $K_{n_1} \cup K_{n_2}$, where $n_1 + n_2 = n$ and $n_1, n_2 \geq 0$ (by definition K_0 is the empty graph).

Show that for any graph G with n vertices the following relation holds:
$$\delta_2(G) \leq [\tfrac{1}{4}(n-1)^2],$$
and this inequality becomes an equality if and only if G is isomorphic to the complete bipartite graph $K_{p,q}$ where $p, q \geq 0$ and $p+q=n$.

10

Coloring Problems

10.1 Show that if each vertex of a graph G has degree at most equal to k, then the chromatic number of the graph G satisfies the inequality

$$\chi(G) \leq k+1.$$

10.2 Let $\bar{G}$ be the complement of a graph G with n vertices. Prove that the following inequalities are satisfied by the chromatic number:

$$2\sqrt{n} \leq \chi(G) + \chi(\bar{G}) \leq n+1,$$

$$n \leq \chi(G)\chi(\bar{G}) \leq \left[\left(\frac{n+1}{2}\right)^2\right].$$

10.3 Suppose that a planar graph G has a Hamiltonian cycle. Show that the faces of all representations of G in the plane can be colored with four colors so that each two faces which have a common edge are colored differently.

10.4 Draw an arbitrary number of lines in the plane so that no three of them are concurrent. One can obtain a planar graph G by considering the points of intersection of the lines as vertices of a graph and the segments between neighboring intersections as edges of the graph. Show that

$$\chi(G) \leq 3.$$

10.5 Show that in a connected planar graph with n vertices and m edges there are $m-n+2$ faces (including the infinite face) in every planar representation (Euler's formula).

10.6 Show that every planar graph with n vertices has at most $3n-6$ edges and every planar graph with n vertices which does not contain triangles has at most $2n-4$ edges.

10.7 Show that the graphs K_5 and $K_{3,3}$ are not planar.

10.8 Prove that every planar graph contains a vertex x with degree $d(x) \leq 5$. Construct a planar graph with the property that $d(x) \geq 5$ for every vertex x.

10.9 A planar graph G with n vertices and m edges has the smallest length of its elementary cycles equal to $g(G) = g \geqslant 3$. Show that

$$m \leqslant \frac{g}{g-2}(n-2).$$

10.10 Show that each planar graph G has chromatic number $\chi(G) \leqslant 5$.

10.11 Construct a graph G_1 with chromatic number $\chi(G_1) = 3$, and a graph G_2 with chromatic number $\chi(G_2) = 4$, which do not contain triangles.

10.12 Consider an infinite graph G defined as follows: The set of vertices of G is $\{(a, b) | a, b \in \mathscr{Z} \text{ and } a > 0, b > 0\}$; every vertex (a, b) is adjacent to all vertices $(a+b, 1), (a+b, 2), \ldots, (a+b, n), \ldots$, and thus to all points having positive integer coordinates on the line $x = a + b$.

Show that G does not contain triangles and that its chromatic number $\chi(G) = \infty$.

10.13 If G is a planar graph with $n \geqslant 4$ vertices of degrees $d_1, d_2, \ldots, d_n$, show that

$$\sum_{i=1}^{n} d_i^2 \leqslant 2(n+3)^2 - 62.$$

Verify that for every $n \geqslant 4$ there is a planar graph with all faces triangular such that the inequality becomes an equality.

10.14 Let G be a graph with vertex set X of cardinality n and set of edges U. A λ-coloring of G is a function

$$f : X \to \{1, \ldots, \lambda\}$$

where $\lambda \geqslant 1$ is a natural number such that if $[x, y] \in U$, then $f(x) \neq f(y)$.

Show that the number of λ-colorings of the graph G can be expressed in the form of a polynomial of degree n in λ [called the chromatic polynomial of the graph G and denoted $P_G(\lambda)$] in the following manner:

$$P_G(\lambda) = \sum_{V \subset U} (-1)^{|V|} \lambda^{c(V)},$$

where $c(V)$ represents the number of connected components of the spanning graph (X, V) of G.

10.15 Let G be a graph and $e = [x, y]$ one of its edges. Denote by $G - e$ the graph obtained from G by suppressing the edge e and by $G | e$ the graph obtained from G by suppressing the vertices x and y and the edges incident with these vertices; replace them with a new vertex z which will be adjacent to all vertices of the graph G which were adjacent to either x or y. Show that

$$P_G(\lambda) = P_{G-e}(\lambda) - P_{G|e}(\lambda).$$

10.16 Denote by K_n the complete graph on n vertices, by T_n a tree with n

vertices, and by C_n an elementary cycle with n vertices and n edges. Verify that:
(a) $P_{K_n}(\lambda) = \lambda(\lambda-1)\cdots(\lambda-n+1)$;
(b) $P_{T_n}(\lambda) = \lambda(\lambda-1)^{n-1}$;
(c) $P_{C_n}(\lambda) = (\lambda-1)^n + (-1)^n(\lambda-1)$.

10.17 If G is a graph with n vertices, then its chromatic polynomial has the form
$$P_G(x) = x^n - a_{n-1}x^{n-1} + a_{n-2}x^{n-2} - \cdots + (-1)^{n-1}a_1 x,$$
where $a_i \geq 0$ for every i. If G is connected, then $a_i \geq \binom{n-1}{i-1}$ for $i = 1, \ldots, n-1$.

10.18 Show that for every graph G it is the case that the chromatic polynomial $P_G(\lambda)$ has no roots in the interval $(0, 1)$ and that
$$P_G(\tau+1) \neq 0, \quad \text{where} \quad \tau = \frac{\sqrt{5}+1}{2}.$$

10.19 The chromatic index of a graph G, denoted $q(G)$, is the smallest number of colors with which the edges of G can be colored so that each two edges with common endpoints have different colors. If D denotes the maximum degree of the vertices of the graph G, show that
$$q(G) = D \quad \text{or} \quad q(G) = D+1.$$
This result is Vizing's theorem.

10.20 Show that the chromatic index of the complete graph K_n is given by:
$$q(K_n) = \begin{cases} n & \text{for } n \text{ odd,} \\ n-1 & \text{for } n \text{ even.} \end{cases}$$

10.21 There are n players participating in a chess tournament. Each player must play one match against each of the other $n-1$ players, and none plays more than one match per day. Determine the minimum number of days necessary to run the tournament.

10.22 A k-coloring of the vertices of a graph G is a partition of the set of vertices into k classes, such that each class contains only pairwise nonadjacent vertices. Show that k^{n-k} is the maximum number of k-colorings of the n vertices of a graph G with chromatic number $\chi(G) = k$. The graph which has this maximum number of colorings is formed from the complete graph with k vertices together with $n-k$ isolated vertices.

10.23 Show that the number of k-colorings of the vertices of a tree with n vertices is equal to $S(n-1, k-1)$ for every $n \geq 2$ and $k \geq 2$.

10.24 Let G be a graph with vertex set X and which does not contain a complete subgraph with $k+1$ vertices. Prove that there exists a k-chromatic graph H with the same vertex set such that
$$d_H(x) \geq d_G(x), \quad x \in X.$$

Coloring Problems

Use this result to prove Turán's theorem.

10.25 Let G be a graph with n vertices, m edges, and chromatic number $\chi(G) = k$ ($1 \leqslant k \leqslant n$). Prove that

$$m \leqslant M(n, k+1)$$

and that equality holds if and only if G is isomorphic to the Turán graph with n vertices, k parts, and $M(n, k+1)$ edges.

10.26 Let G be a graph and $P_G(\lambda)$ its chromatic polynomial. G is said to be chromatically unique if $P_H(\lambda) = P_G(\lambda)$ implies that the graph H is isomorphic to G. Prove that Turán's graph $T(n, k)$, on n vertices and with a maximum number $M(n, k)$ of edges with respect to the property that it does not contain any complete subgraph with k vertices, is chromatically unique for every $2 \leqslant k \leqslant n+1$.

10.27 Prove that the number of k-colorings of the vertices of a graph G is given by

$$\frac{1}{k!} \sum_{i=0}^{k} (-1)^i \binom{k}{i} P_G(k-i),$$

where $P_G(\lambda)$ is the chromatic polynomial of G.

10.28 Let $M(x_1, y_1)$ and $N(x_2, y_2)$ be two points in the Euclidean plane E^2. It is known that the following definitions yield metrics for the Euclidean plane:

$$d_4(M, N) = |x_1 - x_2| + |y_1 - y_2| \quad \text{(city-block distance)},$$

$$d_8(M, N) = \max(|x_1 - x_2|, |y_1 - y_2|) \quad \text{(chessboard distance)}.$$

Define the infinite graphs G_4 and G_8 as follows: The vertex set of these graphs is the set of points of E^2, two vertices being adjacent if and only if their city-block or chessboard distance is equal to 1. Prove that the chromatic number of these graphs is equal to 4, that is,

$$\chi(G_4) = \chi(G_8) = 4.$$

10.29 If a graph G contains no clique K_p and $\chi(G) = p \geqslant 3$, show that G has at least $p+2$ vertices. For any $p \geqslant 3$ construct a graph G with $p+2$ vertices and without p-cliques such that $\chi(G) = p$.

11
Hamiltonian Problems

11.1 Show that the complete bipartite graph $K_{n,n}$ contains
$$\tfrac{1}{2}(n-1)!\,n!$$
Hamiltonian cycles.

11.2 Prove that the number of Hamiltonian cycles in the complete graph K_n which use h given edges (which pairwise have no common vertices) is equal to
$$(n-h-1)!\,2^{h-1}$$
for every $0 \leqslant h \leqslant n/2$.

11.3 Show that for n odd, $n \geqslant 3$, the edges of the complete graph K_n can be covered by $(n-1)/2$ Hamiltonian cycles without common edges.

11.4 Let G be a graph with n vertices $x_1, \ldots, x_n$ whose degrees satisfy the inequality
$$d_1 \leqslant d_2 \leqslant \cdots \leqslant d_n.$$
Show that G contains a Hamiltonian cycle if any one of the following three conditions is satisfied:

(a) $d_1 \geqslant n/2$ (Dirac);
(b) $d_p \leqslant p$, $d_q \leqslant q$ implies that $d_p + d_q \geqslant n$ for every $p \neq q$ (Bondy);
(c) $d_k \leqslant k < n/2$ implies that $d_{n-k} \geqslant n-k$ (Chvátal).

11.5 Let G be a graph with $n \geqslant 2$ vertices for which each vertex has degree greater than $n/2$. Show that each two vertices of G can be joined by a Hamiltonian walk.

11.6 If G is a regular graph of degree n with $2n+1$ vertices, show that G has a Hamiltonian cycle.

11.7 Let G be a k-connected graph which does not contain a subset formed from $k+1$ pairwise nonadjacent vertices ($k \geqslant 2$). Show that G has a Hamiltonian cycle.

11.8 Let G be a graph with $n \geqslant 3$ vertices and m edges. If the inequality
$$m \geqslant \binom{n-1}{2} + 2$$

is satisfied, then G contains a Hamiltonian cycle and there is a graph with $m = \binom{n-1}{2}+1$ edges which does not contain a Hamiltonian cycle.

11.9 Let G be a graph with n vertices of degree greater than or equal to k. Show that:
  (a) G contains an elementary cycle of length greater than or equal to $k+1$;
  (b) If G is 2-connected, then it contains either a Hamiltonian cycle or an elementary cycle of length greater than or equal to $2k$.

11.10 Let G be a graph with n vertices and more than $(n-1)k/2$ edges where $k \geq 2$. Show that G contains an elementary cycle of length at least equal to $k+1$.

11.11 Let G be a digraph with n vertices such that the indegree $d^-(x)$ and the outdegree $d^+(x)$ of every vertex x satisfy the inequality

$$d^-(x) \geq \frac{n}{2} \quad \text{and} \quad d^+(x) \geq \frac{n}{2}.$$

Show that G contains a Hamiltonian circuit.

11.12 Show that a tournament is strongly connected if and only if it contains a Hamiltonian circuit.

11.13 Let $P_j(n, k)$ denote the number of ways of choosing k edges from the set of the $n-1$ edges of a walk P of length n, such that these k edges generate exactly j connected components on P. Prove that the following relation holds:

$$P_j(n, k) = \binom{k-1}{j-1}\binom{n-k}{j}.$$

11.14 Denote by $H(n, k)$ [$DH(n, k)$] the number of Hamiltonian walks [Hamiltonian paths] having k edges in common with a given Hamiltonian walk [Hamiltonian path] in the complete graph K_n [complete digraph K_n^*]. Show that

$$H(n, k) = \sum_{i=k}^{n-1} (-1)^{i-k} \binom{i}{k} \frac{(n-i)!}{2} \sum_{j=1}^{i} \binom{i-1}{j-1}\binom{n-i}{j} 2^j,$$

$$DH(n, k) = \sum_{i=k}^{n-1} (-1)^{i-k} \binom{i}{k} \binom{n-1}{i} (n-i)!.$$

11.15 The cube G^3 of a graph G is defined as follows: G^3 has the same vertex set as G; two vertices are adjacent in G^3 if and only if their distance in G is at most 3. Show that if G is connected, then for any two distinct vertices x, y of G there is a Hamiltonian walk in G^3 having x, y as its endpoints.

12

Permutations

12.1 Let r be the smallest positive integer such that $p^r = e$, where e is the identity permutation. Show that r is equal to the least common multiple of the lengths of the cycles of the permutation p.

12.2 Verify that the number of permutations of n elements which have k cycles is $|s(n, k)|$.

12.3 For a permutation $p \in S_m$, let $c(p)$ denote the number of cycles (including the cycles of length 1) in the representation of p as a product of disjoint cycles. Show that the following equalities hold for all positive integers m and n:

$$\frac{1}{m!} \sum_{p \in S_m} n^{c(p)} = \binom{n+m-1}{m},$$

$$\frac{1}{m!} \sum_{p \in S_m} \operatorname{sgn}(p)\, n^{c(p)} = \binom{n}{m},$$

where $\operatorname{sgn}(p)$ is the signature of the permutation p.

12.4 Show that

$$\sum \frac{1}{x_1 x_2 \cdots x_p} = \frac{p!}{n!} |s(n, p)|,$$

where the sum is taken over all integral solutions of the equation

$$x_1 + \cdots + x_p = n$$

such that $x_i \geq 1$ for $i = 1, \ldots, p$.

12.5 Let $d(n, k)$ denote the number of permutations $p \in S_n$ without fixed points and which contain k cycles. Show that:

(a) $d(n+1, k) = n(d(n, k) + d(n-1, k-1))$, where $d(0, 0) = 1$;

(b) $d(2k, k) = 1 \times 3 \times 5 \times \cdots \times (2k-1)$;

(c) $d(n, k) = \sum_{j=0}^{n} (-1)^j \binom{n}{j} c(n-j, k-j)$,

where $c(n, k) = |s(n, k)| = (-1)^{n+k} s(n, k)$ is the number of permutations $p \in S_n$ which contain k cycles.

Permutations

12.6 Let S_n denote the symmetric group of order n, i.e., the group of permutations of the set $\{1, \ldots, n\}$. Two permutations $s, t \in S_n$ are said to be conjugate if there exists a permutation $g \in S_n$ such that $s = gtg^{-1}$. Show that:

(a) Conjugation is an equivalence relation.

(b) Two permutations s and t are conjugate if and only if they have the same number m of cycles and their cycles have, respectively, the same length n_i for $i = 1, \ldots, m$.

(c) Suppose that the permutation t has in its representation as a product of disjoint cycles: λ_1 cycles of length $1, \ldots, \lambda_k$ cycles of length k ($\lambda_1 + 2\lambda_2 + \cdots + k\lambda_k = n$). Show that the number of permutations conjugate to t, or the number of permutations with the same cycle structure as t, is equal to

$$h(\lambda_1, \ldots, \lambda_k) = \frac{n!}{\lambda_1! \, \lambda_2! \cdots \lambda_k! \, 1^{\lambda_1} 2^{\lambda_2} \cdots k^{\lambda_k}}.$$

This is known as Cauchy's formula.

(d) The number of equivalence classes with respect to conjugation of two permutations is equal to the number $P(n)$ of partitions of the integer n.

12.7 Prove Cauchy's identity:

$$\sum_{\substack{c_1 + 2c_2 + \cdots = n \\ c_i \geq 0}} \frac{1}{c_1! \, c_2! \cdots 1^{c_1} \, 2^{c_2} \cdots} = 1.$$

12.8 Choose at random a permutation of the set $\{1, \ldots, n\}$. What is the probability that the cycle which contains the number 1 has length k? (Suppose that all permutations of these n numbers have equal probability. This assumption also holds for the following two problems.)

12.9 What is the probability that a permutation of the set $\{1, \ldots, n\}$ chosen at random contains the numbers 1 and 2 in the same cycle?

12.10 Select a permutation of the set $\{1, \ldots, n\}$ at random. What is its average number of cycles?

12.11 Denote by P_n the number of permutations $p \in S_n$ with the property that $p^2 = e$. Show that:

(a) $P_n = P_{n-1} + (n-1)P_{n-2}$, where $P_0 = P_1 = 1$;

(b) $P_n = n! \sum_{i+2j=n} (i! \, j! \, 2^j)^{-1}$;

(c) $\sum_{n \geq 0} P_n \frac{t^n}{n!} = \exp\left(t + \frac{t^2}{2}\right).$

12.12 Show that the minimum number of transpositions necessary for writing a permutation $p \in S_n$, $p \neq e$, as a product of transpositions is equal to $n - c(p)$, where $c(p)$ is the number of cycles in the permutation p (including the cycles of length 1).

12.13 A set $T=\{t_1,\ldots,t_{n-1}\}$ consists of $n-1$ transpositions of the set $X=\{1,\ldots,n\}$. Associate with it a graph (X, T) with vertices $1,\ldots,n$ whose edges $[i, j]$ are transpositions in the set T.

Show that the product $f = t_1 t_2 \cdots t_{n-1}$ is a circular permutation of the set X if and only if the graph (X, T) is a tree. Deduce from this that the number of ways in which a circular permutation on n elements can be written as a product of $n-1$ transpositions is equal to n^{n-2}.

12.14 Denote by $p(n, k)$ the number of permutations $p \in S_n$ of the set $\{1,\ldots,n\}$ which have $I(p)=k$ inversions [pairs $i<j$ for which $p(i)>p(j)$]. Show that:

(a) $\sum_{p \in S_n} I(p) = \tfrac{1}{2} n! \binom{n}{2}$;

(b) $p(n, k) = p\left(n, \binom{n}{2} - k\right)$;

(c) $p(n, k) = p(n, k-1) + p(n-1, k)$ for $k<n$;

(d) $p(n+1, k) = p(n, k) + p(n, k-1) + \cdots + p(n, k-n)$, where we define $p(n, i) = 0$ for $i > \binom{n}{2}$ or $i<0$ and $p(n, 0)=1$;

(e) $\sum_{0 \leq k \leq \binom{n}{2}} p(n, k) x^k = (1+x)(1+x+x^2) \cdots (1+x+x^2+\cdots+x^{n-1})$.

12.15 Show that the number of permutations p of the set $\{1,\ldots,n\}$ which have the property that there exist k elements j for which $p(j)>p(i)$ for every $i<j$ is equal to $|s(n, k)|$.

12.16 Show that the expression $d(f, g) = \max_{i=1,\ldots,n} |f(i) - g(i)|$, where f and g are two permutations of the set $\{1,\ldots,n\}$, defines a distance on the set S_n. If one denotes by $F(n, r)$ the number of permutations f with the property that $d(e, f) \leq r$, where e is the identity permutation (i.e., $|f(i)-i| \leq r$ for $1 \leq i \leq n$), show that

$$F(n, 1) = F_n, \quad \text{the Fibonacci number.}$$

12.17 Denote by a_n the number of permutations p of the set $\{1,\ldots,n\}$ which satisfy $|p(i)-i| \leq 2$ for every $i=1,\ldots,n$. Show that a_n is the element in the first row and first column of the matrix A^n, where

$$A = \begin{pmatrix} 1 & 1 & 1 & 0 & 0 \\ 1 & 1 & 0 & 0 & 0 \\ 0 & 1 & 0 & 1 & 0 \\ 1 & 0 & 0 & 0 & 1 \\ 1 & 0 & 0 & 0 & 0 \end{pmatrix}.$$

12.18 Find the number $A(n, p)$ of permutations of the set $\{1, 2,\ldots,n\}$ which satisfy the inequality

$$p(k) \leq k + p - 1$$

for $k = 1,\ldots,n$.

Permutations

12.19 An up–down permutation of the set $\{1, 2, \ldots, n\}$ is a permutation

$$\begin{pmatrix} 1 & 2 & \cdots & n \\ a_1 & a_2 & \cdots & a_n \end{pmatrix}$$

with the property

$$a_1 < a_2, \quad a_2 > a_3, \quad a_3 < a_4, \quad a_4 > a_5, \ldots.$$

If A_n denotes the number of up–down permutations of the set $\{1, 2, \ldots, n\}$, show that the exponential generating function of this number is

$$\sum_{n=0}^{\infty} \frac{A_n x^n}{n!} = \sec x + \tan x,$$

where $A_0 = A_1 = 1$.

12.20 A permutation $p(1)p(2) \cdots p(n)$ of the set $\{1, \ldots, n\}$ is said to be 2-ordered if $p(i) < p(i+2)$ for every $1 \leq i \leq n-2$, and 3-ordered if $p(i) < p(i+3)$ for every $1 \leq i \leq n-3$. Show that the number of permutations of the set $\{1, \ldots, n\}$ which are both 2-ordered and 3-ordered is equal to the Fibonacci number F_n for every $n \geq 1$ ($F_0 = F_1 = 1$ and $F_{n+1} = F_n + F_{n-1}$ for $n \geq 1$).

12.21 Let $f(n)$ denote the number of sequences $u_1, u_2, \ldots, u_n$ formed from n numbers in the set $\{1, \ldots, n\}$ and which satisfy the following inequalities:

$$u_i < u_{i+2} \quad \text{for every} \quad i = 1, \ldots, n-2,$$
$$u_i < u_{i+3} \quad \text{for every} \quad i = 1, \ldots, n-3.$$

Determine $f(6)$.

12.22 A permutation $p(1) p(2) \cdots p(n)$ of the set $\{1, \ldots, n\}$ is said to have a fall at $p(i)$ if $p(i) > p(i+1)$, where $1 \leq i \leq n-1$; by definition, every permutation has a fall at $p(n)$. The Eulerian number $A(n, k)$ is defined as the number of permutations of the set $\{1, \ldots, n\}$ having exactly k falls. Show that:

(a) $A(n, k) = kA(n-1, k) + (n-k+1)A(n-1, k-1)$ for $n \geq 2$, and $A(n, n) = A(n, 1) = 1$ for any $n \geq 1$;

(b) $A(n, k) = A(n, n-k+1)$;

(c) $x^n = \sum_{k=1}^{n} A(n, k) \binom{x+k-1}{n}$ for $n \geq 1$;

(d) $A(n, k) = \sum_{j=0}^{k-1} (-1)^j \binom{n+1}{j} (k-j)^n$.

12.23 Show that the number $N(n)$ of permutations $p \in S_n$ such that

$$p^2 = \begin{pmatrix} 1 & 2 & \cdots & n \\ n & n-1 & \cdots & 1 \end{pmatrix}$$

satisfies $N(4m) = N(4m+1) = (2m)!/m!$ for $m \geq 1$, and $N(4m+2) = N(4m+3) = 0$ for $m \geq 0$.

13

The Number of Classes of Configurations Relative to a Group of Permutations

13.1 A ticket-punching machine of the Bucharest Transit System uses nine perforation prongs arranged in a 3-by-3 array inside a rectangle $ABCD$. What is the number of ways in which a ticket can be punched using all possible patterns? The ticket can be put into the slot along edge AB with either of its faces showing.

13.2 If $G \subset S_n$ is a permutation group on the set $X = \{1, \ldots, n\}$ and $x, y \in X$, let $x \sim y(G)$ if there exists a permutation $f \in G$ such that $y = f(x)$. The relation thus defined is an equivalence relation, whose equivalence classes are called orbits of the group G.

If $\lambda_1(g)$ is the number of cycles of length one of the permutation g or the number of elements of X which are invariant under the permutation g, then the number of orbits of a group $G \subset S_n$ is equal to

$$\frac{1}{|G|} \sum_{g \in G} \lambda_1(g).$$

This theorem is due to W. Burnside.

13.3 How many convex polygons with k vertices can be formed from the vertices of a regular polygon with n vertices? Two polygons with k vertices are to be considered distinct if one cannot be obtained from the other by a rotation.

13.4 In how many ways can one color with k colors the vertices of a regular polygon with n vertices? Two colorings are considered distinct if one is not obtained from the other by a rotation.

13.5 Using Burnside's lemma (Problem 3.2) show that:

(a) The number of pairwise nonisomorphic graphs on n vertices is given by the formula

$$g_n = \sum_{(d)} \frac{2^{G_d}}{N_d},$$

where the sum is taken over all non-negative integer solutions (d) of the equation

$$d_1 + 2d_2 + \cdots + nd_n = n, \tag{1}$$

and

$$G_d = \frac{1}{2}\left\{\sum_{k,l=1}^{n} d_k d_l(k,l) - \sum_{k \text{ odd}} d_k\right\},$$

where (k, l) is the greatest common divisor of k and l, and

$$N_d = 1^{d_1} d_1!\, 2^{d_2} d_2! \cdots n^{d_n} d_n!. \tag{2}$$

(b) The number of nonisomorphic digraphs with n vertices is equal to

$$d_n = \sum_{(d)} \frac{2^{D_d}}{N_d}$$

where the sum is taken over all non-negative integer solutions (d) of equation (1). Thus $D_d = \sum_{k,l=1}^{n} d_k d_l(k,l) - \sum_{k=1}^{n} d_k$, where N_d is given by (2).

(c) A digraph is complete if every two distinct vertices x and y are joined by an arc (x, y) or (y, x) or both arcs. The number of nonisomorphic complete digraphs with n vertices is given by

$$c_n = \sum_{(d)} \frac{3^{C_d}}{N_d},$$

where $C_d = \frac{1}{2}\{\sum_{k,l=1}^{n} d_k d_l(k,l) - \sum_{k=1}^{n} d_k - \sum_{k \text{ even}} d_k\}$. The rest of the notation is as above.

(d) The number of nonisomorphic complete antisymmetric digraphs (tournaments) with n vertices is equal to

$$t_n = \sum_{(d)} \frac{2^{T_d}}{N_d},$$

where $T_d = \frac{1}{2}\{\sum_{k,l=1}^{n}(k,l)d_k d_l - \sum_{k=1}^{n} d_k\}$, and the sum is taken over all non-negative integer solutions (d) of the equation

$$d_1 + 3d_3 + 5d_5 + \cdots = n. \tag{3}$$

13.6 Let X be a set of objects denoted $1, \ldots, n$, and A a set of colors, which will be denoted $a_1, \ldots, a_m$. Every function $f: X \to A$ is called a coloring of the objects in X, the object i being colored by $f(i)$ for $i = 1, \ldots, n$. Let G be a group of permutations of the set X. Set $f_1 \sim f_2$ (where f_1, f_2 are two colorings) if there exists a permutation $g \in G$ such that $f_1 g = f_2$.

Show that the binary relation thus defined is an equivalence relation and the number of classes relative to this equivalence is equal to

$$P(G; m, m, \ldots, m),$$

that is, to the numerical value of the cycle index polynomial of G for all variables equal to m. The cycle index polynomial of the group G is defined by the equation

$$P(G; x_1, \ldots, x_n) = \frac{1}{|G|} \sum_{g \in G} x_1^{\lambda_1(g)} x_2^{\lambda_2(g)} \cdots x_n^{\lambda_n(g)},$$

where $\lambda_i(g)$ represents the number of cycles of length i of the permutation $g \in G$ for $1 \leq i \leq n$. (G. Pólya.)

13.7 The number of ways of coloring the six faces of a cube with m colors (two colorings being considered distinct if and only if they cannot be obtained from each other by a rotation of the cube) is equal to

$$\tfrac{1}{24}(m^6 + 3m^4 + 12m^3 + 8m^2).$$

13.8 Determine the cycle index polynomial for the group of rotations of a regular polygon with n vertices. Use this result and Pólya's theorem to obtain another proof of the result of Problem 13.4.

In similar fashion, solve Problem 13.1 by using Pólya's theorem.

13.9 Let f be a permutation of m objects which has order r, and let $G = \{f, f^2, \ldots, f^r = e\}$ denote the cyclic group generated by f. Show that the cycle index polynomial of G is

$$P(G; x_1, x_2, \ldots) = \frac{1}{r} \sum_{i=1}^{r} \prod_{k=1}^{r} x_{k/(k,i)}^{(k,i)\lambda_k(f)}$$

13.10 Find the cycle index polynomial for the automorphism group of the graph illustrated in Figure 13.1.

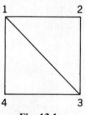

Fig. 13.1

13.11 Prove that the number of nonoriented, pairwise nonisomorphic multigraphs having three vertices and m edges is equal to $[(m+3)^2/12]$ if $m \not\equiv 0 \pmod{6}$ and to $\{(m+3)^2 + 3\}/12$ if $m \equiv 0 \pmod{6}$.

13.12 A Boolean function of n variables is a mapping $f : B^n \to B$ where $B = \{0, 1\}$. A Boolean function is called symmetric if it is invariant under all permutations of its variables, that is, $f(x_1, \ldots, x_n) = f(x_{p(1)}, \ldots, x_{p(n)})$ for any bijection p of the set $\{1, \ldots, n\}$ onto itself. Prove that the number of symmetric Boolean functions of n variables is equal to 2^{n+1}.

14

Problems of Ramsey Type

14.1 If the points of the plane are colored with three colors, show that there will always exist two points of the same color which are 1 unit apart.

14.2 Show that if the points of the plane are colored with two colors, there will always exist an equilateral triangle with all its vertices of the same color. There is, however, a coloring of the points of the plane with two colors for which no equilateral triangle of side 1 has all of its vertices of the same color.

14.3 Show that whenever the points of the plane are colored with two colors, there will always exist an equilateral triangle of side 1 or $\sqrt{3}$ which has all of its vertices of the same color.

14.4 Let T be a 30–60° right triangle with sides 1, $\sqrt{3}$, and 2. Show that for any 2-coloring of the points of the plane there is a triangle congruent to T which has all vertices of the same color.

14.5 Let ABC be an equilateral triangle, and let E be the set of all points contained in the three segments AB, BC, and CA (including A, B and C). Determine whether, in every partition of E into two disjoint subsets, at least one of the two subsets contains the vertices of a right triangle.

14.6 If three distinct integers are chosen, there will always be at least two, say a and b, such that

$$a^3b - ab^3$$

is a multiple of 10.

14.7 Let $a_1, a_2, \ldots, a_{k^2+1}$ be a sequence of numbers. Show that it contains a monotone subsequence with $k+1$ terms.

14.8 Let f be an integer-valued function defined on the set $\{1, \ldots, 2^{n-1}\}$ with the property that for $i=1, \ldots, 2^{n-1}$ one has $1 \leqslant f(i) \leqslant i$. Show that there exists a sequence

$$1 = a_1 < \cdots < a_n \leqslant 2^{n-1}$$

for which $f(a_1) \leq \cdots \leq f(a_n)$. However, this is no longer true if 2^{n-1} is replaced by $2^{n-1}-1$.

14.9 Show that if nine points in the plane are selected so that no three are collinear, then five of the points form the vertices of a convex polygon.

14.10 The graph G is formed from two odd cycles C_n and C_m having vertex sets $A_1, \ldots, A_n$ and $B_1, \ldots, B_m$, respectively. The graph also contains all edges of the form $[A_i, B_j]$ for every $1 \leq i \leq n$ and $1 \leq j \leq m$. Assume that the $mn+m+n$ edges of G are colored red and blue so that no triangle is monochromatic. Show that the $m+n$ edges of the cycles C_m and C_n are either all colored red or all colored blue.

14.11 Show that if the edges of a complete graph with $n_0 = \binom{p+q}{p}$ vertices are colored either red or blue, then there is either a complete subgraph with $p+1$ vertices all of whose edges are red, or a complete subgraph with $q+1$ vertices all of whose edges are blue.

The smallest number $n \leq n_0$ with this property is called the Ramsey number with parameters $p+1$ and $q+1$ and is denoted $R(p+1, q+1)$ for every $p, q \geq 1$ (p, q integers).

14.12 Show that $R(3, 3) = 6$.

14.13 Show that the Ramsey number $R(k, k)$ satisfies the inequalities

$$2^{k/2} \leq R(k, k) \leq 2^{2k-3},$$

for every $k \geq 2$.

14.14 Let $a_1, a_2, \ldots, a_k \geq 1$ be integers and $k \geq 2$. Show that there exists a smallest natural number $n = R_k(a_1, \ldots, a_k)$ called the Ramsey number with parameters $a_1, \ldots, a_k$ with the following property: In any coloring with k colors $c_1, \ldots, c_k$ of the edges of the complete graph K_n, there exists an index i, $1 \leq i \leq k$, and a complete subgraph with a_i vertices which has all of its edges of color c_i.

14.15 If $R_k(3) = R_k(3, \ldots, 3)$ show that

$$R_k(3) \leq [k!e] + 1,$$

where equality holds if $k=2$ and $k=3$.

14.16 Show that $R_k(3) \geq 2^k + 1$.

14.17 Consider an arbitrary partition of the natural numbers $1, 2, \ldots, n$ into k classes. Show that if $n \geq ek!$, then one of the classes will contain three integers x, y, z (not necessarily distinct) such that

$$x + y = z.$$

14.18 Consider a coloring with k colors of all the $2^n - 1$ nonempty subsets of a set with n elements. Show that there exists a natural number $n_0(k)$ such that

Problems of Ramsey Type

for every $n \geq n_0(k)$ there exist two nonvoid disjoint subsets X, Y, such that X, Y, $X \cup Y$ have the same color.

14.19 Denote by K_∞ the complete graph with a countably infinite number of vertices; color its edges with r colors. Show that the graph K_∞ contains a complete infinite monochromatic subgraph.

14.20 Let $(a_n)_{n \in N}$ be an infinite sequence of real numbers. Show that it contains an infinite subsequence which is either strictly increasing, strictly decreasing, or constant.

14.21 Consider an infinite set A of points in space. Show that A contains:
(1) an infinite subset A_1 of collinear points or
(2) an infinite subset A_2 of coplanar points with the property that no three points are collinear or
(3) an infinite subset A_3 of points with the property that no four points are coplanar.

14.22 Show that for every partition of the set of integers $\{1, 2, \ldots, 9\}$ into two classes, at least one of the classes contains an arithmetic progression with three terms.

14.23 Show that in every partition into two classes of the set $M = \{1, 2, \ldots, 256\}$ there is a class containing a geometric progression with three terms.

14.24 Prove or disprove: From the interval $[1, (3^n + 1)/2]$ one can select a set of 2^n integers containing no arithmetic triple (three numbers in arithmetic progression).

14.25 Show that the Ramsey numbers satisfy the inequality

$$R(3, t) \leq \frac{t^2 + 3}{2}$$

for every positive integer $t \geq 2$.

14.26 Show that in any coloring with two colors of the edges of the complete bipartite graph $G = K_{2p+1, 2p+1}$, there exists a monochromatic connected spanning subgraph of G having $2p + 2$ vertices.

14.27 Let $f_r(n)$ be the greatest integer m having the property that any coloring with r colors of the edge set of the complete graph K_n induces a monochromatic connected spanning subgraph with at least m vertices. Prove that:
(a) $f_2(n) = n$;
(b) $f_3(n) = \begin{cases} \left[\frac{n+1}{2}\right] & \text{if } n \not\equiv 2 \pmod 4, \\ \frac{n}{2} + 1 & \text{if } n \equiv 2 \pmod 4. \end{cases}$

14.28 The Ramsey number $R(F_1, F_2)$ for two graphs F_1, F_2 is the minimum p such that every 2-coloring of the edges of K_p contains a green F_1 or a red F_2. Prove that the Ramsey numbers for stars are given by the formula

$$R(K_{1,m}, K_{1,n}) = \begin{cases} m+n & \text{if } m \text{ or } n \text{ is odd,} \\ m+n-1 & \text{if } m \text{ and } n \text{ are both even.} \end{cases}$$

14.29 Let T_m be a tree with m vertices where $m-1$ divides $n-1$. Show that $R(T_m, K_{1,n}) = m+n-1$.

14.30 Show that the Ramsey number $R(K_m, K_{1,n})$ is given by the formula

$$R(K_m, K_{1,n}) = (m-1)n + 1$$

for any $m, n \geq 1$.

14.31 If $m \geq 3$ is a fixed natural number, find the smallest natural number $r(m)$ with the property that every partition into two classes of the set $\{1, 2, \ldots, r(m)\}$ contains a class with m numbers (not necessarily distinct) $x_1, \ldots, x_m$ such that

$$x_1 + \cdots + x_{m-1} = x_m.$$

Part II
SOLUTIONS

Solutions

CHAPTER 1

1.1 (a) One must show that

$$\sum_{k=0}^{[n/2]}\left\{\binom{n}{k}^2+\binom{n}{k-1}^2\right\}-2\sum_{k=0}^{[n/2]}\binom{n}{k}\binom{n}{k-1}=\frac{1}{n+1}\binom{2n}{n}.$$

Let

$$A_n=\sum_{k=0}^{[n/2]}\left\{\binom{n}{k}^2+\binom{n}{k-1}^2\right\} \text{ and } B_n=2\sum_{k=0}^{[n/2]}\binom{n}{k}\binom{n}{k-1}.$$

Starting from the identity $(x+1)^n(x+1)^n=(x+1)^{2n}$ and taking note of standard properties of binomial coefficients, it can be seen that for even n, A_n is the coefficient of x^n in the expansion of $(x+1)^{2n}$, and B_n is the coefficient of x^{n-1} in this expansion. In fact, for even n one can write

$$A_n=\sum_{k=0}^{n}\binom{n}{k}^2=\binom{2n}{n},$$

$$B_n=\sum_{k=1}^{n}\binom{n}{n-k}\binom{n}{k-1}=\binom{2n}{n-1}.$$

Thus in this case

$$A_n-B_n=\binom{2n}{n}\left(1-\frac{n}{n+1}\right)=\frac{1}{n+1}\binom{2n}{n},$$

and the identity is established. For n odd, A_n and B_n are different from these binomial coefficients, and thus

$$A_n=\binom{2n}{n}-\binom{n}{\frac{n+1}{2}}^2 \text{ and } B_n=\binom{2n}{n-1}-\binom{n}{\frac{n-1}{2}}^2.$$

Since

$$\binom{n}{\frac{n+1}{2}}=\binom{n}{\frac{n-1}{2}},$$

71

it follows that the value of $A_n - B_n$ is the same, that is, $\{1/(n+1)\}\binom{2n}{n}$.

(b) Let
$$f(n) = \sum_{k=0}^{n} \binom{n+k}{k} \frac{1}{2^k}.$$

It follows that $f(1) = 2$ and

$$f(n+1) = \sum_{k=0}^{n+1} \binom{n+1+k}{k} \frac{1}{2^k} = \sum_{k=0}^{n+1} \binom{n+k}{k} \frac{1}{2^k} + \sum_{k=0}^{n+1} \binom{n+k}{k-1} \frac{1}{2^k}$$

$$= f(n) + \binom{2n+1}{n+1} \frac{1}{2^{n+1}}$$

$$+ \frac{1}{2} \sum_{k=1}^{n+2} \binom{n+1+k-1}{k-1} \frac{1}{2^{k-1}} - \binom{2n+2}{n+1} \frac{1}{2^{n+2}}$$

$$= f(n) + \tfrac{1}{2} f(n+1).$$

Thus $f(n+1) = 2f(n)$ for every $n \geq 1$, which implies that $f(n) = 2^n$. [D. Beverage, H-283, *Fibonacci Quarterly*, **2** (1978), 16.]

1.2 (a) Recall that the number of strictly increasing words $c_1 c_2 \cdots c_k$ with $c_1 < c_2 < \cdots < c_k$ and $c_i \in \{1, \ldots, n\}$ for every $1 \leq i \leq k$, is equal to $\binom{n}{k}$. It follows that $c_{k-2} \in \{k-2, k-1, \ldots, n-2\}$. Let A_q be the number of strictly increasing words of this form with $c_{k-2} = q$. One can then write

$$\left| \bigcup_{q=k-2}^{n-2} A_q \right| = \sum_{q=k-2}^{n-2} |A_q| = \binom{n}{k}.$$

Since $c_1 c_2 \cdots c_{k-3}$ and $c_{k-1} c_k$ are strictly increasing words with letters from the sets $\{1, \ldots, q-1\}$ and $\{q+1, \ldots, n\}$, respectively, one can conclude that

$$|A_q| = \binom{q-1}{k-3}\binom{n-q}{2}.$$

By setting $p = q - 1$ it is seen that $k - 3 \leq p \leq n - 3$, and (a) follows from this observation.

(b) The desired sum can be written

$$S_n = \sum_{p \geq 0} (-1)^p \binom{2n-p}{p} = \binom{2n}{0} - \binom{2n-1}{1} + \binom{2n-2}{2} - \cdots$$

$$= \binom{2n+1}{0} - \binom{2n}{1} + \binom{2n-1}{0} + \binom{2n-1}{2} - \binom{2n-2}{1} + \cdots$$

$$= \binom{2n+1}{0} - \binom{2n}{1} + \binom{2n-1}{2} - \cdots$$

$$+ \binom{2n-1}{0} - \binom{2n-2}{1} + \binom{2n-3}{2} - \cdots$$

$$= \binom{2n+2}{0} - \binom{2n+1}{1} + \binom{2n}{0} + \binom{2n}{2} - \binom{2n-1}{1} + \cdots$$
$$+ \binom{2n}{0} - \binom{2n-1}{1} + \binom{2n-2}{0} + \cdots$$
$$= S_{n+1} + 2S_n + S_{n-1}.$$

The numbers S_n are therefore seen to satisfy the recurrence relation
$$S_{n+1} = -(S_n + S_{n-1}).$$
Since $S_1 = 0$, $S_2 = -1$, $S_3 = 1$, the proof of the proposition follows by induction on n, and use of the recurrence relation last referred to above.

(c) Differentiate (with respect to x) both sides of the identity
$$(1+x)^n = 1 + x\binom{n}{1} + x^2\binom{n}{2} + \cdots.$$

The result is:
$$n(1+x)^{n-1} = \binom{n}{1} + 2\binom{n}{2}x + 3\binom{n}{3}x^2 + \cdots.$$

Multiplying term by term, one obtains
$$n(1+x)^{2n-1} = \sum_{i=0}^{n}\binom{n}{i}x^i \sum_{j=1}^{n} j\binom{n}{j} x^{j-1}.$$

The proof of (c) is completed by equating the coefficients of x^{n-1} on both sides of this equation and using standard properties of binomial coefficients.

1.3 One can write
$$1 + \sum_{k=0}^{r-1}\binom{r}{k} S_k(n) = 1 + \sum_{k=0}^{r-1}\binom{r}{k}\sum_{p=1}^{n} p^k = 1 + \sum_{p=1}^{n}\sum_{k=0}^{r-1}\binom{r}{k} p^k$$
$$= 1 + \sum_{p=1}^{n}((p+1)^r - p^r) = (n+1)^r.$$

1.4 Expand the expression $(\sqrt{m} + \sqrt{m-1})^n$ by using Newton's binomial formula. If n is even, grouping terms yields
$$(\sqrt{m} + \sqrt{m-1})^n = A_n + B_n\sqrt{m(m-1)}. \tag{1}$$
where A_n, B_n are natural numbers. In the same way, one sees that
$$(\sqrt{m} - \sqrt{m-1})^n = A_n - B_n\sqrt{m(m-1)}, \tag{2}$$
and term-by-term multiplication then gives
$$1 = A_n^2 - m(m-1)B_n^2. \tag{3}$$
Thus be setting $p = A_n^2$ and recalling (3), it can be deduced from (1) that
$$(\sqrt{m} + \sqrt{m-1})^n = \sqrt{p} + \sqrt{p-1}.$$

For n odd one has

$$(\sqrt{m}+\sqrt{m-1})^n = C_n\sqrt{m}+D_n\sqrt{m-1}, \tag{4}$$

and thus C_n and D_n are related by

$$1 = mC_n^2 - (m-1)D_n^2. \tag{5}$$

The desired equality is obtained by setting $p = mC_n^2$.

1.5 (a) To justify the first identity consider two disjoint sets U and V such that $|U|=p$, $|V|=q$, $U \cap V = \varnothing$. The number of k-element subsets of U is equal to $\binom{p}{k}$, and the number of $(m-k)$-element subsets of V is equal to $\binom{q}{m-k}$. Let $A \subset W = U \cup V$ be a subset such that $|A|=m$. Set $|A \cap U|=k$; it follows that $|A \cap V|=m-k$, where $0 \leq k \leq m$. All the m-element subsets A of W are obtained without repetition by carrying out the following procedure for $k=0, 1, \ldots, m$. Take the union of an arbitrary k-element subset of U and an arbitrary $(m-k)$-element subset of V. One finally obtains $\binom{p}{k}\binom{q}{m-k}$ m-element subsets of W. Thus $\sum_{k=0}^{m} \binom{p}{k}\binom{q}{m-k}$ represents the number of m-element subsets of W, that is, $\binom{p+q}{m}$, and this establishes the validity of the first formula.

Another solution starts by expanding the identity

$$(1+x)^p(1+x)^q = (1+x)^{p+q}.$$

Expand by Newton's binomial formula. Observe that identity (a) follows from the identification of the coefficients of x^m in both sides of this polynomial equation.

(b) One can write

$$\binom{k}{m}\binom{n}{k} = \frac{k!}{m!(k-m)!} \cdot \frac{n!}{k!(n-k)!} = \frac{n!}{m!} \cdot \frac{1}{(k-m)!(n-k)!}$$

$$= \frac{n!}{m!(n-m)!} \cdot \frac{(n-m)!}{(k-m)!(n-k)!} = \binom{n}{m}\binom{n-m}{n-k}.$$

Thus the desired sum can be written

$$\sum_{k=m}^{n} \binom{k}{m}\binom{n}{k} = \sum_{k=m}^{n} \binom{n}{m}\binom{n-m}{n-k} = \binom{n}{m} \sum_{k=m}^{n} \binom{n-m}{n-k} = \binom{n}{m} 2^{n-m}.$$

(c) The identity is established by induction on m. For $m=0$ both sides are equal to 1. Assuming that (c) is true for $m=0, 1, \ldots, p-1$, with $p \geq 1$, it follows that

$$\sum_{k=0}^{p} (-1)^k \binom{n}{k} = \sum_{k=0}^{p-1} (-1)^k \binom{n}{k} + (-1)^p \binom{n}{p}$$

$$= (-1)^{p-1} \binom{n-1}{p-1} + (-1)^p \binom{n}{p} = (-1)^p \binom{n-1}{p}.$$

It has thus been shown that (c) is true for $m=p$ and hence it is valid for every $m \geq 0$.

Solutions

(d) It will be shown that both sides of the equation represent an enumeration of the same quantity.

The left side can be interpreted in the following way: Choose k elements from a set M with $|M|=m$ elements in $\binom{m}{k}$ distinct ways. Consider another set N (with $|N|=n$ elements) disjoint from M. Choose m elements from the union of the set N and the k elements previously chosen. One can do this in $\binom{n+k}{m}$ distinct ways. It follows that the number of ordered pairs of sets (X, Y) where $X \subset M$, $Y \subset N \cup X$ with $|X|=k$, $|Y|=m$ is equal to $\binom{m}{k}\binom{n+k}{m}$. If the result is extended to k which varies from 0 to m, the number of all pairs (X, Y) will be equal to

$$\sum_{k=0}^{m} \binom{m}{k}\binom{n+k}{m}.$$

This problem can also be solved by first choosing the sets $Y \cap M$ and $Y \cap N$. If $|Y \cap N|=j$, then $0 \leqslant j \leqslant m$ and it follows that $|Y \cap M|=m-j$. Thus we can choose Y so that $|Y \cap N|=j$ and $|Y \cap M|=m-j$ in $\binom{n}{j}\binom{m}{m-j}$ different ways. It remains to choose the set X so as to satisfy

$$Y \cap M \subset X \subset M,$$

since $Y \subset N \cup X$ and thus $Y \cap M \subset (N \cup X) \cap M = (N \cap M) \cup (X \cap M) = \varnothing \cup X = X$. But the set X can be chosen in 2^j ways, because for each of the j elements of the set $M \setminus (Y \cap M)$ there exist exactly two possibilities: it belongs or does not belong to the set X. It follows that the number of pairs (X, Y) with the stated property is equal to

$$\sum_{j=0}^{m} \binom{n}{j}\binom{m}{j} 2^j,$$

and this observation completes the proof of (d).

(e) One can write identity (d) as an equality between two polynomials in n with rational coefficients. But (d) is valid for every natural number n, and hence for a number of values greater than the degree m of either of the polynomials. Thus the two polynomials in n are identical. It follows that equality results if one replaces n by $-n-1$ on both sides of (d). The left-hand side becomes

$$\sum_{k=0}^{m} \binom{m}{k}\binom{-n-1+k}{m} = \sum_{k=0}^{m} \binom{m}{k}(-1)^m \binom{n+m-k}{m}$$

$$= (-1)^m \sum_{k=0}^{m} \binom{m}{m-k}\binom{n+m-k}{m}$$

$$= (-1)^m \sum_{k=0}^{m} \binom{m}{k}\binom{n+k}{m}.$$

The right-hand side is equal to

$$\sum_{k=0}^{m} \binom{m}{k}\binom{-n-1}{k} 2^k = \sum_{k=0}^{m} \binom{m}{k}(-1)^k \binom{n+k}{k} 2^k = \sum_{k=0}^{m} \binom{m}{k}\binom{n+k}{k}(-2)^k,$$

and this completes the proof of (e).

The notation $\binom{p}{q} = p(p-1)\cdots(p-q+1)/q!$ will also be used for negative values of p.

(f) By using (a) it follows that

$$\sum_{k \geq 0} \binom{p}{k}\binom{q}{k}\binom{n+k}{p+q} = \sum_{k \geq 0} \binom{p}{k}\binom{q}{k} \sum_{j=0}^{k} \binom{k}{j}\binom{n}{p+q-j}$$

$$= \sum_{j \geq 0} \binom{n}{p+q-j} \sum_{k \geq 0} \binom{p}{k}\binom{q}{k}\binom{k}{j}.$$

A further use of (a) yields

$$\sum_{k \geq 0} \binom{p}{k}\binom{q}{k}\binom{k}{j} = \sum_{k \geq 0} \binom{p}{k} \frac{q!}{(q-k)!j!(k-j)!}$$

$$= \sum_{k \geq 0} \binom{p}{k}\binom{q}{j}\binom{q-j}{q-k} = \binom{q}{j}\binom{p+q-j}{q}.$$

Thus the left-hand side of the desired identity becomes

$$\sum_{j \geq 0} \binom{n}{p+q-j}\binom{q}{j}\binom{p+q-j}{q} = \sum_{j \geq 0} \frac{n!}{(n-p-q+j)!(q-j)!(p-j)!j!}$$

$$= \sum_{j \geq 0} \binom{n}{p}\binom{p}{j}\binom{n-p}{q-j}$$

$$= \binom{n}{p} \sum_{j \geq 0} \binom{p}{j}\binom{n-p}{q-j} = \binom{n}{p}\binom{n}{q},$$

again by the use of (a).

(g) Identity (f) is an equality between two polynomials in n, of degree $p+q$. This equality holds for every natural number n, and hence the two polynomials in n are identical. It follows that the two sides are equal if n is replaced by $-1-n$. Since

$$\binom{-1-n+k}{p+q} = (-1)^{p+q} \binom{n+p+q-k}{p+q},$$

$$\binom{-1-n}{p} = (-1)^{p} \binom{n+p}{p}$$

and

$$\binom{-1-n}{q} = (-1)^{q} \binom{n+q}{q},$$

one can deduce (g) from (f). Finally, (h) follows from (g) by setting $q=p$.

1.6 Let $S(l, m, n)$ be the sum of the left-hand side. By replacing

$$\binom{2n+2}{n+k+1} \quad \text{with} \quad \binom{2n+1}{n+k+1} + \binom{2n+1}{n+k},$$

Solutions

one finds that

$$S(l, m, n+1) = 2 \sum_k (-1)^k \binom{2l}{l+k}\binom{2m}{m+k}\binom{2n+1}{n+k+1} \quad (1)$$

On the other hand, by using (1) it follows that

$$(l+n+1)(2m+1)S(l, m, n+1)$$

$$= 2 \sum_k (-1)^k \binom{2l}{l+k}\binom{2m+1}{m+k+1}\binom{2n+1}{n+k+1}$$

$$\times (m+k+1)(l-k+n+k+1)$$

$$= 2 \sum_k (-1)^k \binom{2l-1}{l-k-1}\binom{2m+1}{m+k+1}\binom{2n+1}{n+k+1}(m+k+1)(2l)$$

$$+ 2 \sum_k (-1)^k \binom{2l}{l+k}\binom{2m+1}{m+k+1}\binom{2n+1}{n+k+1}(m+k+1)(n+k+1).$$

For $l > 0$, one has

$$\sum_k (-1)^k \binom{2l-1}{l-k-1}\binom{2m+1}{m+k+1}\binom{2n+1}{n+k+1} = 0,$$

by replacing k with $-k-1$ and observing that the terms of the sum can be grouped in pairs with zero sum, and by using the standard formulas. In this case the last representation of $(l+n+1)(2m+1)S(l, m, n+1)$ is symmetric with respect to m and n, which implies that

$$(l+n+1)(2m+1)S(l, m, n+1) = (l+m+1)(2n+1)S(l, m+1, n). \quad (2)$$

The proof is now completed by induction on n. For $n=0$ both sides have the value $\binom{2l}{l}\binom{2m}{m}$. Assume the formula is true for all triples (l, m, q) with $0 \leq q \leq n$. It then follows that

$$S(l, m, n+1) = \frac{(l+m+1)(2n+1)}{(l+n+1)(2m+1)} \cdot S(l, m+1, n)$$

$$= \frac{(l+m+n+1)!(2l)!(2m)!(2n+2)!}{(l+m)!(m+n+1)!(n+l+1)!l!m!(n+1)!}$$

[P. A. MacMahon, *Quart. J. Pure and Appl. Math.*, **33** (1902), 274–288; J. Dougall, *Proc. Edinburgh Math. Soc.*, **25** (1906), 114–132]. The case $l=m=n$ is due to A. C. Dixon (*Messenger of Math.*, **20** (1891), 79–80).

1.7 Consider the expansion

$$(ax+b)^p(1+x)^q = \left(\sum_{i=0}^{p} \binom{p}{i}(ax)^{p-i}b^i\right)\left(\sum_{j=0}^{q} \binom{q}{j}x^j\right).$$

The coefficient of x^p in the right-hand side is equal to

$$\sum_{i=0}^{p} \binom{p}{i}\binom{q}{i} a^{p-i} b^i.$$

But
$$(ax+b)^p(1+x)^q = \{x(a-b)+b(1+x)\}^p(1+x)^q$$
$$= \sum_{i=0}^{p} \binom{p}{i} (a-b)^{p-i} b^i x^{p-i} (1+x)^{q+i},$$

and hence the coefficient of x^p is also equal to
$$\sum_{i=0}^{p} \binom{p}{i}\binom{q+i}{i} (a-b)^{p-i} b^i,$$
which yields the desired equality.

1.8 First prove the identity
$$\sum_{i=0}^{2n} \binom{2n}{i}^3 x^i = (1+x)^{2n} + \sum_{i=1}^{n} \binom{2i}{i}\binom{2n}{2i}\binom{2n+i}{i} (1+x)^{2n-2i} x^i,$$
and thereby (for $x=-1$) obtain Dixon's formula,
$$\sum_{i=0}^{2n} (-1)^i \binom{2n}{i}^3 = (-1)^n \binom{2n}{n}\binom{3n}{n} = (-1)^n \frac{(3n)!}{(n!)^3}.$$

Equating the coefficients of x^p in the identity which is to be proved, one finds that for $1 \leq p \leq 2n$
$$\binom{2n}{p}^3 = \binom{2n}{p} + \sum_{i=1}^{n} \binom{2i}{i}\binom{2n}{2i}\binom{2n+i}{i}\binom{2n-2i}{p-i}. \tag{1}$$

But
$$\binom{2i}{i}\binom{2n}{2i}\binom{2n-2i}{p-i} = \binom{p}{i}\binom{2n-p}{i}\binom{2n}{p},$$
and hence identity (1) reduces to
$$\binom{2n}{p}^2 = \sum_{i=0}^{n} \binom{p}{i}\binom{2n-p}{i}\binom{2n+i}{i} \tag{2}$$
for every $p=1,\ldots,2n$. In order to prove (2) use the identity of the preceding problem with respect to the following values:

(I) $a=0, b=-1$, and $q=2n$,
(II) $a=x, b=1+x$, and $q=2n-p$.

In case I it can be seen that
$$(-1)^p \binom{2n}{p} = \sum_{i=0}^{p} (-1)^i \binom{p}{i}\binom{2n+i}{i}, \tag{3}$$
while in case II it follows that
$$\sum_{i=0}^{p} \binom{p}{i}\binom{2n-p}{i} x^{p-i}(1+x)^i = \sum_{i=0}^{p} (-1)^{p-i} \binom{p}{i}\binom{2n-p+i}{i} (1+x)^i. \tag{4}$$

Solutions

Multiplying both sides of (4) by $(1+x)^{2n}$ and equating coefficients of x^p, one obtains

$$\sum_{i=0}^{p}\binom{p}{i}\binom{2n-p}{i}\binom{2n+i}{i}=(-1)^p\sum_{i=0}^{p}(-1)^i\binom{p}{i}\binom{2n-p+i}{i}\binom{2n+i}{p}.$$

But

$$\binom{2n-p+i}{i}\binom{2n+i}{p}=\binom{2n+i}{i}\binom{2n}{p},$$

and thus one can write

$$\sum_{i=0}^{p}\binom{p}{i}\binom{2n-p}{i}\binom{2n+i}{i}=(-1)^p\binom{2n}{p}\sum_{i=0}^{p}(-1)^i\binom{p}{i}\binom{2n+i}{i}.$$

By using (3) it is seen that the right-hand side is $\binom{2n}{p}^2$, which establishes (2) and hence Dixon's formula. [See W. Ljunggren, *Nordisk Mat. Tidskr.*, **29** (1947), 35–38.]

Another proof is based on Problem 1.6. By taking $l=m=n$ it can be seen that

$$\sum_{k}(-1)^k\binom{2n}{n+k}^3=\frac{(3n)!}{(n!)^3}.$$

But

$$\sum_{k}(-1)^k\binom{2n}{n+k}^3=(-1)^{-n}\sum_{k=0}^{2n}(-1)^k\binom{2n}{k}^3,$$

from which Dixon's formula follows.

1.9 (a) First note that

$$(1+x+x^2)^n=\{(x+1)+x^2\}^n$$

$$=(x+1)^n+\binom{n}{1}(x+1)^{n+1}x^2$$

$$+\binom{n}{2}(x+1)^{n-2}x^4+\cdots+x^{2n}.$$

This implies that the coefficient of x^n is equal to

$$a_n=1+\binom{n}{1}\binom{n-1}{1}+\binom{n}{2}\binom{n-2}{2}+\cdots+\binom{n}{k}\binom{n-k}{k}+\cdots$$

$$=1+\frac{n(n-1)}{(1!)^2}+\frac{n(n-1)(n-2)(n-3)}{(2!)^2}+\cdots$$

$$+\frac{n(n-1)\cdots(n-2k+1)}{(k!)^2}+\cdots.$$

The exact form of the last term depends on the parity of n. This coefficient can also be written in the form

$$a_n = \sum_{k \geq 0} \frac{(2k)!}{(k!)^2} \binom{n}{2k}.$$

(b) First show that $a_k = a_{2n-k}$. Then set $x = 1/y$ and multiply both sides of the expansion by y^{2n}. It follows that

$$(y^2 + y + 1)^n = a_0 y^{2n} + a_1 y^{2n-1} + \cdots + a_{2n}.$$

Comparison with the first expansion yields $a_k = a_{2n-k}$ for $k = 0, \ldots, 2n$. Now substitute $-x$ for x and obtain

$$(1 - x + x^2)^n = a_0 - a_1 x + a_2 x^2 - \cdots + a_{2n} x^{2n}.$$

By multiplying each side of this expansion by the corresponding side of the expansion in the statement of the problem one sees that

$$(1 + x^2 + x^4)^n = \sum_{k=0}^{4n} (-1)^k \{a_0 a_k - a_1 a_{k-1} + \cdots + (-1)^k a_k a_0\} x^k.$$

The expansion of the right-hand side must contain only even powers of x, because the left-hand side also has this property. Thus the coefficient of x^{2n-1} in the right-hand side is zero or:

$$-(a_0 a_{2n-1} - a_1 a_{2n-2} + a_2 a_{2n-3} - \cdots - a_{2n-1} a_0)$$
$$= -(a_0 a_1 - a_1 a_2 + a_2 a_3 - \cdots - a_{2n-1} a_{2n}) = 0,$$

which is (b).

(c) In order to prove (c) one substitutes x^2 for x in the expansion of the statement of the problem and obtains

$$(1 + x^2 + x^4)^n = a_0 + a_1 x^2 + a_2 x^4 + \cdots + a_{2n} x^{4n}.$$

The coefficient of x^{2n} in this expansion is equal to a_n. Recalling the expansion previously obtained for $(1 + x^2 + x^4)^n$, it can be seen that this coefficient is also equal to

$$a_0 a_{2n} - a_1 a_{2n-1} + a_2 a_{2n-2} - \cdots + a_{2n} a_0 = 2a_0^2 - 2a_1^2 + 2a_2^2 - \cdots + (-1)^n a_n^2,$$

from which (c) follows.

(d) Multiply both sides of the equation in the statement of the problem by $(1-x)^n$ to obtain

$$(1 - x^3)^n = (1 - x)^n (a_0 + a_1 x + a_2 x^2 + \cdots + a_{2n} x^{2n}),$$

and hence

$$1 - \binom{n}{1} x^3 + \binom{n}{2} x^6 - \cdots + (-1)^n \binom{n}{n} x^{3n}$$
$$= \left\{ 1 - \binom{n}{1} x + \binom{n}{2} x^2 - \cdots + (-1)^n \binom{n}{n} x^n \right\} (a_0 + a_1 x + \cdots + a_{2n} x^{2n}).$$

Solutions

If p is not a multiple of 3, the coefficient of x^p on the left-hand side is zero, but if $p=3k$, this coefficient is equal to $(-1)^k \binom{n}{k}$.
On the right-hand side this coefficient is equal to

$$a_p - \binom{n}{1} a_{p-1} + \binom{n}{2} a_{p-2} - \cdots + (-1)^p \binom{n}{p} a_0,$$

which establishes (d).

(e) By taking $x=1$ and $x=-1$ in the expansion of $(1+x+x^2)^n$ one finds that

$$a_0 + a_1 + a_2 + \cdots + a_{2n} = 3^n$$

and

$$a_0 - a_1 + a_2 - \cdots + a_{2n} = 1.$$

Adding and subtracting term by term, (e) is obtained.

(f) To prove (f), let $\alpha = (-1+i\sqrt{3})/2$, one of the cube roots of unity. It follows that $\alpha^2 = (-1-i\sqrt{3})/2$ and $1+\alpha+\alpha^2 = 0$. By replacing x with α one sees that $(1+\alpha+\alpha^2)^n = 0$, and hence

$$a_0 + a_1\alpha + a_2\alpha^2 + \cdots + a_{2n}\alpha^{2n} = 0.$$

After equating the real and imaginary part of the left-hand side to zero and using the notation

$$S_1 = a_0 + a_3 + a_6 + \ldots,$$
$$S_2 = a_1 + a_4 + a_7 + \ldots,$$
$$S_3 = a_2 + a_5 + a_8 + \ldots,$$

it can be shown that

$$S_1 - \frac{S_2 + S_3}{2} = 0,$$

$$\frac{\sqrt{3}}{2}(S_2 - S_3) = 0,$$

whence $S_1 = S_2 = S_3$. But $S_1 + S_2 + S_3 = 3^n$, and hence (f) holds.

(g) Set $x=i$ in the given expansion to obtain

$$i^n = S_1 - S_3 + i(S_2 - S_4),$$

where

$$S_1 = a_0 + a_4 + a_8 + \ldots, \qquad S_2 = a_1 + a_5 + a_9 + \ldots,$$
$$S_3 = a_2 + a_6 + a_{10} + \ldots, \qquad S_4 = a_3 + a_7 + a_{11} + \ldots.$$

By equating real and imaginary parts one obtains, by virtue of (e), the following cases:

(1) $n \equiv 0 \pmod 4$ when $S_2 = S_3 = S_4 = (3^n - 1)/4$ and $S_1 = (3^n + 3)/4$;

(2) $n \equiv 1 \pmod 4$ when $S_1 = S_2 = S_3 = (3^n+1)/4$ and $S_4 = (3^n-3)/4$;
(3) $n \equiv 2 \pmod 4$ when $S_1 = S_2 = S_4 = (3^n-1)/4$ and $S_3 = (3^n+3)/4$;
(4) $n \equiv 3 \pmod 4$ when $S_1 = S_3 = S_4 = (3^n+1)/4$ and $S_2 = (3^n-3)/4$.

Thus three sums are equal and the fourth differs from them by one.

(h) Use induction on n, recalling that $a_k = a_{2n-k}$ for $0 \leq k \leq n$. For $n=2$, it follows that $a_0 = a_4 = 1$; $a_1 = a_3 = 2$; $a_2 = 3$ and the property is satisfied. Suppose that the property is true for every exponent $2 \leq m \leq n$ ($n \geq 2$), and let $(1+x+x^2)^{n+1} = b_0 + b_1 x + b_2 x^2 + \cdots + b_{2n+2} x^{2n+2}$. Since $(1+x+x^2)^{n+1} = (a_0 + a_1 x + \cdots + a_{2n} x^{2n})(1+x+x^2)$, it follows that $b_p = a_p + a_{p-1} + a_{p-2}$ for every $0 \leq p \leq 2n+2$. Take $a_{-1} = a_{-2} = a_{2n+1} = a_{2n+2} = 0$. If $p \leq n$ or $p \geq n+2$, one sees immediately that $b_p > b_{p-1}$ or $b_p > b_{p+1}$ respectively, by using the previously established recurrence relation and induction hypothesis. It remains to show that

$$b_{n+1} > b_n \quad \text{and} \quad b_{n+1} > b_{n+2}.$$

The first of these inequalities can also be written $a_{n+1} + a_n + a_{n-1} > a_n + a_{n-1} + a_{n-2}$ or $a_{n+1} > a_{n-2}$. But $a_{n-2} = a_{n+2}$, and the inequality becomes $a_{n+1} > a_{n+2}$, whose validity follows from the induction hypothesis. The second inequality is true because $b_{n+2} = b_n$.

1.10 Use induction on n. For $n=1$ we note that $S_0 = S_1 = \cdots = S_m = 1$ and $S_{m+1} = 0$ and the formula is verified, since $1+i \equiv 0 \pmod{m+2}$ and $0 \leq i \leq m+1$ imply that $i = m+1$. Assume that the formula is true for every exponent $p \leq n-1$. Consider the expansion

$$(1+x+\cdots+x^m)^{n-1} = b_0 + b_1 x + \cdots + b_{m(n-1)} x^{m(n-1)}.$$

It follows that

$$a_0 + a_1 x + \cdots + a_{mn} x^{mn} = (b_0 + b_1 x + \cdots + b_{m(n-1)} x^{m(n-1)})(1+x+\cdots+x^m),$$

from which the recurrence relation

$$a_i = b_i + b_{i-1} + \cdots + b_{i-m} \tag{1}$$

is obtained. Here $b_j = 0$ for $j < 0$ or for $j > m(n-1)$.

By recalling the definition of the sum S_i, one is led to the recurrence relation

$$S_i = T_i + T_{i-1} + \cdots + T_{i-m}, \tag{2}$$

where

$$T_i = b_i + b_{i+(m+2)} + b_{i+2(m+2)} + \cdots$$

and in (2) the index of T is computed modulo $m+2$. Two cases must be analyzed:

(a) n is even. It follows from the induction hypothesis that the expression for the sum S_i is valid if $p = n-1$, and hence if $n-1+i \equiv 0 \pmod{m+2}$, the sum T_i is smaller by 1 than the other sums, which have the same common value. Since relation (2) holds, it follows that the sum S_j [where $j \equiv i-1 \pmod{m+2}$] is one more than the other sums, which have a common value. Thus $j \equiv i-1 \equiv -n$ (mod $m+2$). This means that $j+n \equiv 0$ (mod $m+2$) and $S_j = \{(m+1)^n - 1\}/(m+2) + 1$. But the other sums are equal to $\{(m+1)^n - 1\}/(m+2)$, since $S_0 + S_1 + \cdots + S_{m+1} = (1+1+\cdots+1)^n = (m+1)^n$. It follows that the property is true for n.

(b) If n is odd, then from the induction hypothesis one can conclude that the sum T_i is greater by one than the sums with a common value. As in (a), one can show that the sum S_j, where $j+n \equiv 0$ (mod $m+2$) is smaller by one than the sums with a common value.

Thus

$$S_j = \frac{(m+1)^n + 1}{m+2} - 1,$$

but the other sums are equal to

$$\frac{(m+1)^n + 1}{m+2}.$$

In both cases the expression for the sum S_i is valid for n and hence the property is true for every natural number n. If $m=1$, then Newton's binomial formula yields

$$S_0 = \binom{n}{0} + \binom{n}{3} + \binom{n}{6} + \cdots,$$

$$S_1 = \binom{n}{1} + \binom{n}{4} + \binom{n}{7} + \cdots,$$

$$S_2 = \binom{n}{2} + \binom{n}{5} + \binom{n}{8} + \cdots.$$

The property just established implies that for every n, two of these sums are equal, and the third differs by one. For $m=2$ one obtains property (g) from the previous problem.

1.11 Recalling the rules for removing parentheses, one sees that the desired coefficient is equal to the number of ways in which k can be written as the sum of two non-negative integers:

$$k = a_1 + a_2$$

where $0 \leq a_1, a_2 \leq n-1$. Two representations will also be considered to be distinct if they differ in the order of their terms. If $0 \leq k \leq n-1$, these $k+1$ representations are $0+k, 1+(k-1), \ldots, k+0$. However, if $n \leq k \leq 2n-2$, the representations $0+k, 1+(k-1), \ldots, (k-n)+n$ and those obtained by permuting two terms do not satisfy the condition $a_1, a_2 \leq n-1$. Thus in this case the number of representations is equal to

$$k+1-2(k-n+1) = 2n-k-1.$$

Note that in both cases the coefficient of x^k can be written in the form

$$n - |n - k - 1|.$$

1.12 The values $a_1, \ldots, a_r$ are obtained in the following manner: Let a_1 be the largest integer x which satisfies the inequality

$$n \geq \binom{x}{r}.$$

Denote by a_2 the largest integer x such that

$$n - \binom{a_1}{r} \geq \binom{x}{r-1}.$$

Finally, a_r will represent the difference

$$n - \left[\binom{a_1}{r} + \binom{a_2}{r-1} + \cdots + \binom{a_{r-1}}{2}\right] = \binom{a_r}{1} = a_r.$$

It will be shown that $a_1 > a_2$; the remaining inequalities $a_2 > a_3 > \cdots > a_r$ can be established similarly. The fact that $a_r \geq 0$ follows from the definition of a_{r-1}. First note that

$$\binom{a_1 + 1}{r} > n \quad \text{and similarly} \quad n - \binom{a_1}{r} \geq \binom{a_2}{r-1}.$$

Thus we can make:

$$\binom{a_1 + 1}{r} - \binom{a_1}{r} > \binom{a_2}{r-1}, \quad \text{that is,} \quad \binom{a_1}{r-1} > \binom{a_2}{r-1}.$$

This inequality implies that $a_1 > a_2$ (by use of the definition of the binomial coefficients).

Suppose that there exists a representation of n as a sum of binomial coefficients different from that given in the statement of the problem

$$\binom{a_1}{r} + \binom{a_2}{r-1} + \cdots + \binom{a_r}{1} = \binom{b_1}{r} + \binom{b_2}{r-1} + \cdots + \binom{b_r}{1}, \quad (1)$$

and with $(a_1, \ldots, a_r) \neq (b_1, \ldots, b_r)$. One can suppose for example that $a_1 > b_1$, since if $a_1 = b_1$, the corresponding terms can be reduced, and this procedure can be repeated for a_2 and b_2.

Since $b_1 > b_2 > \cdots > b_r \geq 0$, it follows that $b_1 - i \geq b_{i+1}$ for $i = 1, \ldots, r-1$ and hence

$$\binom{b_1}{r} + \binom{b_2}{r-1} + \cdots + \binom{b_r}{1} \leq \binom{b_1}{r} + \binom{b_1 - 1}{r-1} + \cdots + \binom{b_1 - (r-1)}{1}.$$

In view of (1) this last inequality implies that

$$\binom{a_1}{r} \leq \binom{b_1}{r} + \binom{b_1 - 1}{r-1} + \cdots + \binom{b_1 - (r-1)}{1}. \quad (2)$$

Solutions

Since $a_1 \geq b_1 + 1$, we see that

$$\binom{a_1}{r} \geq \binom{b_1+1}{r}$$

and thus inequality (2) implies that

$$\binom{b_1+1}{r} \leq \binom{b_1}{r} + \binom{b_1-1}{r-1} + \cdots + \binom{b_1-(r-1)}{1} \tag{3}$$

By applying the recurrence relation for binomial coefficients it follows that

$$\binom{b_1+1}{r} = \binom{b_1}{r} + \binom{b_1}{r-1} = \binom{b_1}{r} + \binom{b_1-1}{r-1} + \binom{b_1-1}{r-2}$$

$$= \cdots = \binom{b_1}{r} + \binom{b_1-1}{r-1} + \cdots + \binom{b_1-(r-1)}{1} + \binom{b_1-(r-1)}{0},$$

which contradicts inequality (3).

Thus the values $a_1, \ldots, a_r$ are uniquely determined and the result is proved.

1.13 The identity can also be written

$$(x+y)^{2n} = \sum_{k=1}^{n} \binom{2n-k-1}{n-1} (x^k + y^k)(xy)^{n-k}(x+y)^k.$$

The coefficient of the term $x^{2n-m}y^m$ obtained by expanding $(x+y)^{2n}$ with the help of Newton's binomial formula is equal to $\binom{2n}{m}$. On the right-hand side, recalling the expansion of $(x+y)^k$ and the standard binomial formulas, one finds that this coefficient is equal to

$$\alpha(n, m) = \sum_{k=1}^{n} \binom{2n-k-1}{n-1} \left[\binom{k}{n-m} + \binom{k}{m-n}\right].$$

If $n = m$, then

$$\alpha(n, n) = 2 \sum_{k=1}^{n} \binom{2n-k-1}{n-1} = 2\left[\binom{2n-2}{n-1} + \binom{2n-3}{n-1} + \cdots + \binom{n-1}{n-1}\right]$$

$$= 2\left[\binom{2n-1}{n} - \binom{2n-2}{n} + \binom{2n-2}{n} - \binom{2n-3}{n} + \cdots \right.$$

$$\left. + \binom{n+1}{n} - \binom{n}{n} + \binom{n-1}{n-1}\right]$$

$$= 2\binom{2n-1}{n} = \binom{2n}{n},$$

and the equality of the coefficients is established.

If for example $m \leq n-1$, then $\binom{k}{m-n} = 0$ and it must be shown that

$$\sum_{k=1}^{n} \binom{2n-k-1}{n-1}\binom{k}{n-m} = \binom{2n}{m} = \binom{2n}{2n-m}. \tag{1}$$

But $\binom{2n}{2n-m}$ is the number of strictly increasing words $c_1 c_2 \cdots c_{2n-m}$ ($c_1 < c_2 < \cdots < c_{2n-m}$) formed with letters of the alphabet $\{1, 2, \ldots, 2n\}$. Denote by A_k the set of those words with the property that $c_{n-m+1} = k+1$ for $k = 1, \ldots, n$. Since $m \leq n-1$, we obtain $n-m+1 \geq 2$ and thus $k+1 \geq 2$, from which it follows that $k \geq 1$.

Similarly $2n - m - (n-m+1) = n-1$, and thus the maximum value of $k+1$ is $2n - m - (n-1) = n - m + 1 \leq n$, since the word is strictly increasing and $m \geq 1$. But if $m = 0$, the coefficient of x^{2n} in both sides is equal to 1 and the equality is thus verified. It follows that the set of strictly increasing words $c_1 c_2 \cdots c_{2n-m}$ formed with letters of the alphabet $\{1, \ldots, 2n\}$ can be written $\bigcup_{k=1}^{n} A_k$. Thus

$$\binom{2n}{2n-m} = \left| \bigcup_{k=1}^{n} A_k \right| = \sum_{k=1}^{n} |A_k|, \qquad (2)$$

where the sets A_k are pairwise disjoint.

If $c_{n-m+1} = k+1$, then the letters $c_1, \ldots, c_{n-m}$ are chosen smaller than c_{n-m+1}, and thus $c_1 \cdots c_{n-m}$ is a strictly increasing word formed from letters of the alphabet $\{1, \ldots, k\}$. The word $c_{n-m+2} \cdots c_{2n-m}$ is a strictly increasing word of length $n-1$ formed with the letters of the set $\{k+2, \ldots, 2n\}$ of cardinality $2n - k - 1$. Thus the word $c_1 \cdots c_{n-m}$ can be chosen in $\binom{k}{n-m}$ ways, and the word $c_{n-m+2} \cdots c_{2n-m}$ can be chosen in $\binom{2n-k-1}{n-1}$ distinct ways, which implies that

$$|A_k| = \binom{2n-k-1}{n-1} \binom{k}{n-m}.$$

Identity (1) now follows by the application of (2).

If $m \geq n+1$, one obtains $\binom{k}{n-m} = 0$, and thus, by making the change of variable $p = 2n - m \leq n - 1$, (1) can be written

$$\sum_{k=1}^{n} \binom{2n-k-1}{n-1} \binom{k}{n-p} = \binom{2n}{2n-p},$$

or

$$\sum_{k=1}^{n} \binom{2n-k-1}{n-1} \binom{k}{m-n} = \binom{2n}{m}.$$

One also sees that $\alpha(n, m) = \binom{2n}{m}$ in this case. [L. Toscano, *Boll. Soc. Math. Calabrese*, **16** (1965), 1–8.]

1.14 One can write

$$\binom{n-1}{k}(k+1)! = \frac{(n-1)!}{(n-k-1)!}(k+1)$$

$$= \frac{(n-1)!}{(n-k-1)!}\{n - (n-k-1)\} = \frac{n!}{(n-k-1)!} - \frac{(n-1)!}{(n-k-2)!}$$

for $0 \leq k \leq n-2$. The desired sum then becomes

Solutions

$$\sum_{k=0}^{n-1}\binom{n-1}{k}n^{n-1-k}(k+1)! = \sum_{k=0}^{n-2}\left[\frac{n!}{(n-k-1)!} - \frac{(n-1)!}{(n-k-2)!}\right]n^{n-1-k} + n!$$

$$= \sum_{k=0}^{n-1}\frac{n!}{(n-k-1)!}n^{n-1-k} - \sum_{k=0}^{n-2}\frac{(n-1)!}{(n-k-2)!}n^{n-1-k}$$

$$= n^n + \sum_{k=1}^{n-1}\frac{(n-1)!}{(n-k-1)!}n^{n-k} - \sum_{k=0}^{n-2}\frac{(n-1)!}{(n-k-2)!}n^{n-1-k}$$

$$= n^n.$$

[J. Riordan, *Ann. Math. Statist.*, **33** (1962), 178–185.]

1.15 The n_1 objects in the first box can be chosen in $\binom{n}{n_1}$ ways, the n_2 objects in the second box can be chosen from the $n-n_1$ remaining objects in $\binom{n-n_1}{n_2}$ ways, etc. The total number of arrangements is equal to

$$\binom{n}{n_1}\binom{n-n_1}{n_2}\binom{n-n_1-n_2}{n_3}\cdots\binom{n-n_1-\cdots-n_{p-1}}{n_p} = \frac{n!}{n_1!n_2!\ldots n_p!}.$$

1.16 Consider the product

$$(a_1^1 + a_2^1 + \cdots + a_p^1)(a_1^2 + a_2^2 + \cdots + a_p^2)\cdots(a_1^n + a_2^n + \cdots + a_p^n)$$

$$= \sum_{\substack{n_1,\ldots,n_p \geq 0 \\ n_1 + \cdots + n_p = n}} (a_1^{i_1}a_1^{i_2}\cdots a_1^{i_{n_1}})(a_2^{j_1}a_2^{j_2}\cdots a_2^{j_{n_2}})\cdots(a_p^{k_1}a_p^{k_2}\cdots a_p^{k_{n_p}}),$$

where the upper indices do not indicate powers, and

$$\{i_1, i_2, \ldots, i_{n_1}\} \cup \cdots \cup \{k_1, k_2, \ldots, k_{n_p}\}$$

is an ordered partition of the set $\{1, \ldots, n\}$, which may contain empty classes. By taking note of the upper indices of each parenthesis in the last sum indicated, one sees that it represents an arrangement of the set of objects $\{1, \ldots, n\}$ in p boxes $y_1, y_2, \ldots, y_p$, such that y_k contains n_k objects for $k = 1, \ldots, p$. In fact, if one takes objects $i_1, \ldots, i_{n_1}$ in box $y_1, \ldots,$ objects $k_1, \ldots, k_{n_p}$ in box y_p, then such an arrangement is also obtained. By using the rules for removing parentheses, it turns out that these are altogether

$$\frac{n!}{n_1!n_2!\ldots n_p!}$$

arrangements of n objects in p boxes which contain $n_1, \ldots, n_p$ objects respectively.

Take $a_i^1 = a_i^2 = \cdots = a_i^n = a_i$ for every $1 \leq i \leq p$; the sum under consideration becomes

$$\sum_{\substack{n_1,\ldots,n_p \geq 0 \\ n_1 + \cdots + n_p = n}} \binom{n}{n_1, \ldots, n_p} a_1^{n_1} a_2^{n_2} \ldots a_p^{n_p},$$

where the upper indices now indicate powers. This follows from the fact that for each representation of n in the form $n = n_1 + \cdots + n_p$, there are

$$\binom{n}{n_1, \ldots, n_p}$$

terms equal to $a_1^{n_1} a_2^{n_2} \cdots a_p^{n_p}$.

1.17 According to Problem 1.19 there are $\binom{m-1}{h-1}$ representations of m as the sum of h positive integers. Let s_i denote the number of terms equal to i in a sum of this type. It follows that $s_1 + \cdots + s_k = h$ and $s_1 + 2s_2 + \cdots + ks_k = m$, where k represents the largest term of the sum.

If the numbers s_i having this property are fixed, then the number of ways in which m can be written as a sum of h positive terms such that s_i terms are equal to i for $i \geq 1$ is equal to the number of arrangements of a set of h objects into k boxes, such that the kth box contains s_k objects. This last number is given by

$$\binom{h}{s_1, \ldots, s_k}$$

Thus the proposed identity is true, since it has been shown that both sides are equal to the number of different representations of m as the sum of h positive integers.

1.18 Let $N = \{1, \ldots, n\}$, $R = \{1, \ldots, r\}$, and $C(N, R)$ the set of increasing functions $f : N \to R$. Identify the increasing function f with the increasing word $b_1 b_2 \cdots b_n$, where $b_i = f(i)$ for $i = 1, \ldots, n$. Since f is increasing, it follows that $1 \leq b_1 \leq b_2 \leq \cdots \leq b_n \leq r$ for $1 \leq i \leq n$.

Consider a mapping

$$F : C(N, R) \to P_n(X),$$

where $P_n(X)$ is the family of n-element subsets of the set $X = \{1, 2, \ldots, r+n-1\}$. The function F is defined by the equation

$$F(b_1 b_2 \cdots b_n) = \{b_1, b_2 + 1, b_3 + 2, \ldots, b_n + (n-1)\}.$$

Because $b_1 \leq b_2 \leq \cdots \leq b_n$, it follows that $b_1 < b_2 + 1 < b_3 + 2 < \cdots < b_n + (n-1)$, and hence the image of a function f under F is in fact an n-element subset of X.

If $b_1 b_2 \cdots b_n \neq c_1 c_2 \cdots c_n$ are both increasing words, then there is an index i such that, for example, $b_1 = c_1, \ldots, b_{i-1} = c_{i-1}$ and $b_i < c_i$ where $1 \leq i \leq n$. This implies that $b_i + (i-1) < c_i + (i-1)$ and $b_i + (i-1) \notin F(c_1 \cdots c_n)$. One can conclude that $F(b_1 \cdots b_n) \neq F(c_1 \cdots c_n)$ and thus the mapping F is injective. If $Y \subset X$ satisfies $|Y| = n$, take $Y = \{y_1, \ldots, y_n\}$ and $1 \leq y_1 < y_2 < \cdots < y_n \leq r + n - 1$.

Let $b_i = y_i - (i-1)$ for every $1 \leq i \leq n$. Then $b_1 \leq b_2 \leq \cdots \leq b_n$ and $b_i \in R$ for every $1 \leq i \leq n$. Using the definition of the function F, one finds that $F(b_1 \cdots b_n) = \{y_1, \ldots, y_n\}$, from which it follows that F is surjective.

Since F is bijective, it can be inferred that the number of increasing functions

$$f : N \to R$$

Solutions

is equal to $|P_n(X)| = \binom{n+r-1}{n} = [r]^n/n!$.

1.19 Define the partial sum

$$S_p = \sum_{k=1}^{p} u_k$$

for every $1 \leq p \leq n-1$. Each representation of m as a sum of n integers corresponds to a word $s_1 s_2 \cdots s_{n-1}$, since $u_1 = s_1$, $u_n = m - s_{n-1}$, and $u_k = s_k - s_{k-1}$ for $2 \leq k \leq n-1$.

If $u_i \geq 0$ for every i, it follows that

$$0 \leq s_1 \leq s_2 \leq \cdots \leq s_{n-1} \leq m;$$

hence the desired number is equal to the number of increasing words of length $n-1$ formed from letters of the set $\{0, 1, \ldots, m\}$ of cardinality $m+1$, that is, the number of combinations of $m+1$ things taken $n-1$ at a time with replacement (see Problem 1.18), and hence to

$$\frac{[m+1]^{n-1}}{(n-1)!} = \binom{m+n-1}{m}.$$

If $u_i > 0$ for every i, then the word $s_1 \cdots s_{n-1}$ is strictly increasing, since

$$1 \leq s_1 < s_2 < \cdots < s_{n-1} \leq m-1.$$

It follows that the number of ways in which m can be expressed as a sum of n positive integers is equal to the number of strictly increasing words formed from $n-1$ letters of the alphabet $\{1, 2, \ldots, m-1\}$ and thus has the numerical value $\binom{m-1}{n-1}$.

1.20 The multinomial formula implies that

$$(x_1 + x_2 + \cdots + x_p)^n = \sum_{\substack{n_1, \ldots, n_p \geq 0 \\ n_1 + \cdots + n_p = n}} \binom{n}{n_1, \ldots, n_p} x_1^{n_1} \cdots x_p^{n_p}.$$

Hence the number of monomials in the expansion of the polynomial $(x_1 + \cdots + x_p)^n$ is equal to the number of representations of n in the form

$$n = n_1 + \cdots + n_p,$$

where each n_i is a non-negative integer, and the representations are considered different if they differ at least in the order of their terms. From the previous problem, one can conclude that this number is also equal to the number of combinations $n+1$ take $p-1$ with replacement. It follows that the number of terms is equal to

$$\frac{(n+1)(n+2)\cdots(n+p-1)}{(p-1)!} = \binom{n+p-1}{p-1}.$$

For $p=2$ there are $\binom{n+1}{1} = n+1$ terms; this can also be shown by using Newton's binomial formula.

1.21 Let

$$f(x) = 2\left\{\binom{x-1}{0} + \binom{x-1}{1} + \cdots + \binom{x-1}{n}\right\}.$$

If $m < n$, it is the case that $\binom{m}{n} = 0$. This together with the formula for the sum of binomial coefficients implies that $f(x) = 2^x$ for every $1 \leq x \leq n+1$, x an integer.

The highest-degree term of the polynomial $f(x)$ is contained in the expansion $2\binom{x-1}{n} = 2\{(x-1)\cdots(x-n)\}/n!$, and thus has degree n. Since the polynomials $f(x)$ and $P(x)$ are both of degree n and take the same values for $n+1$ distinct values of x, it follows that they are identical. Thus

$$P(n+2) = 2\left\{\binom{n+1}{0} + \binom{n+1}{1} + \cdots + \binom{n+1}{n}\right\} = 2(2^{n+1} - 1) = 2^{n+2} - 2.$$

[M. Klamkin, *Pi Mu Epsilon*, **4** (1964), 77, Problem 158.]

1.22 Let the left-hand side be denoted a_n. Then

$$\sum_{n=0}^{\infty} a_n t^n = \left(\sum_{m=0}^{\infty} m t^m\right)^k = \left\{t \frac{d}{dt}\left(\frac{1}{1-t}\right)\right\}^k = \frac{t^k}{(1-t)^{2k}} = t^k (1-t)^{-2k}.$$

Expanding $(1-t)^{-2k}$ by Newton's generalized binomial formula, one sees that the coefficient of t^n in the expression $t^k(1-t)^{-2k}$ is equal to

$$(-1)^{n-k}\binom{-2k}{n-k} = \frac{(n+k-1)(n+k-2)\cdots(2k+1)2k}{(n-k)!} = \frac{(n+k-1)!}{(2k-1)!(n-k)!}$$

$$= \frac{n(n^2-1^2)\cdots\{n^2-(k-1)^2\}}{(2k-1)!}.$$

1.23 The coefficient of x^r in the expansion of $(1 + x + x^2 + \cdots)^n$ will be equal to the number of ways in which r can be represented as a sum of n nonnegative integers:

$$r = r_1 + r_2 + \cdots + r_n$$

It follows from Problem 1.19 that this number is equal to $\binom{n+r-1}{r}$.

1.24 The formula

$$\log\frac{1}{1-x} = x + \frac{x^2}{2} + \cdots + \frac{x^n}{n} + \cdots$$

implies that $1/n$ is the coefficient of x^n in this expansion. But $\log\{1/(1-x)\} = \log(1 + x + x^2 + \cdots)$, and hence the coefficient of x^n in the expansion of this logarithm coincides with the coefficient of x^n in the expansion

$$\log\{1 + (x + x^2 + \cdots + x^n)\} = (x + x^2 + \cdots + x^n) - \frac{(x + \cdots + x^n)^2}{2} + \cdots$$

$$+ (-1)^{n-1}\frac{(x + \cdots + x^n)^n}{n} + \cdots.$$

Solutions

Now consider powers smaller than or equal to n. It turns out that the coefficient of x^n in the expansion $(-1)^{p+1}(x+\cdots+x^n)^p/p$ is equal to

$$\sum (-1)^{j_1+\cdots+j_n+1} \binom{j_1+\cdots+j_n}{j_1,\ldots,j_n} \frac{1}{p},$$

where the sum is taken for $j_i \geq 0$ with $j_1+\cdots+j_n=p$ and $j_1+2j_2+\cdots+nj_n=n$. (This deduction requires the use of the multinomial formula. See Problem 1.16.)
However, this coefficient is equal to

$$\sum (-1)^{j_1+\cdots+j_n+1} \frac{(j_1+j_2+\cdots+j_n-1)!}{j_1!j_2!\cdots j_n!}.$$

By substituting for p the values $p=1,\ldots,n$, the sum on the left-hand side of the statement of the problem is obtained. [J. Sheehan, *Amer. Math. Monthly*, **77** (1970), 168.]

1.25 For natural numbers n_i and n_j such that $n_i \geq n_j+2$, one can show that

$$\binom{n_i}{h}+\binom{n_j}{h}>\binom{n_i-1}{h}+\binom{n_j+1}{h},$$

or

$$\binom{n_i}{h}-\binom{n_i-1}{h}>\binom{n_j+1}{h}-\binom{n_j}{h}.$$

The recurrence relation for binomial coefficients further implies that this is equivalent to $\binom{n_i-1}{h-1}>\binom{n_j}{h-1}$. The last inequality follows from the fact that $n_i-1>n_j$. It implies that the desired minimum is attained when the numbers $n_1,\ldots,n_k$ satisfy the inequality $-1 \leq n_i-n_j \leq 1$ for every $i,j=1,\ldots,k$, i.e., r numbers are equal to $t+1$ and $k-r$ numbers are equal to $t=[n/k]$. Similarly, in order to obtain the maximum one must show that there do not exist two numbers n_i, n_j which are both greater than 1, since, replacing them by n_i+1 and n_j-1 respectively, one would obtain a larger sum. Thus the maximum is attained for a representation of n in the form $n=(n-k+1)+1+\cdots+1$.

1.26 Consider the parts $m_1, m_2, \ldots, m_k$ in the representation of n in the form

$$n=m_1+\cdots+m_k$$

which maximize (for fixed k) the products indicated in (a) and (b). We show that these are as nearly equal as possible, that is,

$$-1 \leq m_i-m_j \leq 1 \tag{1}$$

for every $i,j=1,\ldots,k$. In fact, if the contrary is assumed, then there are two indices i and j such that $m_i \geq m_j+2$. In this case, one can write

$$(m_i-1)(m_j+1)=m_im_j+m_i-(m_j+1)>m_im_j,$$

which contradicts the maximality of the product $m_1m_2\ldots m_k$. In the same way

it can be shown that
$$\binom{m_i-1}{2}\binom{m_j+1}{2} > \binom{m_i}{2}\binom{m_j}{2},$$
and this is equivalent to $(m_i-2)(m_j+1) > m_i(m_j-1)$. By reducing terms in the same way one sees that $m_i > m_j+1$ and the inequality is true by hypothesis. Thus the product
$$\binom{m_1}{2}\binom{m_2}{2}\cdots\binom{m_k}{2}$$
cannot be a maximum, and this observation completes the proof of inequalities (1).

Now consider the case (a) for an arbitrary value k. One finds that max $(m_1, \ldots, m_k) \leq 4$, since in the contrary case one obtains $5 < 3 \times 2$, and this contradicts the maximality of the product under consideration. In the same way one cannot have $m_i = m_j = 4$ since $4 \times 4 = 16 < 3 \times 3 \times 2 = 18$. There do not exist three numbers m_i equal to 2, since $2 \times 2 \times 2 = 8 < 3 \times 3 = 9$. Note that $2 \times 2 = 4$, and thus almost all numbers $m_1, \ldots, m_k$ are equal to 3. It may be that a single number $m_i = 2$, two numbers are equal to 2, or a single number is equal to 4. Thus, denoting by $A(n)$ the maximum value in case (a), one can deduce that

$$A(n) = \begin{cases} 3^{n/3} & \text{for } n \equiv 0 \pmod{3}, \\ 4 \times 3^{(n-4)/3} & \text{for } n \equiv 1 \pmod{3}, \\ 2 \times 3^{(n-2)/3} & \text{for } n \equiv 2 \pmod{3}. \end{cases}$$

Now we proceed in a similar manner for case (b), by first showing that max $(m_1, \ldots, m_k) \leq 7$. In fact, if there is an $m_i \geq 8$, then
$$\binom{m}{2} < \binom{5}{2}\binom{m-5}{2},$$
which is equivalent to $m^2 - m < 10(m-5)(m-6)$, or $9m^2 - 109m + 300 > 0$. This trinomial has two real roots in the interval $(3, 8)$, which establishes the given inequality. Thus if there exists an $m_i \geq 8$, then the product of the binomial coefficients from (b) cannot be a maximum, since one can replace the part equal to m_i by two parts equal to 5 and $m_i - 5$ respectively, and the product of the binomial coefficients therefore increases. Since $\binom{5}{2} > \binom{3}{2}\binom{2}{2}$, $\binom{6}{2} > \binom{3}{2}\binom{3}{2}$, and $\binom{7}{2} > \binom{3}{2}\binom{4}{2}$, it follows that for $n \leq 7$ the desired maximum is equal to $\binom{n}{2}$.

If $n \geq 8$, in the case of the maximum it follows that max $(m_1, \ldots, m_k) \leq 6$, since in the opposite case there exists an $m_i = 7$. But relation (1) implies that there must exist at least one number m_j which is equal to 6 or to 7. For $m_j = 6$,
$$315 = \binom{7}{2}\binom{6}{2} < \binom{5}{2}\binom{4}{2}\binom{4}{2} = 360.$$
Similarly one finds that
$$441 = \binom{7}{2}\binom{7}{2} < \binom{5}{2}\binom{5}{2}\binom{4}{2} = 600,$$

and thus the maximum cannot be obtained for (b).

Further observe that at most two of the numbers m_i can be equal to 4 or to 6 in view of the inequalities

$$3375 = \binom{6}{2}\binom{6}{2}\binom{6}{2} < \binom{5}{2}\binom{5}{2}\binom{4}{2}\binom{4}{2} = 3600,$$

$$216 = \binom{4}{2}\binom{4}{2}\binom{4}{2} < \binom{6}{2}\binom{6}{2} = 225.$$

If $n \geqslant 8$, neither of the numbers m_i can be equal to 2 or to 3. For example, suppose that there exists an $m_i = 3$. It follows from (1) that there must also exist a number $m_j \in \{2, 3, 4\}$ and hence

$$3 = \binom{2}{2}\binom{3}{2} < \binom{5}{2} = 10, \qquad 9 = \binom{3}{2}\binom{3}{2} < \binom{6}{2} = 15,$$

$$18 = \binom{3}{2}\binom{4}{2} < \binom{7}{2} = 21.$$

In conclusion, almost all numbers m_i are equal to 5, but at most two of them are equal to 4 or to 6, so that relation (1) is satisfied.

Let $B(n)$ denote the maximum in case (b). Then:

$$B(n) = \binom{n}{2} \qquad \text{for} \quad n \leqslant 7,$$

and for $n \geqslant 8$ the following identities hold:

(1) $B(n) = 10^{n/5}$ for $n \equiv 0 \pmod 5$ and $m_1 = \cdots = m_k = 5$;

(2) $B(n) = 15 \times 10^{(n-6)/5}$ for $n \equiv 1 \pmod 5$ and $m_1 = 6$, $m_2 = \cdots = m_k = 5$;

(3) $B(n) = 225 \times 10^{(n-12)/5}$ for $n \equiv 2 \pmod 5$ and $m_1 = m_2 = 6$, $m_3 = \cdots = m_k = 5$;

(4) $B(n) = 36 \times 10^{(n-8)/5}$ for $n \equiv 3 \pmod 5$ and $m_1 = m_2 = 4$, $m_3 = \cdots = m_k = 5$;

(5) $B(n) = 6 \times 10^{(n-4)/5}$ for $n \equiv 4 \pmod 5$ and $m_1 = 4$, $m_2 = \cdots = m_k = 5$.

Consider

$$\max_{1 \leqslant k \leqslant n} \max_{n_1 + \cdots + n_k = n} \prod_{i=1}^{k} \binom{n_i}{h}.$$

It can be shown that for n sufficiently large, the maximum is attained for a representation of n which satisfies (1). Also almost all the numbers m_i are equal to $2h + o(h)$. The function $o(h)$ satisfies the relation

$$\lim_{h \to \infty} \frac{o(h)}{\ln h} = \frac{1}{2}.$$

[I. Tomescu, *Discrete Mathematics*, **37** (1981), 263–277.]

For example, for $h=3$ almost all numbers m_i are equal to 7 for n sufficiently large.

1.27 Using the standard formulas for binomial coefficients, the inequality becomes
$$\binom{n_1-x}{h-x}\binom{n_2-y}{h-y} \geqslant \binom{2h}{x+y}.$$
Since $n_1 \geqslant 3h$ and $n_2 \geqslant 3h$, it is sufficient to show that
$$\binom{3h-x}{h-x}\binom{3h-y}{h-y} \geqslant \binom{2h}{x+y}.$$
Let $x+y=k$ (constant) where $0 \leqslant k \leqslant 2h$. Then
$$\binom{3h-x}{h-x}\binom{3h-y}{h-y} = \binom{3h-x}{2h}\binom{3h-y}{2h}$$
$$= \frac{(3h-x)(3h-y)(3h-x-1)(3h-y-1)\cdots(h-x+1)(h-y+1)}{((2h)!)^2}.$$

By grouping the factors two by two in the order in which they are written it can be observed that their sum is constant and equal respectively to $6h-k$, $6h-k-2, \ldots, 2h-k+2$. Thus the products become minimal when the factors differ to the largest extent possible among themselves.

One can distinguish two cases:

(a) $k \leqslant h$. In this case the products of the two factors become minimal simultaneously for $x=0, y=k$ or $x=k, y=0$. Let $x=0, y=k$. In this case it is sufficient to show that
$$\binom{3h}{h}\binom{3h-k}{h-k} \geqslant \binom{2h}{k}.$$
But
$$\binom{3h}{h} > \binom{2h}{h} = \max_k \binom{2h}{k} \geqslant \binom{2h}{k},$$
and hence the inequality is established in this case, since $\binom{3h-k}{h-k} \geqslant 1$.

(b) $k>h$. Here the products of the two factors become minimal simultaneously for $x=k-h, y=h$ (or $x=h$ and $y=k-h$), when one shows that
$$\binom{4h-k}{2h} \geqslant \binom{2h}{k}.$$
Recall the standard formulas for binomial coefficients. It suffices to verify that
$$\binom{4h-k}{2h-k} \geqslant \binom{2h}{2h-k}.$$

But $4h-k \geqslant 2h$, so $k \leqslant 2h$, and the inequality follows from the monotonicity of the binomial coefficients. The proof also establishes that equality holds only when $k=2h$, that is, for $x=y=h$.

1.28 Let X be a set with nk elements, and

$$X = A_1 \cup \cdots \cup A_k$$

a partition of X in which the sets $A_1, \ldots, A_k$ each contain n elements. The number of p-element subsets of X is equal to $\binom{nk}{p}$.

We show that the left-hand side of the equality under consideration also represents the number of p-element subsets of X. Consider a p-element subset $Y \subset X$ and the sets $Y \cap A_1, Y \cap A_2, \ldots, Y \cap A_k$. It follows that $0 \leqslant |Y \cap A_i| \leqslant n$ for every $i=1,\ldots,k$. Let α_i denote the cardinality of the set of numbers equal to i in the sequence $|Y \cap A_1|, |Y \cap A_2|, \ldots, |Y \cap A_k|$. It follows that

$$\alpha_1 + 2\alpha_2 + \cdots + n\alpha_n = |Y| = p.$$

If the numbers $\alpha_1, \alpha_2, \ldots, \alpha_n \geqslant 0$ are fixed, then it is possible to select j elements in α_j sets among $A_1, \ldots, A_k$ in $\binom{n}{j}^{\alpha_j}$ ways for $j=1, 2, \ldots, n$. The union of these elements is the set Y. On the other hand, one can select α_1 sets from among $A_1, \ldots, A_k$, α_2 sets from the remainder, and so on, in

$$\binom{k}{\alpha_1}\binom{k-\alpha_1}{\alpha_2}\binom{k-\alpha_1-\alpha_2}{\alpha_3}\cdots\binom{k-\alpha_1-\cdots-\alpha_{n-1}}{\alpha_n}$$

$$= \frac{k!}{\alpha_1!(k-\alpha_1)!} \cdot \frac{(k-\alpha_1)!}{\alpha_2!(k-\alpha_1-\alpha_2)!} \cdots \frac{(k-\alpha_1-\cdots-\alpha_{n-1})!}{\alpha_n!\{k-(\alpha_1+\cdots+\alpha_n)\}!}$$

$$= \frac{k!}{\alpha_1!\ldots\alpha_n!\{k-(\alpha_1+\cdots+\alpha_n)\}!}$$

distinct ways.

Thus the number of p-element subsets of X can be written as the sum of the numbers

$$\frac{k!}{\alpha_1!\ldots\alpha_n!\{k-(\alpha_1+\cdots+\alpha_n)\}!}\binom{n}{1}^{\alpha_1}\cdots\binom{n}{n}^{\alpha_n}$$

over all representations $p=\alpha_1+2\alpha_2+\cdots+n\alpha_n$ where the α_i are non-negative integers. If $\alpha_1=p$ and $\alpha_2=\cdots=\alpha_n=0$, then $k \geqslant \alpha_1+\cdots+\alpha_n$, or $k \geqslant p$. For $p=k-1$ one obtains an identity due to I. M. Voloshin (*J. Combinatorial Theory*, **A12** (1972), 202–216).

1.29 (a) We prove equality (a) by induction on n. For $n=1$ both sides are equal to $x+y+1$. Differentiating with respect to y, one sees that

$$\frac{\partial}{\partial y}(x+y+n)^n = n(x+y+n)^{n-1}$$

$$= n\{x+(y+1)+(n-1)\}^{n-1}$$

$$\times \frac{\partial}{\partial y} \sum_{k=0}^{n} \binom{n}{k} x(x+k)^{k-1}(y+n-k)^{n-k}$$

$$= \sum_{k=0}^{n-1} \binom{n}{k} x(x+k)^{k-1}(n-k)(y+n-k)^{n-k-1}$$

$$= n \sum_{k=0}^{n-1} \binom{n-1}{k} x(x+k)^{k-1}\{(y+1)+(n-1)-k\}^{(n-1)-k}.$$

By the induction hypothesis the two right-hand members of these equalities are equal.

In order to prove the first Abel identity it is sufficient to show that (a) holds for a particular value of y, such that $(y+n-k)^{n-k}$ is defined for $k=n$, and hence for $y \neq k-n$. Choose $y=-x-n$. The right-hand side vanishes, but the left-hand side is equal to

$$\sum_{k=0}^{n} \binom{n}{k} x(x+k)^{k-1}(-x-k)^{n-k} = x \sum_{k=0}^{n} \binom{n}{k} (-1)^{n-k}(x+k)^{n-1}$$

$$= \sum_{k=0}^{n} (-1)^{n-k} \binom{n}{k} \sum_{j=0}^{n-1} \binom{n-1}{j} k^j x^{n-j}$$

$$= \sum_{j=0}^{n-1} \binom{n-1}{j} x^{n-j} \sum_{k=0}^{n} (-1)^{n-k} k^j \binom{n}{k}$$

$$= \sum_{j=0}^{n-1} \binom{n-1}{j} x^{n-j} n! S(j, n) = 0,$$

since the Stirling number of the second kind $S(j, n) = 0$ for $j = 0, \ldots, n-1$.

(b) The left-hand side of identity (a) can be written as follows:

$$\sum_{k=0}^{n} \binom{n}{k} x(x+k)^{k-1}(y+n-k)^{n-k-1}(y+n-k)$$

$$= \sum_{k=0}^{n} \binom{n}{k} x(x+k)^{k-1} y(y+n-k)^{n-k-1}$$

$$+ \sum_{k=0}^{n} \binom{n}{k} x(x+k)^{k-1}(y+n-k)^{n-k-1}(n-k).$$

Again by using (a), the second term of this sum is seen to be equal to

$$\sum_{k=0}^{n-1} n \binom{n-1}{k} x(x+k)^{k-1}((y+1)+(n-1)-k)^{(n-1)-k}$$

$$= n(x+(y+1)+(n-1))^{n-1} = n(x+y+n)^{n-1}.$$

Solutions

Thus

$$\sum_{k=0}^{n} \binom{n}{k} x(x+k)^{k-1} y(y+n-k)^{n-k-1} = (x+y+n)^n - n(x+y+n)^{n-1}$$

$$= (x+y)(x+y+n)^{n-1},$$

from which (b) follows by dividing both sides by xy.

(c) If $(1/x)(y+n)^{n-1} + (1/y)(x+n)^{n-1}$ is subtracted from both sides of identity (b), then

$$\sum_{k=1}^{n-1} \binom{n}{k} (x+k)^{k-1}(y+n-k)^{n-k-1} = \frac{1}{x}\{(x+y+n)^{n-1} - (y+n)^{n-1}\}$$

$$+ \frac{1}{y}\{(x+y+n)^{n-1} - (x+n)^{n-1}\}.$$

By taking the limit on both sides as $x \to 0$ and $y \to 0$, (c) is obtained.

1.30 Let $y=0$ in the first Abel identity, and equate the coefficient of x in both sides.

1.31 If $f_1(n)$ and $f_2(n)$ are solutions of the recurrence relation

$$f(n+2) = af(n+1) + bf(n), \tag{1}$$

it will follow that for every two real numbers C_1 and C_2 the function $h(n) = C_1 f_1(n) + C_2 f_2(n)$ is also a solution of equation (1). In fact, since $f_1(n)$ and $f_2(n)$ satisfy (1), one can conclude that

$$C_1 f_1(n+2) + C_2 f_2(n+2) = C_1\{af_1(n+1) + bf_1(n)\} + C_2\{af_2(n+1) + bf_2(n)\}$$

$$= a\{C_1 f_1(n+1) + C_2 f_2(n+1)\} + b\{C_1 f_1(n) + C_2 f_2(n)\},$$

or $h(n+2) = ah(n+1) + bh(n)$. This is equivalent to saying that $h(n)$ satisfies relation (1).

Now we show that if r_1 is a root of the quadratic equation

$$r^2 = ar + b, \tag{2}$$

then the sequence $1, r_1, r_1^2, \ldots, r_1^n, \ldots$ is a solution of equation (1). Let $f(n) = r_1^n$; then $f(n+1) = r_1^{n+1}$ and $f(n+2) = r_1^{n+2}$. Substituting these values in (1), one sees that

$$r_1^{n+2} = ar_1^{n+1} + br_1^n,$$

since a simplification of r_1^n yields $r_1^2 = ar_1 + b$. The assumption $b \neq 0$ implies that $r_1 \neq 0$. If the characteristic equation (2) has two distinct roots $r_1 \neq r_2$, then $f_1(n) = r_1^n$ and $f_2(n) = r_2^n$ are solutions of equation (1).

Recall that for all real numbers C_1 and C_2 the function $f(n) = C_1 r_1^n + C_2 r_2^n$ is a solution of the recurrence relation (1). Now we show that every solution of (1) has this form. First observe that every solution of (1) is uniquely determined by its initial values $f(0)$ and $f(1)$. It is thus sufficient to show that the system of equations

$$C_1 + C_2 = f(0),$$
$$C_1 r_1 + C_2 r_2 = f(1),$$

has a solution for every choice of $f(0) = b_0$ and $f(1) = b_1$. The solution of the system, when $r_1 \neq r_2$, is given by

$$C_1 = \frac{f(1) - r_2 f(0)}{r_1 - r_2}, \quad C_2 = \frac{r_1 f(0) - f(1)}{r_1 - r_2},$$

and thus the proof of case (a) is established.

However if $r_1 = r_2$, then from the system of equations one finds that $C_1 + C_2 = f(0)$ and $C_1 + C_2 = f(1)/r_1$. This latter system is, in general, incompatible. Now we show that in this case equation (1) also has the solution $f_2(n) = nr_1^n$ in addition to $f_1(n) = r_1^n$. In fact, if the quadratic equation

$$r^2 = ar + b$$

has a double root, it follows from Viète's relations that $a = 2r_1$, $b = -r_1^2$, and thus equation (2) can also be written

$$r^2 = 2r_1 r - r_1^2. \tag{3}$$

The recurrence relation (1) thus has the form

$$f(n+2) = 2r_1 f(n+1) - r_1^2 f(n). \tag{4}$$

Verify that $f_2(n) = nr_1^n$ is a solution of (4). Observe that

$$f_2(n+2) = (n+2)r_1^{n+2} \quad \text{and} \quad f_2(n+1) = (n+1)r_1^{n+1},$$

and thus (4) becomes

$$(n+2)r_1^{n+2} = 2(n+1)r_1^{n+2} - nr_1^{n+2},$$

which is an identity.

Thus $f_2(n) = nr_1^n$ is a solution of recurrence relation (4). One can conclude by similar reasoning that $f(n) = C_1 f_1(n) + C_2 f_2(n) = r_1^n (C_1 + C_2 n)$ is a solution of (4). The constants C_1 and C_2 can be chosen so that $f(n)$ satisfies arbitrary initial conditions for $n = 0$ and $n = 1$. In fact one obtains the system

$$C_1 = f(0),$$
$$r_1(C_1 + C_2) = f(1),$$

which has the solution $C_1 = f(0)$ and $C_2 = \{f(1) - r_1 f(0)\}/r_1$. Thus the general solution in case (b) has the desired form. In general, for a linear recurrence relation with constant coefficients of the form

$$f(n+k) = a_1 f(n+k-1) + \cdots + a_k f(n) \tag{5}$$

Solutions

the characteristic equation can be written

$$r^k = a_1 r^{k-1} + \cdots + a_k.$$

If the roots of this equation are $r_1, \ldots, r_s$ and their multiplicities are respectively $m_1, \ldots, m_s$, where $m_1 + \cdots + m_s = k$, one can show similarly that the general solution of (5) has the form

$$f(n) = \sum_{i=1}^{s} (C_{i,1} + C_{i,2} n + \cdots + C_{i,m_i} n^{m_i - 1}) r_i^n.$$

The constants $C_{i,1}, \ldots, C_{i,m_i}$ for $1 \leq i \leq s$ are uniquely determined by assuming that the general solution $f(n)$ satisfies the initial conditions $f(0) = b_0, \ldots, f(k-1) = b_{k-1}$.

1.32 Let a_n equal the number of ways of making the purchases described in the statement of the problem. In the first instance there are three possibilities: the student buys a bun and thus there still exist a_{n-1} possible purchases with the $n-1$ remaining dollars. He buys an ice cream, and thus he can spend the rest of the money in a_{n-2} ways. Similarly there are a_{n-2} ways of spending the rest of the money if he has bought a pastry. It follows that

$$a_n = a_{n-1} + 2a_{n-2}$$

for initial values $a_1 = 1$, $a_2 = 3$ (two buns on two consecutive days, an ice cream or a pastry on the first day).

In order to solve this recurrence relation one must use the characteristic equation (see Problem 1.31)

$$r^2 - r - 2 = 0,$$

which has solutions $r_1 = 2$, $r_2 = -1$. The general solution has the form

$$a_n = C_1 2^n + C_2 (-1)^n,$$

where C_1 and C_2 are determined by the system

$$2C_1 - C_2 = 1,$$
$$4C_1 + C_2 = 3.$$

Thus $C_1 = \frac{2}{3}$, $C_2 = \frac{1}{3}$, and

$$a_n = \tfrac{1}{3}\{2^{n+1} + (-1)^n\}.$$

1.33 Cover the rectangle with dominoes by starting from the side AD of length 3. One obtains one of the cases of Figure 1.1. In the last two cases, a

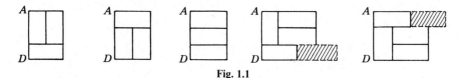

Fig. 1.1

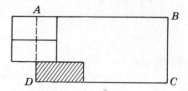

Fig. 1.2

square with side 1 remains uncovered in the 3×3 square with side AD. It can be covered in a unique way by the striped domino. Thus

$$U(2n+2) = 3U(2n) + 2V(2n-2), \qquad (1)$$

where $V(2n)$ denotes the number of ways in which one can cover the figure consisting of a rectangle ABCD of sides 3 and $2n$ to which one has added a rectangle with sides 2 and 1, by attaching its long side to the side AD of the rectangle.

From Figure 1.2 it follows that the additional domino can be covered in two ways, and hence

$$V(2n) = U(2n) + V(2n-2). \qquad (2)$$

Now in (1) use the expression for $V(2n-2)$ given by (2) to obtain $U(2n+2) = 3U(2n) + 2U(2n-2) + 2V(2n-4)$. From (1) it follows that $2V(2n-4) = U(2n) - 3U(2n-2)$, and thus the relation becomes $U(2n+2) = 3U(2n) + 2U(2n-2) + U(2n) - 3U(2n-2)$. This establishes the linear recurrence relation

$$U(2n+2) = 4U(2n) - U(2n-2). \qquad (3)$$

By a simple argument one can deduce that $U(2) = 3$ and $U(4) = 11$. The characteristic equation of the recurrence (3) is $r^2 - 4r + 1 = 0$, with roots $r_{1,2} = 2 \pm \sqrt{3}$, and thus the solution of the recurrence (3) is

$$U(2n) = C_1 r_1^n + C_2 r_2^n$$

The constants C_1 and C_2 are determined by recalling the values $U(2)$ and $U(4)$:

$$U(2n) = C_1(2+\sqrt{3})^n + C_2(2-\sqrt{3})^n,$$

where

$$C_1 = \frac{\sqrt{3}+1}{2\sqrt{3}} \quad \text{and} \quad C_2 = \frac{\sqrt{3}-1}{2\sqrt{3}}$$

[I. Tomescu, E 2417, *American Mathematical Monthly*, **80**(5) (1973), 559–560.]

1.34 Consider a word of length $n-1$ formed from the letters of the alphabet A, such that a and b are not adjacent letters. One can then form a word of length n which satisfies this condition by placing in front of the word:

(1) the letter c or d or b if the first letter is a;
(2) the letter c or d or a if the first letter is b;
(3) any of the four letters a, b, c, d if the word begins with c or with d.

Solutions

Each word of length n which satisfies the conditions of the problem can be obtained from a word of length $n-1$ by performing these operations. Let x_n denote the number of words of length n which start with a or with b, and let y_n denote the number of words with first letter c or d which satisfy the given conditions. The following recurrence relations hold:

$$x_n = x_{n-1} + 2y_{n-1},$$
$$y_n = 2x_{n-1} + 2y_{n-1}.$$

By adding the equations term by term one sees that

$$x_n + y_n = 3(x_{n-1} + y_{n-1}) + y_{n-1} = 3(x_{n-1} + y_{n-1}) + 2(x_{n-2} + y_{n-2}).$$

Now calculate $z_n = x_n + y_n$, which satisfies the recurrence relation

$$z_n = 3z_{n-1} + 2z_{n-2}.$$

One obtains $z_1 = 4$ and $z_2 = 14$, since the words of length 2 which satisfy the given conditions are precisely the $4^2 = 16$ words of length 2 with the exception of the words ab and ba.

We use the general method given in Problem 1.31 to solve this recurrence. The characteristic equation $r^2 - 3r - 2 = 0$ has roots $r_{1,2} = (3 \pm \sqrt{17})/2$, and hence there is a solution of the form

$$z_n = C_1 r_1^n + C_2 r_2^n,$$

where the constants C_1 and C_2 are determined by the initial conditions:

$$z_1 = 4 \quad \text{and} \quad z_2 = 14.$$

Solving the system

$$C_1 \frac{3 + \sqrt{17}}{2} + C_2 \frac{3 - \sqrt{17}}{2} = 4,$$

$$C_1 \frac{13 + 3\sqrt{17}}{2} + C_2 \frac{13 - 3\sqrt{17}}{2} = 14,$$

yields

$$C_1 = \frac{\sqrt{17} + 5}{2\sqrt{17}} \quad \text{and} \quad C_2 = \frac{\sqrt{17} - 5}{2\sqrt{17}}.$$

Thus the solution of the problem can be written in the form

$$z_n = \frac{\sqrt{17} + 5}{2\sqrt{17}} \left(\frac{3 + \sqrt{17}}{2} \right)^n + \frac{\sqrt{17} - 5}{2\sqrt{17}} \left(\frac{3 - \sqrt{17}}{2} \right)^n.$$

1.35 By utilizing the recurrence relation for binomial coefficients and observing that $\binom{n-k}{k} = 0$ for $k > [n/2]$ one can deduce that

$$a_n = \sum_{k \geq 0} \left[\binom{n-k-1}{k} + \binom{n-k-1}{k-1} \right] z^k$$

$$= \sum_{k \geq 0} \binom{n-1-k}{k} z^k + z \sum_{k \geq 1} \binom{(n-2)-(k-1)}{k-1} z^{k-1}.$$

This yields the recurrence relation

$$a_n = a_{n-1} + z a_{n-2} \tag{1}$$

with initial values $a_0 = a_1 = 1$. The characteristic equation of this recurrence is

$$r^2 - r - z = 0$$

with solutions $r_1 = (1 + \sqrt{1+4z})/2$ and $r_2 = (1 - \sqrt{1+4z})/2$. If $z \neq -\frac{1}{4}$, the general solution of recurrence relation (1) is

$$a_n = C_1 r_1^n + C_2 r_2^n \tag{2}$$

where C_1 and C_2 are determined by the system

$$C_1 + C_2 = 1,$$

$$C_1 \frac{1+\sqrt{1+4z}}{2} + C_2 \frac{1-\sqrt{1+4z}}{2} = 1.$$

Therefore

$$C_1 = \frac{1+\sqrt{1+4z}}{2\sqrt{1+4z}} \quad \text{and} \quad C_2 = \frac{\sqrt{1+4z}-1}{2\sqrt{1+4z}},$$

and thus from (2) it follows that

$$a_n = \frac{1}{\sqrt{1+4z}} \left\{ \left(\frac{1+\sqrt{1+4z}}{2} \right)^{n+1} - \left(\frac{1-\sqrt{1+4z}}{2} \right)^{n+1} \right\}.$$

If $z = -\frac{1}{4}$ one has $r_1 = r_2 = \frac{1}{2}$ and hence, according to Problem 1.31, the general solution of recurrence relation (1) is

$$a_n = r_1^n (C_1 + C_2 n), \tag{3}$$

where $C_1 = a_0 = 1$ and $C_2 = (a_1 - r_1 a_0)/r_1 = 1$. After making these substitutions together with $r_1 = \frac{1}{2}$ in (3), one finds that

$$a_n = \frac{n+1}{2^n} \quad \text{for} \quad z = -\tfrac{1}{4}.$$

1.36 Let $S(n, k; x_1, \ldots, x_k) = (x_1 + \cdots + x_k)^n - \sum (x_1 + \cdots + x_{k-1})^n + \sum (x_1 + \cdots + x_{k-2})^n - \cdots$; then the desired sum is $S_n(x_1, \ldots, x_n) = a_0 S(n, n; x_1, \ldots, x_n) + a_1 S(n-1, n; x_1, \ldots, x_n) + a_2 S(n-2, n; x_1, \ldots, x_n) + \cdots + a_n S(0, n; x_1, \ldots, x_n)$. Since $e^x = 1 + x/1! + x^2/2! + \cdots$, it can be seen that the expression for the exponential generating function is

$$\sum_{n=0}^{\infty} S(n, k; x_1, \ldots, x_k) \frac{z^n}{n!} = e^{(x_1 + \cdots + x_k)z} - \sum e^{(x_1 + \cdots + x_{k-1})z} + \sum e^{(x_1 + \cdots + x_{k-2})z}$$

$$- (e^{x_1 z} - 1)(e^{x_2 z} - 1) \cdots (e^{x_k z} - 1)$$

$$= \left(\frac{x_1 z}{1!} + \frac{(x_1 z)^2}{2!} + \cdots \right) \left(\frac{x_2 z}{1!} + \frac{(x_2 z)^2}{2!} + \cdots \right) \cdots \left(\frac{x_k z}{1!} + \frac{(x_k z)^2}{2!} + \cdots \right).$$

By equating the coefficients of z^n on the two sides of this equation for $n=1,\ldots,k$, one can conclude that $S(n,k;x_1,\ldots,x_k)=0$ for $0\leqslant n<k$ and $S(n,n;x_1,\ldots,x_n)=n!x_1\cdots x_n$. Thus $S_n(x_1,\ldots,x_n)=a_0 S(n,n;x_1,\ldots,x_n)=a_0 n!x_1\cdots x_n$. [L. Carlitz, *Fibonacci Quarterly*, **18**(1) (1980), 85.]

1.37 Let
$$p_i(x_1,\ldots,x_{i-1},x_{i+1},\ldots,x_n)=p(x_1,\ldots,x_{i-1},0,x_{i+1},\ldots,x_n)$$
be the polynomial obtained from p by replacing the variable x_i with 0. It follows that the value of the polynomial $z^k p$ for $x_i=0$ is equal to
$$z^k p|_{x_i=0}=z^{k-1}p_i+z^k p_i \tag{1}$$
for $k\geqslant 1$, where $z^0 p_i=p_i$ by definition.

In fact, the $\binom{n}{k}$ combinations of the n variables taken k at a time can be written as the union of the $\binom{n-1}{k-1}$ combinations of the n variables taken k at a time which contain x_i, and those $\binom{n-1}{k}$ combinations of n variables taken k at a time which do not contain x_i. Thus one can write
$$z^k p = z_1^k p + z_2^k p, \tag{2}$$
where $z_1^k p$ is the sum of all the polynomials which can be obtained from p by replacing k of the variables (including x_i) $x_1,\ldots,x_n$ with zero, and $z_2^k p$ is the sum of all those polynomials which can be obtained from p by replacing k of the variables $x_1,\ldots,x_{i-1},x_{i+1},\ldots,x_n$ with zero, in all possible ways.

From (2) it follows that
$$z^k p|_{x_i=0}=z_1^k p|_{x_i=0}+z_2^k p|_{x_i=0}=z_1^k p+z_2^k p_i=z^{k-1}p_i+z^k p_i,$$
which is equivalent to (1). Thus $(p-z^1 p+z^2 p-\cdots)|_{x_i=0}=p_i-(p_i+z^1 p_i)+(z^1 p_i+z^2 p_i)-(z^2 p_i+z^3 p_i)+\cdots+(-1)^{n-1}z^{n-2}p_i=0$, since $z^{n-1}p_i=z^n p_i=\cdots=0$. Therefore the monomial $x_1 x_2 \cdots x_n$ divides the polynomial $p-z^1 p+\cdots$, since the latter vanishes for $x_i=0$ when $1\leqslant i\leqslant n$. Since $p-z^1 p+\cdots$ has degree at most equal to m, it follows that
$$p-z^1 p+z^2 p-\cdots=\begin{cases}0 & \text{for } m<n \\ cx_1\ldots x_n & \text{for } m=n\end{cases}$$

However, the polynomial $z^k p$ with $k\geqslant 1$ does not contain any term of the form $x_1\cdots x_n$, from which it can be inferred that c is the coefficient of $x_1\cdots x_n$ in the representation of the polynomial p.

If $p(x_1,\ldots,x_n)=(x_1+\cdots+x_n)^k$, then the coefficient of $x_1\cdots x_n$ for $n=k$ is equal to $c=n!/(1!)^n=n!$ by the multinomial formula (Problem 1.16). On the other hand, the numerical value of the polynomial $p-z^1 p+\cdots$ for $x_1=\cdots=x_n=1$ is equal to
$$\sum_{n-i=0}^{n}(-1)^{n-i}\binom{n}{n-i}i^k=(-1)^n\sum_{i=0}^{n}(-1)^i\binom{n}{i}i^k.$$

Using the results previously obtained, one can show that this expression is equal to 0 for $0 \leqslant k < n$ and is equal to $c = n!$ for $k = n$, from which Euler's identity follows.

1.38 Consider the left-hand side of the identity as a polynomial in n. The constant term is then equal to

$$-\binom{p}{1}(-1)^p + \binom{p}{2}(-2)^p - \cdots + (-1)^p \binom{p}{p}(-p)^p$$

$$= p^p - \binom{p}{1}(p-1)^p + \binom{p}{2}(p-2)^p - \cdots + (-1)^{p-1}\binom{p}{p-1} = p!.$$

(Use Euler's formula, Problem 1.37.)

Similarly it can be shown that the coefficient of n^p is equal to

$$1 - \binom{p}{1} + \binom{p}{2} - \cdots + (-1)^p \binom{p}{p} = 0,$$

the coefficient of n^k for $1 \leqslant k \leqslant p-1$ is equal to:

$$(-1)^{p-k}\binom{p}{k}\left\{-\binom{p}{1} + \binom{p}{2}2^{p-k} - \binom{p}{3}3^{p-k} + \cdots + (-1)^p p^{p-k}\right\}.$$

If $p - k = m$, then $1 \leqslant m \leqslant p - 1$ and thus the coefficient of n^k is equal to

$$(-1)^k \binom{p}{m}\left\{p^m - \binom{p}{1}(p-1)^m + \binom{p}{2}(p-2)^m - \cdots + (-1)^{p-1}\binom{p}{p-1}\right\}$$

$$= (-1)^k \binom{p}{m} S_{m,p} = 0,$$

since $m < p$.

1.39 The identity will be established by mathematical induction on n. For $n = 1$ it reduces to the identity $1/x - 1/(x+1) = 1/x(x+1)$. Suppose that the proposed identity is true. Replacing x by $x+1$ yields

$$\frac{\binom{n}{0}}{x+1} - \frac{\binom{n}{1}}{x+2} + \cdots + (-1)^n \frac{\binom{n}{n}}{x+n+1} = \frac{n!}{(x+1)\cdots(x+n+1)}.$$

Subtract this relation from the original equation to obtain

$$\frac{\binom{n+1}{0}}{x} - \frac{\binom{n+1}{1}}{x+1} + \cdots + (-1)^{n+1}\frac{\binom{n+1}{n+1}}{x+n+1} = \frac{n!}{(x+1)\cdots(x+n)}\left(\frac{1}{x} - \frac{1}{x+n+1}\right)$$

$$= \frac{(n+1)!}{x(x+1)\cdots(x+n+1)}$$

by virtue of the recurrence relation for binomial coefficients, that is, the identity in the statement of the problem for $n+1$.

Solutions

CHAPTER 2

2.1 Subtract from the total of 40 students the number who prefer mathematics, physics, and chemistry respectively:

$$40-14-16-11.$$

The students who prefer mathematics and physics are subtracted twice, and thus they must also be added once.

A similar procedure for the two other pairs of subjects yields the number

$$40-14-16-11+7+8+5$$

The students who prefer all three subjects were subtracted three times and after were added three times. In order to obtain the number of students who did not prefer any of the three subjects it is necessary to subtract once the four students who had a preference for all three subjects. The final result is

$$40-14-16-11+7+8+5-4=15.$$

2.2 The proof uses induction on $q \geqslant 2$. For $q=2$ the formula becomes

$$|A_1 \cup A_2| = |A_1| + |A_2| - |A_1 \cap A_2|,$$

which can easily be verified.

Suppose the formula is true for each union of at most $q-1$ sets. It follows that

$$|A_1 \cup \cdots \cup A_q| = |A_1 \cup \cdots \cup A_{q-1}| + |A_q| - |(A_1 \cup \cdots \cup A_{q-1}) \cap A_q|.$$

Applying the distributive property for intersections of sets, one has

$$(A_1 \cup \cdots \cup A_{q-1}) \cap A_q = (A_1 \cap A_q) \cup (A_2 \cap A_q) \cup \cdots \cup (A_{q-1} \cap A_q),$$

and from the inductive hypothesis it follows that

$$|A_1 \cup \cdots \cup A_q| = \sum_{1 \leqslant i < q} |A_i| - \sum_{1 \leqslant i < j < q} |A_i \cap A_j| + \cdots (-1)^q \left| \bigcap_{i=1}^{q-1} A_i \right|$$

$$+ |A_q| - \sum_{1 \leqslant i < q} |A_i \cap A_q| + \cdots + (-1)^{q+1} \left| \bigcap_{i=1}^{q} A_i \right|,$$

by using the idempotent property of intersection in the form

$$(A_i \cap A_q) \cap (A_j \cap A_q) = A_i \cap A_j \cap A_q, \ldots, \quad \bigcap_{i=1}^{q-1}(A_i \cap A_q) = \bigcap_{i=1}^{q} A_i.$$

By regrouping terms, one obtains the formula of inclusion and exclusion for unions of q sets. The formula is called the Principle of Inclusion and Exclusion because of the alternating signs of the right-hand side.

2.3 Let $P \subset Q = \{1, 2, \ldots, q\}$ be a fixed set with $|P| = p$. The number of elements which belong to all the sets A_i with $i \in P$ and do not belong to any of the sets A_j with $j \in Q \setminus P$ coincides with the set of elements which belong to the set $\bigcap_{i \in P} A_i$ and do not belong to any set

$$A_j \cap \bigcap_{i \in P} A_i \subset \bigcap_{i \in P} A_i \quad \text{for} \quad j \in Q \setminus P.$$

By applying the Principle of Inclusion and Exclusion one finds that the number of elements which belong to all A_i with $i \in P$ and do not belong to any A_j with $j \in Q \setminus P$ is equal to

$$\left|\bigcap_{i \in P} A_i\right| - \left|\bigcup_{j \in Q \setminus P}\left(A_j \cap \bigcap_{i \in P} A_i\right)\right| = \left|\bigcap_{i \in P} A_i\right| - \sum_{\substack{K \supset P \\ |K|=p+1}} \left|\bigcap_{i \in K} A_i\right| + \sum_{\substack{K \supset P \\ |K|=p+2}} \left|\bigcap_{i \in K} A_i\right| - \cdots$$

$$= \sum_{K \supset P} (-1)^{|K|-|P|} \left|\bigcap_{i \in K} A_i\right|.$$

Sum these numbers with respect to all the subsets $P \subset Q$ with $|P|=p$ elements:

$$\sum_{|P|=p} \sum_{K \supset P} (-1)^{|K|-|P|} \left|\bigcap_{i \in K} A_i\right| = \sum_{\substack{K \subset Q \\ |K| \geq p}} \sum_{\substack{P \subset K \\ |P|=p}} (-1)^{|K|-|P|} \left|\bigcap_{i \in K} A_i\right|$$

by changing the order of the summation.

The index set $P \subset K$ with $|P|=p$ and $|K|=k$ can be chosen in $\binom{k}{p}$ ways for each choice of K, and since $\left|\bigcap_{i \in K} A_i\right|$ does not depend on P, it follows that the desired number is equal to

$$\sum_{k=p}^{q} (-1)^{k-p} \binom{k}{p} \sum_{\substack{K \subset Q \\ |K|=k}} \left|\bigcap_{i \in K} A_i\right|.$$

2.4 Let A_i denote the set of natural numbers less than or equal to n which are multiples of p_i. It follows that

$$|A_i| = \frac{n}{p_i}; \quad |A_i \cap A_j| = \frac{n}{p_i p_j},$$

since the numbers p_i and p_j being prime are also relatively prime.

The natural numbers which are less than or equal to n and which are prime to n are numbers in the set $X = \{1, \ldots, n\}$ which do not belong to any of the sets A_i for $1 \leq i \leq q$. Thus

$$\varphi(n) = n - |A_1 \cup \cdots \cup A_q|$$

$$= n - \sum_{i=1}^{q} |A_i| + \sum_{1 \leq i < j \leq q} |A_i \cap A_j| - \cdots + (-1)^q \left|\bigcap_{i=1}^{q} A_i\right|$$

$$= n - \sum_{i=1}^{q} \frac{n}{p_i} + \sum_{1 \leq i < j \leq q} \frac{n}{p_i p_j} - \cdots + (-1)^q \frac{n}{p_1 p_2 \cdots p_q}$$

$$= n \left(1 - \frac{1}{p_1}\right)\left(1 - \frac{1}{p_2}\right) \cdots \left(1 - \frac{1}{p_q}\right).$$

Solutions

2.5 Let A_i denote the set of the $(n-1)!$ permutations which have i as a fixed point, and apply the Principle of Inclusion and Exclusion to find the number of permutations which have at least one fixed point. This number is equal to

$$|A_1 \cup \cdots \cup A_n| = \sum_{i=1}^{n} |A_i| - \sum_{1 \leq i < j \leq n} |A_i \cap A_j| + \cdots + (-1)^{n-1} \left|\bigcap_{i=1}^{n} A_i\right|.$$

But $|A_{i_1} \cap A_{i_2} \cap \cdots \cap A_{i_k}| = (n-k)!$, since a permutation of the set $A_{i_1} \cap \cdots \cap A_{i_k}$ has fixed points in the positions $i_1, i_2, \ldots, i_k$, and the other positions contain a permutation of the $n-k$ remaining elements. The latter set of permutations has cardinality $(n-k)!$. But k positions $i_1, \ldots, i_k$ can be chosen from the set of n positions in $\binom{n}{k}$ ways, and thus

$$\left|\bigcup_{i=1}^{n} A_i\right| = \binom{n}{1}(n-1)! - \binom{n}{2}(n-2)! + \cdots + (-1)^{n-1}\binom{n}{n},$$

$$D(n) = n! - \left|\bigcup_{i=1}^{n} A_i\right| = n! - \binom{n}{1}(n-1)! + \cdots + (-1)^n \binom{n}{n},$$

from which the given expression for $D(n)$ follows.

Thus $\lim_{n \to \infty} D(n)/n! = 1/e$, and hence for large n the probability that a permutation of n elements chosen at random has no fixed points is approximately equal to $1/2.7$.

Since the p fixed points $(0 \leq p \leq n)$ can be chosen in $\binom{n}{p}$ ways and the other $n-p$ points are no longer fixed, it follows that the number of permutations in S_n with p fixed points is equal to $\binom{n}{p} D(n-p)$. One uses the fact that for every choice of p fixed points there are $D(n-p)$ permutations of the remaining objects without fixed points, if, by definition, one takes $D(0) = 1$.

2.6 By Problem 2.5, $D(n)$ can be expressed as

$$D(n) = n!\left(1 - \frac{1}{1!} + \cdots + \frac{(-1)^n}{n!}\right).$$

In order to obtain the expression for $E(n)$, let A_i be the set of even permutations $p \in S_n$ such that $p(i) = i$. Since there are $\frac{1}{2} n!$ even permutations in S_n, it follows that

$$E(n) = \tfrac{1}{2} n! - \left|\bigcup_{i=1}^{n} A_i\right|$$

$$= \tfrac{1}{2} n! - \sum_{i=1}^{n} |A_i| + \sum_{1 \leq i < j \leq n} |A_i \cap A_j| - \cdots + (-1)^n \left|\bigcap_{i=1}^{n} A_i\right|.$$

Using $|A_{i_1} \cap A_{i_2} \cap \cdots \cap A_{i_k}| = \tfrac{1}{2}(n-k)!$, one can show analogously that

$$E(n) = \tfrac{1}{2} n! - \binom{n}{1}\tfrac{1}{2}(n-1)! + \binom{n}{2}\tfrac{1}{2}(n-2)! - \cdots + (-1)^{n-1}\binom{n}{n-1} + (-1)^n$$

$$= \tfrac{1}{2}\{D(n) + (-1)^{n-1}(n-1)\}.$$

2.7 Let $n = p_1^{i_1} p_2^{i_2} \cdots p_q^{i_q}$, where $p_1, \ldots, p_q$ are pairwise distinct primes. The result can be proved by induction on $i_1 + i_2 + \cdots + i_q$. If $i_1 = 1$ and $i_2 = \cdots = i_q = 0$, then n is a prime and the sum of the left hand side of the equation becomes $\varphi(1) + \varphi(n) = 1 + n - 1 = n$, and the equality is satisfied if, by definition, $\varphi(1) = 1$. Suppose that the property is true for all numbers for which $i_1 + \cdots + i_q \leq r - 1$, and let n be a natural number such that $i_1 + \cdots + i_q = r$. Let

$$D_1 = \{ p_1^{j_1} p_2^{j_2} \cdots p_q^{j_q} \mid 0 \leq j_1 \leq i_1 - 1, 0 \leq j_2 \leq i_2, \ldots, 0 \leq j_q \leq i_q \},$$

$$D_2 = \{ p_1^{i_1} p_2^{j_2} \cdots p_q^{j_q} \mid 0 \leq j_2 \leq i_2, \ldots, 0 \leq j_q \leq i_q \}.$$

It follows that $D_1 \cup D_2$ represents a partition of the set D of divisors of n. Thus, using the inductive hypothesis, one can write

$$\sum_{d \mid n} \varphi(d) = \sum_{d \in D_1} \varphi(d) + \sum_{d \in D_2} \varphi(d) = \frac{n}{p_1} + p_1^{i_1} \frac{n}{p_1^{i_1}} \left(1 - \frac{1}{p_1}\right) = n,$$

since, if a and b are relatively prime, then $\varphi(ab) = \varphi(a) \varphi(b)$. (Gauss).

2.8 Suppose that a square matrix of order 3 satisfies the given conditions. By adding 1 to all its elements one obtains the matrix

$$\begin{pmatrix} a_1 & b_1 & c_1 \\ a_2 & b_2 & c_2 \\ a_3 & b_3 & c_3 \end{pmatrix},$$

where $a_1 + b_1 + c_1 = r + 3$, $a_2 + b_2 + c_2 = r + 3$, and so on. Using the result of Problem 1.19, $r + 3$ can be written as a sum of three positive numbers in $\binom{r+2}{2}$ ways. Thus by completing the first two rows of the matrix in all possible ways one obtains $\binom{r+2}{2}^2$ matrices. The elements of the third row are now determined from the condition that the sum of the elements of each column must be equal to $r + 3$. However, one must eliminate the case when there are negative or zero elements on the third line. Observe that the following relation holds:

$$b_1 + c_1 + b_2 + c_2 = r + 3 + a_3. \tag{1}$$

If $a_3 \leq 0$, the minimum value for a_3 is $r + 3 - 2(r+1) = 1 - r$, and thus by using (1), it follows that the sum of any two of the numbers b_1, c_1, b_2, c_2 is smaller than $r + 3$. It follows that the matrix will be completed correctly in row 3 (except for the element a_3), with elements that are greater than zero. Suppose that $a_3 \leq 0$. In this case one can generate all matrices of the indicated form with row and column sum equal to $r + 3$, for which the unique nonpositive element is a_3, by writing the number $r + 3 + a_3$ as a sum of four positive numbers. Again by using Problem 1.19, it can be seen that for $a_3 = 0, -1, \ldots, -r+1$, the number of solutions of equation (1) is equal to

$$\binom{r+2}{3} + \binom{r+1}{3} + \cdots + \binom{3}{3} = \binom{r+3}{4} - \binom{r+2}{4} + \binom{r+2}{4} - \binom{r+1}{4}$$

$$+ \cdots + \binom{5}{4} - \binom{4}{4} + \binom{3}{3} = \binom{r+3}{4}.$$

Solutions

Thus if one of a_3, b_3, c_3 is negative or zero (in this case two of them cannot be zero), the number of matrices which do not satisfy the given condition is equal to $\binom{r+3}{4}$. Thus the desired number of matrices is equal to $\binom{r+2}{2}^2 - 3\binom{r+3}{4}$.

These matrices are also called magic squares. Sometimes one also requires that the sum of the diagonal elements be equal to r. [P. A. MacMahon, *Combinatory Analysis II*, Cambridge University Press, Cambridge, 1916.]

2.9 Since
$$\frac{D(n)}{n!} = 1 - \frac{1}{1!} + \frac{1}{2!} - \cdots + \frac{(-1)^n}{n!},$$
it can be seen that this expression is the coefficient of t^n in the expansion
$$\frac{e^{-t}}{1-t} = \left(1 - \frac{t}{1!} + \frac{t^2}{2!} - \cdots\right)(1 + t + t^2 + \cdots).$$
These two recurrence relations can be obtained by a straightforward calculation.

2.10 Let $Y = \{y_1, \ldots, y_m\}$, and for every $1 \leq i \leq m$ let A_i be the set of functions from X to Y for which y_i is not the image of any element in X:
$$A_i = \{f : X \to Y \mid y_i \notin f(X)\}.$$
It follows that $s_{n,m} = m^n - |A_1 \cup A_2 \cup \cdots \cup A_m|$, since the total number of functions from X to Y is equal to m^n. By using the Principle of Inclusion and Exclusion one can verify that
$$s_{n,m} = m^n - \sum_{i=1}^{m} |A_i| + \sum_{1 \leq i < j \leq m} |A_i \cap A_j| - \cdots + (-1)^m \left|\bigcap_{i=1}^{m} A_i\right|.$$
But A_i is the set of functions defined on X with values in $Y \setminus \{y_i\}$, and thus $|A_i| = (m-1)^n$, and $A_i \cap A_j$ is the set of functions defined on X with values in $Y \setminus \{y_i, y_j\}$. Thus $|A_i \cap A_j| = (m-2)^n$. In general,
$$|A_{i_1} \cap A_{i_2} \cap \cdots \cap A_{i_k}| = (m-k)^n, \quad \text{where } 1 \leq i_1 < i_2 < \cdots < i_k \leq m.$$
Since there are $\binom{m}{k}$ subsets of indices $K \subset \{1, \ldots, m\}$ with $|K| = k$, it follows that each sum
$$\sum_{\substack{K \subset \{1, \ldots, m\} \\ |K| = k}} \left|\bigcap_{i \in K} A_i\right|$$
contains $\binom{m}{k}$ terms, each of which is equal to $(m-k)^n$. This implies that
$$s_{n,m} = m^n - \binom{m}{1}(m-1)^n + \binom{m}{2}(m-2)^n - \cdots + (-1)^{m-1}\binom{m}{m-1}.$$
For $m = n$, $s_{n,m}$ represents the number of bijections $f : X \to Y$ with $|X| = |Y| = n$. Hence $s_{n,n} = n!$, or
$$n! = n^n - \binom{n}{1}(n-1)^n + \binom{n}{2}(n-2)^n - \cdots + (-1)^{n-1} n.$$

If $n < m$, there does not exist a surjection from X to Y and thus $s_{n,m} = 0$.

2.11 Let $Z = \{y_1, \ldots, y_r\}$. Then, in the notation of the preceding problem,

$$s_{n,m,r} = m^n - |A_1 \cup \cdots \cup A_r|$$
$$= m^n - \binom{r}{1}(m-1)^n + \binom{r}{2}(m-2)^n - \cdots + (-1)^r(m-r)^n.$$

For $r = m$ one can conclude that $s_{n,m,m} = s_{n,m}$, since $Z = Y$.

2.12 The number of words which use all $2n$ letters of the alphabet A is equal to

$$\frac{(2n)!}{(2!)^n} = \frac{(2n)!}{2^n}.$$

[Identical letters can be permuted among themselves in $(2!)^n = 2^n$ distinct ways, to yield the same word formed with the $2n$ letters of A.] Let A_i denote the set of words formed with the $2n$ letters of A for which the two letters denoted a_i are adjacent. It follows that the desired number is equal to

$$\frac{(2n)!}{2^n} - |A_1 \cup A_2 \cup \cdots \cup A_n|. \tag{1}$$

In order to evaluate $|A_1 \cup \cdots \cup A_n|$, apply the Principle of Inclusion and Exclusion:

$$|A_1 \cup \cdots \cup A_n| = \sum_{i=1}^n |A_i| - \sum_{1 \leq i < j \leq n} |A_i \cap A_j| + \cdots$$

$$+ (-1)^{k-1} \sum_{1 \leq i_1 < \cdots < i_k \leq n} |A_{i_1} \cap A_{i_2} \cap \cdots \cap A_{i_k}| + \cdots$$

$$+ (-1)^{n-1} \left| \bigcap_{i=1}^n A_i \right|. \tag{2}$$

One proceeds to calculate in the general case the number of elements in $A_{i_1} \cap A_{i_2} \cap \cdots \cap A_{i_k}$ and show that this number does not depend on the choice of indices $1 \leq i_1 < i_2 < \cdots < i_k \leq n$. If a word belongs to this set, it means that it belongs to each of the sets $A_{i_1}, A_{i_2}, \ldots, A_{i_k}$, and thus the letters $a_{i_1}, a_{i_1}; a_{i_2}, a_{i_2}; \ldots; a_{i_k}, a_{i_k}$ are adjacent. The words for which these k pairs of letters are adjacent are obtained in the following manner: Form all words having $2n - k$ letters taken from an alphabet obtained from A by suppressing one copy of each letter from $a_{i_1}, a_{i_2}, \ldots, a_{i_k}$. Then in each word thus formed, repeat the letters $a_{i_1}, \ldots, a_{i_k}$ by adding the letter a_{i_j} immediately after itself for $j = 1, \ldots, k$. It follows that

$$|A_{i_1} \cap \cdots \cap A_{i_k}| = \frac{(2n-k)!}{(2!)^{n-k}} = \frac{2^k(2n-k)!}{2^n}. \tag{3}$$

Since the indices $i_1, \ldots, i_k$ satisfy

$$1 \leq i_1 < \cdots < i_k \leq n,$$

Solutions

they can be chosen in $\binom{n}{k}$ ways, and hence the number of words which do not contain two identical adjacent letters can be written using (1), (2), and (3) as

$$\frac{(2n)!}{2^n} - \binom{n}{1}\frac{2(2n-1)!}{2^n} + \binom{n}{2}\frac{2^2(2n-2)!}{2^n} - \cdots + \frac{(-1)^n 2^n n!}{2^n}.$$

2.13 Every digraph which does not contain a circuit has at least one vertex x at which no arc begins [i.e., $d^+(x)=0$] and at least one vertex y in which no arc terminates [i.e., $d^-(y)=0$]. Denote by A_i the set of digraphs with n vertices, which have labels from the set $\{1,\ldots,n\}$ which do not contain a circuit, and which have the property $d^+(i)=0$. It follows that

$$a_n = |A_1 \cup A_2 \cup \cdots \cup A_n|$$

$$= \sum_{i=1}^n |A_i| - \sum_{1 \le i<j \le n} |A_i \cap A_j| + \cdots + (-1)^{n-1}\left|\bigcap_{i=1}^n A_i\right|.$$

But

$$\sum_{(i_1,\ldots,i_k)} |A_{i_1} \cap \cdots \cap A_{i_k}| = \binom{n}{k} 2^{k(n-k)} a_{n-k},$$

since the number of subgraphs with $n-k$ vertices labeled from the set $\{1,\ldots,n\} \setminus \{i_1,\ldots,i_k\}$ and without circuits is equal to a_{n-k}, and since the vertices $i_1, \ldots, i_k$ have the property $d^+(i_1) = \cdots = d^+(i_k) = 0$, that is, they can be joined by arcs having a uniquely determined orientation to the other $n-k$ vertices in $(2^{n-k})^k = 2^{k(n-k)}$ ways.

Thus each term of the given sum is equal to $2^{k(n-k)} a_{n-k}$, and the sum contains $\binom{n}{k}$ such terms. This observation yields the recurrence relation for the numbers a_n.

Using the given formula one finds that $a_1 = 1$, $a_2 = 3$, $a_3 = 25$. These values can also be obtained by direct computation. [R. W. Robinson, *New Directions in the Theory of Graphs*, Academic Press, New York–London, 1973, 239–273].

2.14 Let $A_\varnothing(n; 1^m)$ denote the number of arrangement schemes of n identical objects in m pairwise distinct cells; it follows that $A_\varnothing(n; 1^m) = \binom{n+m-1}{n}$. In fact, if the empty cell is included, this number represents the number of ways in which n can be written as a sum of m non-negative integers. Using Problem 1.19 one can show that this number is also equal to $\binom{n+m-1}{n}$.

In order to obtain the expression in the statement of the problem, one should first arrange the λ_1 groups of one object, then the λ_2 groups of two identical objects, and so on. The result is

$$A_\varnothing(1^{\lambda_1} 2^{\lambda_2} \cdots n^{\lambda_n}; 1^m) = A_\varnothing(1; 1^m)^{\lambda_1} A_\varnothing(2; 1^m)^{\lambda_2} \cdots A_\varnothing(n; 1^m)^{\lambda_n}$$

$$= \binom{m}{1}^{\lambda_1} \binom{m+1}{2}^{\lambda_2} \cdots \binom{m+n-1}{n}^{\lambda_n}.$$

In order to obtain the expression for the number of arrangement schemes which do not leave any cell empty, apply the Principle of Inclusion and Exclusion. Denoting by S_i the set of arrangement schemes which leave the cell i

empty, it follows that

$$A(1^{\lambda_1} \cdots n^{\lambda_n}; 1^m) = A_\emptyset(1^{\lambda_1} \cdots n^{\lambda_n}; 1^m) - \left| \bigcup_{i=1}^{m} S_i \right|$$

$$= A_\emptyset(1^{\lambda_1} \cdots n^{\lambda_n}; 1^m) - \sum_{i=1}^{m} |S_i| + \sum_{1 \le i < j \le m} |S_i \cap S_j| - \cdots$$

$$+ (-1)^m \left| \bigcap_{i=1}^{m} S_i \right|$$

$$= A_\emptyset(1^{\lambda_1} \cdots n^{\lambda_n}; 1^m) - \binom{m}{1} A_\emptyset(1^{\lambda_1} \cdots n^{\lambda_n}; 1^{m-1})$$

$$+ \binom{m}{2} A_\emptyset(1^{\lambda_1} \cdots n^{\lambda_n}; 1^{m-2}) - \cdots.$$

This is equivalent to the expression sought, since $\bigcap_{i=1}^{m} S_i = \emptyset$ and $\binom{m}{k} = \binom{m}{m-k}$.
[P. A. MacMahon, *Combinatory Analysis I*, Cambridge University Press, Cambridge, 1915.]

2.15 The number under investigation is of the form $A(1^{\lambda_1} 2^{\lambda_2} \cdots n^{\lambda_n}; 1^m)$. For example, for the number $150 = 2 \times 3 \times 5^2$ one has

$$A(1^2 2^1; 1^3) = \binom{3}{1}\binom{1}{1}^2\binom{2}{2} - \binom{3}{2}\binom{2}{1}^2\binom{3}{2} + \binom{3}{3}\binom{3}{1}^2\binom{4}{2} = 21$$

factorizations as a product of three natural numbers. If some factors equal to 1 are allowed, then q admits $A_\emptyset(1^{\lambda_1} \cdots n^{\lambda_n}; 1^m)$ factorizations as a product of m factors.

2.16 We show that the number $M(p, q)$ represents the number of edge coverings of the set of vertices of the complete bipartite graph $K_{p,q}$. Suppose that the union of two sets which generate the graph $K_{p,q}$ is $X \cup Y$ where $X = \{x_1, \ldots, x_p\}$ and $Y = \{y_1, \ldots, y_q\}$. Denote by A_p the family of sets of edges of the graph which cover the vertices of X, and by A_p^i the family of sets of edges which cover the vertices of X and do not cover the vertex y_i for $i = 1, \ldots, q$.

Each vertex in X can be covered by edges in $2^q - 1$ ways, since the endpoints of these edges in the set Y form a nonempty subset of Y. It follows that

$$|A_p| = (2^q - 1)^p.$$

Thus $M(p, q) = |A_p| - |\bigcup_{i=1}^{q} A_p^i|$. By using the Principle of Inclusion and Exclusion it can be shown that

$$\left| \bigcup_{i=1}^{q} A_p^i \right| = \sum_{i=1}^{q} |A_p^i| - \sum_{1 \le i < j \le q} |A_p^i \cap A_p^j| + \cdots$$

$$= \binom{q}{1}(2^{q-1} - 1)^p - \binom{q}{2}(2^{q-2} - 1)^p + \cdots + (-1)^q \binom{q}{1},$$

and thus $M(p, q)$ has the given meaning. Since the graphs $K_{p,q}$ and $K_{q,p}$ are isomorphic, it follows that $M(p, q) = M(q, p)$. [I. Tomescu, *J. Combinatorial Theory*, **B28**(2) (1980), 127–141.]

2.17 From Newton's binomial formula one obtains

$$x^k = (x-1+1)^k = \sum_{i=0}^{k} \binom{k}{i}(x-1)^i$$

and

$$(x-1)^k = \sum_{i=0}^{k} \binom{k}{i}(-1)^{k-i}x^i.$$

It follows that for $k \leq n$ one has

$$x^k = \sum_{i=0}^{n} \binom{k}{i}(x-1)^i = \sum_{i=0}^{n} \binom{k}{i} \sum_{j=0}^{n} \binom{i}{j}(-1)^{i-j}x^j$$

$$= \sum_{j=0}^{n} \left\{ \sum_{i=0}^{n} \binom{k}{i}\binom{i}{j}(-1)^{i-j} \right\} x^j.$$

Since the polynomials x^k for $k = 0, \ldots, n$ are of different degrees, it follows that they are linearly independent and therefore the above representation is unique. Thus

$$\sum_{i=0}^{n} \binom{k}{i}\binom{i}{j}(-1)^{i-j} = \delta_{k,j}, \tag{1}$$

where the Kronecker symbol is defined by $\delta_{k,j} = 1$ for $k = j$ and $\delta_{k,j} = 0$ for $k \neq j$. Hence the numbers $a_{ki} = \binom{k}{i}$ and $b_{ki} = (-1)^{k-i}\binom{k}{i}$, defined for $0 \leq i, k \leq n$, form two square matrices A and B of order $n+1$. Let a denote the column vector with $n+1$ components $a_0, \ldots, a_n$, and b the column vector with $n+1$ components $b_0, \ldots, b_n$. The problem states that $a = Ab$ implies $b = Ba$. But (1) shows that the matrices A and B are inverse to each other, that is, $AB = I_{n+1}$. Thus by left-multiplying the equation $a = Ab$ by B, one finds that $Ba = (BA)b = I_{n+1}b = b$.

2.18 If $m = u_1 + \cdots + u_n$ and $u_i \geq 2$ for $1 \leq i \leq n$, it follows that $m - n = (u_1 - 1) + \cdots + (u_n - 1)$ and $u_i - 1 \geq 1$ for $1 \leq i \leq n$. Thus by the Problem 1.19 the number of such representations is equal to $\binom{m-n-1}{n-1}$. Let A_i denote the set of representations $m = u_1 + \cdots + u_n$ where $u_1 \geq 1, \ldots, u_{i-1} \geq 1, u_i = 1, u_{i+1} \geq 1, \ldots, u_n \geq 1$. Because the number of representations for which $u_s \geq 1$ for $1 \leq s \leq n$ is equal to $\binom{m-1}{n-1}$, it follows that

$$\binom{m-n-1}{n-1} = \binom{m-1}{n-1} - \left| \bigcup_{i=1}^{n} A_i \right|$$

$$= \binom{m-1}{n-1} + \sum_{i=1}^{n} \sum_{\substack{K \subset \{1,\ldots,n\} \\ |K| = i}} (-1)^i \left| \bigcap_{i \in K} A_i \right|$$

by applying the Principle of Inclusion and Exclusion. Since $\bigcap_{i \in K} A_i$ is the set of representations of m as a sum of n parts $u_1, \ldots, u_n$ such that $u_k = 1$ for any $k \in K$, it follows that

$$\left| \bigcap_{i \in K} A_i \right| = \binom{m-i-1}{n-i-1}$$

and the set K may be chosen in $\binom{n}{i}$ ways such that $|K| = i$. One thus finds that $|\bigcap_{i \in K} A_i| = 0$ if $K = \{1, \ldots, n\}$, because $m \geq n+1$, and hence one can assume that $i \leq n-1$. The last equality becomes

$$\binom{m-n-1}{n-1} = \sum_{i=0}^{n-1} (-1)^i \binom{n}{i} \binom{m-i-1}{n-i-1}.$$

2.19 Let $A = (a_{ij})_{i,j=1,\ldots,n}$ and $B = (b_{ij})_{i,j=1,\ldots,n}$ be two compatible n-by-n matrices and suppose that $a_{ij} + b_{ij} \neq 0$. In this case at least one of the terms is nonzero, say a_{ij}, and thus $x_i \leq x_j$ for $1 \leq i, j \leq n$. It follows that $A + B$ is a compatible matrix.

Let $AB = C = (c_{ij})_{i,j=1,\ldots,n}$, and suppose that $c_{ij} \neq 0$. Since

$$c_{ij} = \sum_{k=1}^{n} a_{ik} b_{kj},$$

there exists an index k such that $a_{ik} \neq 0$ and $b_{kj} \neq 0$. Thus by definition $x_i \leq x_k$ and $x_k \leq x_j$, and by transitivity $x_i \leq x_j$. It follows that the matrix C is compatible.

Let A be a compatible nonsingular matrix of order n. By renumbering the elements of V one can assume that $x_i \leq x_j$ implies $i \leq j$. Thus A is upper triangular, which implies that $\det A = \prod_{i=1}^{n} a_{ii} \neq 0$. It follows that $a_{ii} \neq 0$ for $i = 1, \ldots, n$.

Let $B = (b_{ij})_{i,j=1,\ldots,n} = A^{-1}$, and suppose that B is not compatible, that is, there is an element x_i which is not less than or equal to x_j and such that $b_{ij} \neq 0$. Choose an index i which is maximal relative to this property for a fixed index j.

Since x_i is not less than or equal to x_j, we have $i > j$ and thus

$$\sum_{k=1}^{n} a_{ik} b_{kj} = 0.$$

But $a_{ii} b_{ij} \neq 0$, which implies the existence of an index $k \neq i$ such that $a_{ik} b_{kj} \neq 0$ or $a_{ik} \neq 0$ and $b_{kj} \neq 0$. Since A is compatible, it follows that $x_i \leq x_k$ and thus $i < k$. From the convention for choosing the index i, it follows that $b_{kj} \neq 0$ implies $x_k \leq x_j$. By transitivity one finds that $x_i \leq x_j$, which contradicts the hypothesis. Thus $B = A^{-1}$ is compatible.

2.20 Let $Z = (z_{ij})_{i,j=1,\ldots,n}$ be an n-by-n matrix such that

$$z_{ij} = \begin{cases} 1 & \text{if } x_i \leq x_j, \\ 0 & \text{otherwise.} \end{cases}$$

The conditions imposed on the function μ can be written in matrix form as $MZ = I$, where $M = (m_{ij})_{i,j=1,\ldots,n}$ and $m_{ij} = \mu(x_i, x_j)$, and I is the identity matrix of order n. In fact, one can write

Solutions

$$S_{ii} = \sum_{k=1}^{n} m_{ik}z_{ki} = m_{ii}z_{ii} + 0 = 1,$$

since $m_{ii} = z_{ii} = 1$ and $m_{ik} \neq 0$ implies $x_i \leqslant x_k$, and hence $z_{ki} = 0$ for $k \neq i$. Furthermore, $s_{ij} = \sum_{k=1}^{n} m_{ik}z_{kj} = 0$ for x_i which is not less than or equal to x_j, since $m_{ik} \neq 0$ and $z_{kj} \neq 0$ imply that $x_i \leqslant x_k$ and $x_k \leqslant x_j$; thus $x_i \leqslant x_j$, which is contrary to the hypothesis.

For $x_i < x_j$ one has $s_{ij} = \sum_{x_i \leqslant x_k \leqslant x_j} m_{ik} = 0$ by use of the third condition imposed upon the Moebius function.

Since $z_{ii} = 1$ for $i = 1, \ldots, n$, it follows from the solution of the preceding problem that the matrix Z is nonsingular and similarly that $M = Z^{-1}$ is a compatible matrix. Thus the desired function $\mu(x, y)$ is defined by the matrix Z^{-1}.

Since M is compatible with the partial order relation defined in V, it follows that $\mu(x, y) = 0$ if x is not less than or equal to y. Thus, from the fact that $s_{ii} = 1$ and $s_{ij} = 0$ for every $i \neq j$, one can conclude that the function μ defined by the matrix Z^{-1} satisfies the last two conditions imposed on the Moebius function.

2.21 Observe that the function μ can be obtained inductively as follows: $\mu(a, a) = 1$, and if $\mu(a, y)$ is defined for every y such that $a \leqslant y < b$, then from the equation

$$\sum_{a \leqslant y \leqslant b} \mu(a, y) = 0$$

one obtains the value

$$\mu(a, b) = - \sum_{a \leqslant y < b} \mu(a, y).$$

Thus $\mu(a, x)$ can be calculated for every x which satisfies $a \leqslant x \leqslant b$, using the previously calculated values for elements which lie between a and x.

(a) Now we show that $\mu(X, Y) = (-1)^{|Y|-|X|}$ for every $X \subset Y \subset S$. By definition set $\mu(X, Y) = 0$ if $X \not\subset Y$. In fact, the function thus defined satisfies

$$\sum_{A \subset Y \subset B} (-1)^{|Y|-|A|} = \sum_{k=0}^{|B|-|A|} \binom{|B|-|A|}{k} (-1)^k$$

$$= \begin{cases} 1 & \text{for } A = B, \\ 0 & \text{for } B \supset A \text{ and } B \neq A. \end{cases}$$

(b) We show that

$$\mu(x, y) = \mu\left(\frac{y}{x}\right),$$

where $\mu(x)$ is the number-theoretic Moebius function defined as follows: If the decomposition of k into prime factors has the form $k = p_1^{\alpha_1} p_2^{\alpha_2} \cdots p_r^{\alpha_r}$, where

$p_1; \ldots, p_r$ are pairwise distinct primes, then

$$\mu(k) = \begin{cases} (-1)^r & \text{if } \alpha_1 = \alpha_2 = \cdots = \alpha_r = 1, \\ 0 & \text{otherwise.} \end{cases}$$

We see that $\mu(a, a) = (-1)^0 = 1$. If $a \neq b$ and $a \mid b$ it follows that

$$\sum_{a \mid y \mid b} \mu(a, y) = \sum_{a \mid y \mid b} \mu\left(\frac{y}{a}\right) = \sum_{d \mid b/a} \mu(d).$$

Let $b/a = k = p_1^{\alpha_1} \cdots p_r^{\alpha_r}$. Then

$$\sum_{d \mid k} \mu(d) = \sum_{1 \leq i_1 < \cdots < i_s \leq r} \mu(p_{i_1} p_{i_2} \cdots p_{i_s});$$

the rest of the terms are zero by definition of the function $\mu(k)$. Thus

$$\sum_{d \mid k} \mu(d) = \sum_{s=0}^{r} \binom{r}{s}(-1)^s = 0.$$

Also, it follows from the definition of the function $\mu(k)$ that $\mu(a, b) = 0$ if a does not divide b.

(c) In this case every interval of the form

$$[a, b] = \{z \mid a \leq z \leq b\}$$

is a chain (totally ordered set). Now calculate $\mu(a_1, a_n)$ for the chain

$$a_1 < a_2 < \cdots < a_n.$$

By definition $\mu(a_1, a_1) = 1$. Considering the interval $[a_1, a_2]$, one sees that $\mu(a_1, a_1) + \mu(a_1, a_2) = 0$ and thus $\mu(a_1, a_2) = -1$. For the interval $[a_1, a_3]$ the following equation is obtained:

$$\mu(a_1, a_1) + \mu(a_1, a_2) + \mu(a_1, a_3) = 0,$$

from which it follows that $\mu(a_1, a_3) = 0$. For every $i \geq 3$ one can write

$$\mu(a_1, a_1) + \mu(a_1, a_2) + \cdots + \mu(a_1, a_i) = 0,$$

and hence

$$\mu(a_1, a_3) = \cdots = \mu(a_1, a_n) = 0.$$

Thus the values of the Moebius function are given by

$$\mu(x, y) = \begin{cases} 1 & \text{if } x = y, \\ -1 & \text{if } (x, y) \text{ is an arc,} \\ 0 & \text{otherwise.} \end{cases}$$

2.22 If $y < x$, then $\sum_{y \leq z \leq x} \mu(z, x) = 0$. It follows that

$$\sum_{z \leq x} \mu(z, x) g(z) = \sum_{z \leq x} \sum_{y \leq z} \mu(z, x) f(y) = \sum_{y \leq x} f(y) \sum_{y \leq z \leq x} \mu(z, x) = f(x).$$

Solutions

In fact, in terms of the matrices M and Z introduced in the solution of Problem 2.20, it remains to show that

$$ZM = I,$$

but this equation is satisfied because $M = Z^{-1}$.

Another solution is the following: Let $f = (f(x_1), \ldots, f(x_n))$ and $g = (g(x_1), \ldots, g(x_n))$. The definition of g can also be written

$$g = fZ;$$

this implies that $f = gZ^{-1} = gM$, which is the desired relation.

2.23 For every function $f : B \to C$ where $|C| = x$ and x is a natural number, let the kernel of the function f be the partition p of B defined as follows: The elements b_i and b_j of B belong to the same class of p if and only if $f(b_i) = f(b_j)$. The number $f(p)$ of functions $f : B \to C$ which have the same kernel p is equal to

$$f(p) = x(x-1) \cdots (x-k+1) = [x]_k,$$

where $k = c(p)$ is the number of classes in the partition p. In fact, from the definition of the kernel it follows that the desired number is equal to the number of injective functions defined on a set with k elements and taking its values in a set with x elements, that is, $[x]_k$.

Let

$$F(p) = \sum_{q \leq p} f(q).$$

It follows that $F(p)$ represents the number of functions $f : B \to C$ which have as their kernel a partition q of B such that $q \leq p$. Thus q can be obtained by taking the union of some classes of p, which implies that

$$F(p) = x^k,$$

since the desired number is equal to the number of functions on a k-element set and with values in an x-element set. It is thus the case that

$$\sum_{q \leq p} [x]_{c(q)} = x^{c(p)}.$$

By using the Moebius inversion formula one finds that

$$\sum_{q \leq p} x^{c(q)} \mu(q, p) = [x]_{c(p)}.$$

Let $p = 1$ in this formula, for which the number of classes $c(p) = n$. This equation can also be written

$$\sum_{k=1}^{n} x^k \sum_{c(q) = k} \mu(q, 1) = [x]_n.$$

For $k = 1$ the coefficient of x in the left-hand side is equal to $\mu(0, 1)$, since the only partition q with $c(q)=1$ is the partition 0, while the coefficient of x in the polynomial

$$[x]_n = x(x-1)\cdots(x-n+1)$$

is equal to $(-1)^{n-1}(n-1)!$. Since this identity holds for all positive integral values of x, it follows that the two polynomials in x are equal and hence $\mu(0, 1) = (-1)^{n-1}(n-1)!$. [R. W. Frucht, Gian-Carlo Rota, *Scientia*, 122 (1963), 111–115.]

2.24 Let $U = Z - I = (u_{ij})_{i,j=1,\ldots,n}$, where

$$u_{ij} = \begin{cases} 1 & \text{if } x_i < x_j, \\ 0 & \text{otherwise.} \end{cases}$$

The element in row i and column j of the matrix U^n is of the form

$$\sum_{(k_1,\ldots,k_{n-1})} u_{ik_1} u_{k_1 k_2} \cdots u_{k_{n-1} j} = 0$$

for every $1 \leq i, j \leq n$, since a nonzero element in the given summation would correspond to a walk of elements of V of the form

$$x_i < x_{k_1} < x_{k_2} < \cdots < x_{k_{n-1}} < x_j.$$

But V has only n elements, and hence such a walk does not exist. One thus obtains the matrix identity

$$(I+U)\{I - U + U^2 - U^3 + \cdots + (-1)^{n-1} U^{n-1}\} = I + (-1)^{n-1} U^n = I,$$

and therefore

$$I - U + U^2 - U^3 + \cdots = (I+U)^{-1} = Z^{-1} = M,$$

that is, M is the matrix which defines the Moebius function of the set V. It follows that

$$\mu(0, 1) = \sum_{s=1}^{n} (-1)^s u_{0,1}^{(s)},$$

where $u_{0,1}^{(s)}$ is the element of the matrix U^s located in the row which corresponds to 0 and in the column which corresponds to 1 of V. Finally one finds that

$$u_{0,1}^{(s)} = \sum_{(i_s,\ldots,i_s)} u_{0 i_1} u_{i_1 i_2} \cdots u_{i_{s-1} 1} = \sum_{0 < x_{i_1} < \cdots < x_{i_{s-1}} < 1} 1 = p_s,$$

since the remaining terms are zero according to the definition of the matrix U.

2.25 Apply induction on $n \geq 2$. If $n = 2$ then $S_1 - S_2 \leq |A_1 \cup A_2| \leq S_1$, or

$$|A_1| + |A_2| - |A_1 \cap A_2| \leq |A_1 \cup A_2| \leq |A_1| + |A_2|.$$

The right-hand side of this relation is obvious, and the left-hand side is a consequence of the Principle of Inclusion and Exclusion.

Solutions

Suppose that Bonferroni inequalities are valid for each $n-1$ subsets of X. One can see that

$$\left|\bigcup_{i=1}^{n} A_i\right| = \left|\bigcup_{i=1}^{n-1} A_i\right| + |A_n| - \left|\bigcup_{i<n} (A_i \cap A_n)\right|.$$

By the induction hypothesis, if h is even one can write

$$\left|\bigcup_{i=1}^{n-1} A_i\right| \leq \sum_{k=1}^{h-1} (-1)^{k-1} \sum_{\substack{K \subset \{1,\ldots,n-1\} \\ |K|=k}} \left|\bigcap_{i \in K} A_i\right|$$

and

$$-\left|\bigcup_{i<n} (A_i \cap A_n)\right| \leq \sum_{k=1}^{h-2} (-1)^k \sum_{\substack{K \subset \{1,\ldots,n-1\} \\ |K|=k}} \left|\bigcap_{i \in K} A_i \cap A_n\right|.$$

One can thus conclude

$$\left|\bigcup_{i=1}^{n} A_i\right| \leq \sum_{k=1}^{h-1} (-1)^{k-1} S_k.$$

If h is odd one can deduce in a similar manner that

$$\left|\bigcup_{i=1}^{n} A_i\right| \geq \sum_{k=1}^{h-1} (-1)^{k-1} S_k,$$

which completes the proof.

CHAPTER 3

3.1 In order to obtain the terms which contain x^k in the expansion of $[x]_n$ one must multiply the factor x from $k-1$ parentheses among $(x-1), (x-2), \ldots, (x-n+1)$ by $n-k$ constant terms in these parentheses.

Let $K = \{i_1, i_2, \ldots, i_s\}$ be a set of numbers, and use the notation

$$P(K) = \prod_{j=1}^{s} i_j.$$

It follows that the numbers $s(n, k)$ can be written in the form

$$s(n, k) = (-1)^{n-k} \sum_{\substack{P \subset \{1,\ldots,n-1\} \\ |P|=n-k}} P(K), \tag{1}$$

and this sum contains $\binom{n-1}{n-k}$ terms.

In order to obtain the coefficient of x^k in the expansion of $[x]^n$ one must multiply $n-k$ constant terms from the factors $(x+1), (x+2), \ldots, (x+n-1)$ and add the numbers obtained from all $\binom{n-1}{n-k}$ choices of the constant terms. By using

(1) one finds that the coefficient of x^k in the expansion of $[x]^n$ is equal to

$$\sum_{\substack{K \subset \{1,\ldots,n-1\} \\ |P|=n-k}} P(K),$$

that is, it is $|s(n, k)|$.

3.2 Part (a) is called Vandermonde's formula. It can be proved by induction on n. It is clear that the assertion is true for $n=1$. If (a) holds for $n \leq m-1$, where $m \geq 2$, then

$$[x+y]_m = [x+y]_{m-1}(x+y-m+1)$$

$$= \sum_{k=0}^{m-1} \binom{m-1}{k} [x]_k [y]_{m-k-1}(x-k+y-m+k+1)$$

$$= \sum_{k=0}^{m-1} \binom{m-1}{k} [x]_{k+1}[y]_{m-k-1} + \sum_{k=0}^{m-1} \binom{m-1}{k} [x]_k [y]_{m-k}$$

$$= \sum_{k=0}^{m} \left[\binom{m-1}{k-1} + \binom{m-1}{k} \right] [x]_k [y]_{m-k}$$

$$= \sum_{k=0}^{m} \binom{m}{k} [x]_k [y]_{m-k}.$$

Part (b) is called Nörlund's formula and can be proved analogously.

3.3 In order to prove (a) let X and Y be two sets having n and m elements respectively. Every function $f : X \to Y$ can be considered to be surjective if one changes the codomain, that is, if $f: X \to f(X) = \{f(x) | x \in X\} \subset Y$. Thus the total number m^n of functions from X to Y is equal to the number of functions in the union of the sets

$$A_k = \{f : X \to Y \mid |f(X)| = k\}$$

for $k=1, \ldots, m$; these sets are pairwise disjoint. It can be seen that $|A_k| = \binom{m}{k} s_{n,k} = \binom{m}{k} k! S(n, k)$ (Problem 3.4), since the set $f(X)$ with k elements can be selected from Y in $\binom{m}{k}$ ways. Hence

$$m^n = \sum_{k=1}^{n} \binom{m}{k} k! S(n, k) = \sum_{k=1}^{m} m(m-1) \cdots (m-k+1) S(n, k)$$

$$= \sum_{k=1}^{n} [m]_k S(n, k) \quad \text{for } n \geq m,$$

since $[m]_k = 0$ for $m+1 \leq k \leq n$. It must be shown that the polynomial $x^n - \sum_{k=1}^{n} S(n, k)[x]_k$ is identically zero.

But this polynomial has degree at most equal to $n-1$, since $[x]_n S(n, n) = x(x-1) \cdots (x-n+1)$ contains the term x^n, but the other terms in the sum do not contain x^n. The resulting equality, which is valid for $m=1, 2, \ldots, n$, shows that this polynomial of degree at most $n-1$ has at least n distinct roots, and hence is the zero polynomial.

Solutions

The same argument can also be used to prove (b) by considering monotone functions $f : X \to R$ where $|X|=n$, $|R|=r$, and the set R is totally ordered. Every monotone function $f : X \to R$ is in one-to-one correspondence with an increasing word of length n: $f(x_1)f(x_2) \cdots f(x_n)$, where $X = \{x_1, \ldots, x_n\}$ and $f(x_1) \leq f(x_2) \leq \cdots \leq f(x_n)$, and the number of these increasing words—called combinations with replacement of r take n—is given by the formula

$$\frac{[r]^n}{n!} = \frac{r(r+1)\cdots(r+n-1)}{n!} \qquad \text{(Problem 1.18)}.$$

If the number of distinct letters in the word $f(x_1) \cdots f(x_n)$ is k $(1 \leq k \leq n)$, then these k letters can be chosen from the set R in $\binom{r}{k}$ ways. The number of increasing words of length n with exactly k letters is equal to $\binom{n-1}{k-1}$. In fact, if these k letters are $a_1 < a_2 < \cdots < a_k$, then the increasing words have the form $c_1 c_2 \cdots c_n$ where $c_1 = c_2 = \cdots = c_{i_1-1} = a_1, c_{i_1} = c_{i_1+1} = \cdots = c_{i_2-1} = a_2, \ldots, c_{i_{k-1}} = c_{i_{k-1}+1} = \cdots = a_k$, and $2 \leq i_1 < i_2 < \cdots < i_{k-1} \leq n$. Thus the number of increasing words of a length n with exactly k letters is equal to the number of sequences $2 \leq i_1 < i_2 < \cdots < i_{k-1} \leq n$, that is, it is equal to $\binom{n-1}{k-1}$. It follows that

$$\frac{[r]^n}{n!} = \sum_{k=1}^{n} \binom{r}{k}\binom{n-1}{k-1},$$

or

$$[r]^n = \sum_{k=1}^{n} \frac{n!}{k!}\binom{n-1}{k-1}[r]_k.$$

This equality is valid for every $r \geq 1$, which implies (b).

3.4 To every surjection f of the set $X = \{x_1, \ldots, x_n\}$ onto the set $Y = \{y_1, \ldots, y_m\}$ there corresponds a partition of the set X into m classes, namely

$$f^{-1}(y_1) \cup f^{-1}(y_2) \cup \cdots \cup f^{-1}(y_m).$$

The order of writing the classes is not taken into consideration, and hence there are $m!$ surjective functions from X to Y which generate the same partition of the set X. In other words, if the elements of Y are permuted in $m!$ different ways, then one obtains $m!$ different surjections by starting with the given surjection f. However, all of them will generate the same partition of X into m classes. Two surjections which differ by a permutation of the elements of Y cannot generate different partitions, and furthermore every partition of X into m classes can be obtained in this manner. It follows that

$$S(n,m) = \frac{1}{m!} s_{n,m} = \frac{1}{m!} \sum_{k=0}^{m-1} (-1)^k \binom{m}{k}(m-k)^n.$$

Consider the set consisting of the $S(n, m-1)$ partitions into $m-1$ classes of a set with n elements $x_1, \ldots, x_n$. One can obtain $S(n, m-1)$ partitions into m classes of a set with $n+1$ elements $x_1, \ldots, x_{n+1}$ by adding to each partition a new class consisting of only the element x_{n+1}. The element x_{n+1} can be added to each of the already existing m classes of a partition of $\{x_1, \ldots, x_n\}$ in m distinct ways. These two procedures yield, without repetitions, all the partitions of the set $\{x_1, \ldots, x_n\}$ into m classes. It follows that $S(n+1, m) = S(n, m-1) + mS(n, m)$. This recurrence relation allows one to compute the numbers $S(n, m)$ line by line, by using the values $S(n, 1) = S(n, n) = 1$ for every n and $S(n, m) = 0$ for $m > n$. The values for $n \leqslant 5$ are given in the following table:

$S(n, m)$	$m=1$	2	3	4	5
$n=1$	1	0	0	0	0
2	1	1	0	0	0
3	1	3	1	0	0
4	1	7	6	1	0
5	1	15	25	10	1

3.5 Consider the set of the $S(n+1, m)$ partitions of a set X with $n+1$ elements into m classes. For every such partition suppress the class which contains the $(n+1)$st element. One thereby obtains a partition of a k-element set K into $m-1$ classes where $m-1 \leqslant k \leqslant n$. In fact $k \geqslant m-1$, since the $m-1$ remaining classes each contain at least one element.

The partitions into $m-1$ classes thereby obtained are pairwise distinct, since otherwise the partitions of the $(n+1)$-element set X into m classes would not be pairwise distinct. In this way one obtains all the partitions of all subsets $K \subset X$ ($|K| \geqslant m-1$) into $m-1$ classes. In fact the partition $(K_i)_{1 \leqslant i \leqslant m-1}$ of K was obtained from the partition $K_1 \cup K_2 \cup \cdots \cup K_{m-1} \cup (X \setminus K)$ of X by suppressing the class $X \setminus K$ which contains the $(n+1)$st element.

The $n-k$ elements which form a class, together with the $(n+1)$st element, can be chosen in $\binom{n}{n-k} = \binom{n}{k}$ different ways. It follows that the number of partitions into m classes of an $(n+1)$-element set is equal to

$$S(n+1, m) = \sum_{k=m-1}^{n} \binom{n}{k} S(k, m-1).$$

The recurrence relation for the numbers B_n can be obtained analogously by suppressing the class which contains the $(n+1)$st element in a partition of an $(n+1)$-element set. One can find in the same manner a partition of a k-element set ($0 \leqslant k \leqslant n$). The $n-k$ elements which are contained in the same class, together with the $(n+1)$st element, can be chosen from the set $X \setminus \{x_{n+1}\}$ in $\binom{n}{k}$ distinct ways.

The term 1, corresponding to the case $k=0$, occurs when the partition of X consists of a single class.

3.6 Starting from a partition of a set X of type $1^{k_1} 2^{k_2} \cdots n^{k_n}$, one can obtain a permutation of X by writing the elements of X in the order in which they appear in the classes of the partition. One first writes the classes with one element, then the classes with two elements, etc. Since the order of the elements

Solutions

in a class and the order of the classes in a partition are not significant, it follows that the same partition generates $(1!)^{k_1} (2!)^{k_2} \cdots (n!)^{k_n}$ different permutations of the set X by permuting the elements in each class. For each of these permutations one also obtains $k_1! k_2! \ldots k_n!$ permutations by permuting the classes with the same number of elements among themselves. It is easy to see that in this way one generates without repetitions all the $n!$ permutations of the set X, from which it follows that

$$\text{Part}(1^{k_1}\ 2^{k_2} \cdots n^{k_n};\ n)(1!)^{k_1} k_1! \cdots (n!)^{k_n} k_n! = n!.$$

Let p be a permutation of the set $X = \{1, \ldots, n\}$ of type $1^{k_1} \cdots n^{k_n}$ written in increasing order of the length of its cycles. By suppressing the parentheses which enclose the cycles of p one obtains a word of length n formed from all the letters of the alphabet $\{1, \ldots, n\}$. The number of such words is $n!$.

But the same permutation generates $k_1! k_2! \cdots k_n! 1^{k_1} 2^{k_2} \cdots n^{k_n}$ different words, since the k_i cycles of length i ($1 \leq i \leq n$) can be permuted in $k_1! \cdots k_n!$ different ways. On the other hand, a cycle of length i can be written in i different ways by taking as the first element of the cycle each of its i elements. In this way one obtains the $1^{k_1} \cdots n^{k_n}$ other possibilities of generating distinct words of length n. One thus finds (without repetition) all $n!$ words of length n formed from the numbers $1, \ldots, n$. It follows that

$$\text{Perm}(1^{k_1}\ 2^{k_2} \cdots n^{k_n};\ n)\ 1^{k_1} k_1! \cdots n^{k_n} k_n! = n!.$$

3.7 Use the identities of Problems 3.2 and 3.3 to express the polynomial $[x+y]_n$ in two different ways:

$$[x+y]_n = \sum_{k=0}^{n} s(n, k)(x+y)^k = \sum_{k=0}^{n} s(n, k) \sum_{i=0}^{k} \binom{k}{i} x^i y^{k-i},$$

$$[x+y]_n = \sum_{k=0}^{n} \binom{n}{k} [x]_k [y]_{n-k} = \sum_{k=0}^{n} \binom{n}{k} \sum_{i=0}^{k} s(k, i) x^i \sum_{j=0}^{n-k} s(n-k, j) y^j.$$

Equating the coefficients of $x^i y^j$ in these two expressions yields the first identity.

In order to obtain the second recurrence relation one must expand $(x+y)^n$:

$$(x+y)^n = \sum_{k=0}^{n} S(n, k)[x+y]_k = \sum_{k=0}^{n} S(n, k) \sum_{i=0}^{k} \binom{k}{i} [x]_i [y]_{k-i},$$

$$(x+y)^n = \sum_{k=0}^{n} \binom{n}{k} x^k y^{n-k} = \sum_{k=0}^{n} \binom{n}{k} \sum_{i=0}^{k} S(k, i)[x]_i \sum_{j=0}^{n-k} S(n-k, j)[y]_j.$$

In this case equating the coefficients of $[x]_i [y]_j$ produces the desired result.

3.8

$$[x]_n = \sum_{k=0}^{n} s(n, k) x^k = \sum_{k=0}^{n} s(n, k) \sum_{m=0}^{k} S(k, m)[x]_m$$

and

$$x^n = \sum_{k=0}^{n} S(n, k)[x]_k = \sum_{k=0}^{n} S(n, k) \sum_{m=0}^{k} s(k, m)x^m.$$

Equating the coefficients of $[x]_n$ and x^n respectively in these expressions gives the identities in the statement of the problem. One can equate these coefficients since the families of polynomials $(x^n)_{n \geq 0}$ and $([x]_n)_{n \geq 0}$ each form a basis for the vector space of polynomials with real coefficients, and thus every polynomial can be uniquely expressed as a linear combination of these polynomials.

3.9 The proof uses induction on n. For $n=1$ the property is immediate. From the table of Stirling numbers (Problem 3.4) it can be seen that the property holds for small values of n, by referring to each line of the table.

Suppose that the property holds for every $i \leq n$. Thus $M(i) \leq M(j)$ for every $1 \leq i \leq j \leq n$. Let k be an index such that $2 \leq k \leq M(n)$. It follows from the recurrence relation

$$S(n+1, k) = S(n, k-1) + kS(n, k)$$

that

$$S(n+1, k) - S(n+1, k-1) = \{S(n, k-1) - S(n, k-2)\}$$
$$+ k\{S(n, k) - S(n, k-1)\} + S(n, k-1).$$

The right-hand side of this equation is positive by the induction hypothesis.

Suppose that $M(n) + 2 \leq k \leq n+1$. By using the recurrence relation

$$S(n+1, k) = \sum_{j=1}^{n} \binom{n}{j} S(j, k-1),$$

one can conclude that

$$S(n+1, k) - S(n+1, k-1) = \sum_{j=1}^{n} \binom{n}{j} \{S(j, k-1) - S(j, k-2)\}.$$

The induction hypothesis now yields $S(j, k-1) < S(j, k-2)$, since $M(j) \leq M(n)$ for every $j \leq n$. It follows that $S(n+1, k) < S(n+1, k-1)$ for every $M(n) + 2 \leq k \leq n+1$. Thus the sequence $(S(n, k))_{k=1,\ldots,n}$ is unimodal, and $M(n+1) = M(n)$ or $M(n+1) = M(n) + 1$. No examples are known of sequences $(S(n, k))_k$ which have two equal maxima for $n \geq 3$.

3.10 The proof of (a) follows from equating the coefficients of x^k in the expansions:

$$\sum_{k=0}^{n} s_a(n, k)x^k = (x \mid a)_n = (x \mid a)_{n-1}(x - a_{n-1})$$

$$= \left[\sum_{k=0}^{n-1} s_a(n-1, k)x^k\right](x - a_{n-1}).$$

In the case of (b) one can equate the coefficients of $(x \mid a)_k$ in the expansions

Solutions

$$\sum_{k=0}^{n} S_a(n,k)(x\,|\,a)_k = x^n = x^{n-1} \cdot x = \sum_{k=0}^{n-1} S_a(n-1,k)(x\,|\,a)_k(x - a_k + a_k),$$

in view of the fact that $(x\,|\,a)_k(x - a_k) = (x\,|\,a)_{k+1}$.

For (c) equate the coefficients of x^k in the expansion

$$(x\,|\,a)_n = \frac{1}{x - a_n}(x\,|\,a)_{n+1} = \left(\frac{1}{x} + \frac{a_n}{x^2} + \frac{a_n^2}{x^3} + \cdots\right)(x\,|\,a)_{n+1}.$$

The proof of (d) is obtained by iterating recurrence relation (b) for different values of n.

Recurrence relation (e) follows from repeated application of (a) for decreasing values of n.

In order to establish (f) one can write

$$x^n = \sum_{k=0}^{n} S_a(n,k)(x\,|\,a)_k = \sum_{k=0}^{n} S_a(n,k) \sum_{m=0}^{k} s_a(k,m)x^m$$

$$= \sum_{m=0}^{n} \left(\sum_{k=0}^{n} S_a(n,k) s_a(k,m)\right) x^m.$$

[L. Comtet, *C. R. Acad. Sci. Paris*, **A275** (1972), 747–750.]

3.11 Let $f_k = \sum_{n=k}^{\infty} S_a(n,k) t^n$ and observe that

$$f_k = \sum_{n \geq k} \{S_a(n-1, k-1) + a_k S_a(n-1, k)\} t^n = t f_{k-1} + a_k t f_k.$$

Thus

$$f_k = \frac{f_{k-1} t}{1 - a_k t} = \cdots = \frac{f_0 t^k}{(1 - a_1 t) \cdots (1 - a_k t)} \quad \text{and} \quad f_0 = \sum_{n=0}^{\infty} a_0^n t^n = \frac{1}{1 - a_0 t}.$$

3.12 Let $X = \{x_1, \ldots, x_n\}$. The set of partitions of X into k classes which contain at least i elements can be written in the form $P_1 \cup P_2$, where P_1 represents the set of partitions of X into k classes which contain at least i elements for which the element x_n belongs to a class with more than i elements, and P_2 represents the set of partitions with the same property for which x_n belongs to a class with exactly i elements.

It follows that $S_i(n,k) = |P_1| + |P_2| = k S_i(n-1, k) + \binom{n-1}{i-1} S_i(n-i, k-1)$, since the partitions in P_1 are generated, without repetition, by starting from the partitions of the set $\{x_1, \ldots, x_{n-1}\}$ into k classes of cardinality greater than or equal to i. One then adds, in turn, x_n in k ways to each class. The partitions in P_2 can be obtained by considering all the $\binom{n-1}{i-1}$ subsets of X of cardinality i which contain x_n as a class of the partition. To this one adds, in turn, the $S_i(n-i, k-1)$ partitions of the $n-i$ remaining elements in $k-1$ classes of cardinality greater than or equal to i.

In order to justify (b), consider the set of all functions $f : \{x_1, \ldots, x_n\} \to \{1, \ldots, k\}$ such that $|f^{-1}(s)| = j_s$ for every $1 \leq s \leq k$. It is clear that the number of such functions is equal to the number of arrangements of a set of n objects in

k boxes such that the sth box contains j_s objects for $s = 1, \ldots, k$, which by Problem 1.15 is equal to $n!/j_1! \cdots j_k!$. By summing these numbers for all solutions of the equation $j_1 + \cdots + j_k = n$ with $j_s \geq i$ for $s = 1, \ldots, k$, one obtains the number of surjections f which satisfy $|f^{-1}(s)| \geq i$ for every $s = 1, \ldots, k$. This number must be divided by $k!$, since the order of classes in a partition of X is immaterial.

3.13 Every partition into two classes $X = A \cup B$ is completely determined by the set A, which is taken to be different from X and $\emptyset$. Thus the number of choices for A is equal to the number of subsets of the n-element set X minus two. It follows that $S(n, 2) = (2^n - 2)/2 = 2^{n-1} - 1$, since the order of the classes in a partition is not taken into consideration.

Analogously, every partition of an n-element set X into $n-1$ classes contains a class with two elements and $n-2$ classes with one element. The two elements can be chosen in $\binom{n}{2}$ ways.

In order to prove (c), first show that the exponential generating function of the Stirling numbers of the second kind is equal to

$$\sum_{n=k}^{\infty} \frac{S(n, k)}{n!} x^n = \frac{1}{k!} (e^x - 1)^k.$$

In fact one has

$$e^{px} = 1 + \frac{px}{1!} + \frac{p^2 x^2}{2!} + \cdots \quad \text{and} \quad \frac{1}{k!}(e^x - 1)^k = \frac{1}{k!} \sum_{p=0}^{k} (-1)^{k-p} \binom{k}{p} e^{px},$$

and thus the coefficient of x^n in this expansion is equal to

$$\frac{1}{k!} \sum_{p=0}^{k} (-1)^{k-p} \binom{k}{p} \frac{p^n}{n!} = \frac{S(n, k)}{n!},$$

in view of the known expression for the Stirling numbers of the second kind (Problem 3.4).

Thus $k! S(n, k)$ is the coefficient of x^n in the expansion of $(e^x - 1)^k$, multiplied by $n!$. It follows from this that $(1/n!)\{1 - 1!S(n, 2) + 2!S(n, 3) - 3!S(n, 4) + \cdots\}$ is the coefficient of x^n in the expansion of the sum of the series

$$e^x - 1 - \tfrac{1}{2}(e^x - 1)^2 + \tfrac{1}{3}(e^x - 1)^3 - \tfrac{1}{4}(e^x - 1)^4 + \cdots. \tag{1}$$

Since $x - \tfrac{1}{2}x^2 + \tfrac{1}{3}x^3 - \tfrac{1}{4}x^4 + \cdots = \ln(1+x)$, one can conclude that the sum (1) is equal to $\ln(1 + e^x - 1) = x$, which establishes the identity for $n \geq 2$.

3.14 In order to obtain an expression for $c(n, 3)$ one must determine the number of ways in which one can express $X = A \cup B \cup C$ where A, B, C are nonempty and each element of X belongs to two of the sets A, B, C. In fact each element $x \in X$ can belong to the sets A, B, or C in $\binom{3}{2} = 3$ ways. In this way one obtains 3^n expressions for X in the form $A \cup B \cup C$, three of which do not satisfy the given conditions, since they correspond to situations in which either A, B or C is empty. Thus the number of bicoverings of X with three classes is equal to $(3^n - 3)/3! = \tfrac{1}{2}(3^{n-1} - 1)$, since the order of the sets A, B, and C is not taken into consideration. [L. Comtet, *Studia Sci. Math. Hung.*, **3** (1968), 137–152.]

Solutions

3.15 Consider a new element $z \notin X$, and in each partition of the set $X \cup \{z\}$ suppress the class which contains the element z. One obtains in this way all the partial partitions of X without repetitions, except for the case in which the partition of $X \cup \{z\}$ consists of a single class. The desired number is thus equal to $B_{n+1} - 1$.

3.16 Let $(a_n)_{n \in N}$ be a sequence of real numbers, and denote by $f(x)$ the sum of the series $\sum_{n=0}^{\infty} a_n [x]_n$ for those values of x for which the series is convergent. Now define the operator L by the relation $L(f(x)) = \sum_{n=0}^{\infty} a_n$, if this series is convergent.

We show that if $g(x) = \sum_{n=0}^{\infty} a_n x^n$ and if the operator L is defined for the function g, then $L(g(x)) = \sum_{n=0}^{\infty} a_n B_n$, where $B_0 = 1$. In fact, in view of Problem 3.3 one obtains the partial sums

$$S_m = \sum_{n=0}^{m} a_n x^n = \sum_{n=0}^{m} a_n \sum_{k=1}^{n} [x]_k S(n, k) = \sum_{k=0}^{m} \left(\sum_{n=k}^{m} a_n S(n, k) \right) [x]_k,$$

where $[x]_0 = 1$. Thus

$$L(g(x)) = \lim_{m \to \infty} \sum_{k=0}^{m} \left(\sum_{n=k}^{m} a_n S(n, k) \right) = \lim_{m \to \infty} \sum_{n=0}^{m} a_n \left(\sum_{k=1}^{n} S(n, k) \right)$$

$$= \lim_{m \to \infty} \sum_{n=0}^{m} a_n B_n = \sum_{n=0}^{\infty} a_n B_n.$$

In order to obtain the desired formula consider the Taylor series expansion

$$\exp(tx) = e^{tx} = 1 + \frac{tx}{1!} + \frac{t^2 x^2}{2!} + \cdots + \frac{t^n x^n}{n!} + \cdots.$$

Substituting $e^t = u + 1$ and using Newton's generalized binomial formula yields

$$e^{tx} = (u+1)^x = \sum_{n=0}^{\infty} \frac{[x]_n}{n!} u^n.$$

It follows that

$$L(e^{tx}) = \sum_{n=0}^{\infty} \frac{u^n}{n!} = e^u = \exp(e^t - 1)$$

and it can be seen from the previous proof that

$$L(e^{tx}) = \sum_{n=0}^{\infty} \frac{B_n}{n!} t^n.$$

By equating the two expressions for $L(e^{tx})$ one obtains the expression for the exponential generating function of the numbers B_n.

3.17 The Bell number B_n represents, by definition, the total number of partitions of an n-element set and thus can be written in the form

$$B_n = \sum_{k=1}^{n} S(n, k) = \sum_{k=1}^{\infty} S(n, k),$$

since $S(n, k) = 0$ for $k > n$. In view of the expression for the Stirling number of the second kind (Problem 3.4) it follows that

$$B_n = \sum_{k=1}^{\infty} \frac{1}{k!} \sum_{j=1}^{k} (-1)^{k-j} \binom{k}{j} j^n = \sum_{j=1}^{\infty} \frac{j^n}{j!} \sum_{k=j}^{\infty} \frac{(-1)^{k-j}}{(k-j)!} = \frac{1}{e} \sum_{j=1}^{\infty} \frac{j^n}{j!}.$$

The difference between the number of partitions of an n-element set into an even and an odd number of classes is equal to

$$\sum_{k=1}^{n} (-1)^k S(n, k) = \sum_{k=1}^{\infty} (-1)^k S(n, k) = \sum_{k=1}^{\infty} \frac{(-1)^k}{k!} \sum_{j=1}^{k} (-1)^{k-j} \binom{k}{j} j^n$$

$$= \sum_{j=1}^{\infty} \frac{(-1)^j j^n}{j!} \sum_{k=j}^{\infty} \frac{1}{(k-j)!} = e \sum_{j=1}^{\infty} \frac{(-1)^j j^n}{j!}.$$

3.18 A subset $S \subset X$ can be put into correspondence with a binary word $a_1 a_2 \cdots a_n$ with $a_i = 1$ if $i \in S$ and $a_i = 0$ otherwise. If S does not contain two consecutive integers, then the binary word associated with the set S will not contain two neighboring 1's. In this way a bijection has been defined between the k-element subsets of X which do not contain two consecutive integers and the set of binary words of length n, k of whose letters are 1 and $n-k$ of which are 0, and which do not contain two neighboring 1's. In order to count the number of elements in the latter set one can consider $n-k$ digits equal to 0 and indexed from 1 to $n-k$. To these one must add k digits equal to 1 so that no two 1's are adjacent. Each digit 1 can be characterized by the index of the 0 which precedes it. Thus one must choose k pairwise distinct integers from the set $\{0, 1, \ldots, n-k\}$. It is possible to do this in $f(n, k) = \binom{n-k+1}{k}$ distinct ways. To the set $\{1, \ldots, n-k\}$ of indices of digits equal to zero the number 0 has been added, which corresponds to the case in which the digit 1 occurs in the first position of the word $a_1 a_2 \cdots a_n$.

It follows that F_{n+1} represents the number of subsets of X which do not contain two consecutive integers. This includes the empty set, which corresponds to a word with n positions equal to 0. The summation which defines F_{n+1} contains nonzero terms for $n-k+1 \geqslant k$ or $k \leqslant [(n+1)/2]$.

In order to prove the recurrence relation satisfied by the Fibonacci numbers, observe that every binary word of length n which does not contain two consecutive ones has 0 in its last position or 01 in the last two positions. The words which remain after eliminating 0 or 01 have length $n-1$ or $n-2$ respectively and do not contain two consecutive 1's. Thus there exists a bijection from the set of binary words of length n which do not contain two consecutive 1's onto the union of the disjoint sets formed from binary words of length $n-1$ or $n-2$, respectively, which do not contain two consecutive 1's. Thus $F_n = F_{n-1} + F_{n-2}$ for every $n \geqslant 2$, and $F_0 = F_1 = 1$.

Solutions

3.19 In order to find the expression for the number $f^*(n, k)$, observe that the subsets which contain the number n cannot contain either $n-1$ or 1. The number of such subsets is equal to $f(n-3, k-1)$, and there are $f(n-1, k)$ which do not contain n. In both cases it is assumed that the sets do not contain two consecutive integers modulo $n+1$.

For example, the subsets which contain the number n are obtained from the $f(n-3, k-1)$ subsets of $X \setminus \{1, n-1, n\}$ which have $k-1$ elements and do not contain two consecutive integers. In this case the number n is added. In view of the fact that every k-element subset of X which does not contain two consecutive integers modulo $n+1$ belongs to the union of these two disjoint sets of k-element subsets which do not contain two consecutive integers, it follows that

$$f^*(n, k) = f(n-3, k-1) + f(n-1, k) = \binom{n-k-1}{k-1} + \binom{n-k}{k}$$

$$= \left(\frac{k}{n-k} + 1\right)\binom{n-k}{k} = \frac{n}{n-k}\binom{n-k}{k}.$$

The Lucas numbers can be expressed as a function of the Fibonacci numbers as follows:

$$L_n = \sum_{k \geq 0} f^*(n, k) = \sum_{k \geq 0} \{f(n-1, k) + f(n-3, k-1)\}$$

$$= \sum_{k \geq 0} f(n-1, k) + \sum_{k \geq 1} f(n-3, k-1) = F_n + F_{n-2}.$$

Thus one can write

$$L_{n+1} = F_{n+1} + F_{n-1} = F_n + F_{n-1} + F_{n-2} + F_{n-3} = L_n + L_{n-1}$$

for $n \geq 2$. One finds that $L_1 = 1$ and $L_2 = 3$. With $L_0 = 2$, by definition, the recurrence relation for the Lucas numbers yields the fact that $L_2 = 3$.

3.20 First we show by induction on n that

$$\begin{pmatrix} 1 & 1 \\ 1 & 0 \end{pmatrix}^{n+1} = \begin{pmatrix} F_{n+1} & F_n \\ F_n & F_{n-1} \end{pmatrix} \quad (1)$$

for $n \geq 1$. For $n=1$ one finds that $F_0 = F_1 = 1$ and $F_2 = 2$ and thus (1) is satisfied.
Supposing that (1) holds, one can show that

$$\begin{pmatrix} 1 & 1 \\ 1 & 0 \end{pmatrix}^{n+2} = \begin{pmatrix} F_{n+1} & F_n \\ F_n & F_{n-1} \end{pmatrix}\begin{pmatrix} 1 & 1 \\ 1 & 0 \end{pmatrix} = \begin{pmatrix} F_{n+1}+F_n & F_{n+1} \\ F_n+F_{n-1} & F_n \end{pmatrix} = \begin{pmatrix} F_{n+2} & F_{n+1} \\ F_{n+1} & F_n \end{pmatrix},$$

and thus (1) is true for every n. By equating the determinants of the matrices in both sides of (1) one obtains the desired recurrence relation.

3.21 It is clear that the number u_n is also equal to the number of representations of n as a sum of the numbers 1 or 2. Two representations are considered to be distinct even if they differ only in the order of their terms. For example:

$$2=1+1,$$
$$3=2+1=1+2=1+1+1,$$
$$4=2+2=2+1+1=1+2+1=1+1+2=1+1+1+1.$$

The first term is 1 or 2. In the first case the number of representations is equal to u_{n-1}, since the remainder of the terms are equal to 1 or 2, and their sum is equal to $n-1$. In the second case the number of representations is equal to u_{n-2}. It follows that

$$u_n = u_{n-1} + u_{n-2} \tag{1}$$

with initial values $u_1 = 1$, $u_2 = 2$. If one takes $u_0 = 1$, the result is the sequence of Fibonacci numbers, and hence $u_n = F_n$.

In order to solve equation (1) one can use the characteristic equation

$$r^2 = r + 1,$$

which has solutions $r_1 = (1+\sqrt{5})/2$ and $r_2 = (1-\sqrt{5})/2$, and thus the general solution has the form

$$u_n = C_1 r_1^n + C_2 r_2^n,$$

where the constants C_1 and C_2 are determined from the initial conditions $u_0 = 1$ and $u_1 = 1$. One thus obtains the system

$$C_1 + C_2 = 1,$$
$$\frac{1+\sqrt{5}}{2} C_1 + \frac{1-\sqrt{5}}{2} C_2 = 1.$$

It follows that $C_1 = (\sqrt{5}+1)/2\sqrt{5}$ and $C_2 = (\sqrt{5}-1)/2\sqrt{5}$, and hence

$$u_n = F_n = \frac{\sqrt{5}+1}{2\sqrt{5}} \left(\frac{\sqrt{5}+1}{2}\right)^n + \frac{\sqrt{5}-1}{2\sqrt{5}} \left(\frac{1-\sqrt{5}}{2}\right)^n$$

$$= \frac{1}{\sqrt{5}} \left\{ \left(\frac{1+\sqrt{5}}{2}\right)^{n+1} - \left(\frac{1-\sqrt{5}}{2}\right)^{n+1} \right\}$$

for every $n \geq 0$. If one sets $f(x) = \sum_{n=0}^{\infty} F_n x^n$, then it follows in turn that

$$xf(x) = \sum_{n=1}^{\infty} F_{n-1} x^n \quad \text{and} \quad x^2 f(x) = \sum_{n=2}^{\infty} F_{n-2} x^n,$$

and in view of (1) one has

$$f(x) - xf(x) - x^2 f(x) = F_0 + (F_1 - F_0)x + \sum_{n=2}^{\infty} (F_n - F_{n-1} - F_{n-2})x^n = F_0 = 1,$$

from which it follows that

$$f(x) = \frac{1}{1-x-x^2}.$$

Solutions 131

3.22 Since the sequence of Fibonacci numbers contains arbitrarily large terms, it follows that there exists an index m such that $F_m \leqslant n < F_{m+1}$. Continuing in this way one finds that

$$0 \leqslant n - F_m < F_{m+1} - F_m = F_{m-1}.$$

If $n - F_m > 0$, then in view of the fact that 1 is a term of the Fibonacci sequence, one can find an index $s \leqslant m-2$ such that

$$F_s \leqslant n - F_m < F_{s+1}.$$

As before, it follows that $0 \leqslant n - F_m - F_s < F_{s-1}$.

If $n - F_m - F_s = 0$, the proof is finished. Otherwise repeat the same argument, and after a finite number of steps the desired representation is obtained

$$n = F_m + F_s + F_p + \cdots + F_q,$$

where the indices of two consecutive terms of the sum differ by at least 2.

3.23 The Catalan number C_n is defined as the number of ways in which one can insert parentheses into a nonassociative product of n factors written in the order $x_1, x_2, \ldots, x_n$. If there exists a unique pair of parentheses which is not contained in other parentheses, then this pair contains in its interior the product of the factors $x_2, \ldots, x_n$ which remain outside of the factor x_1, or it contains in its interior the product of the factors $x_1, \ldots, x_{n-1}$, which remain outside of the factor x_n. If there are two pairs of parentheses which are not contained in other parentheses, then these pairs contain the product of the factors $x_1, \ldots, x_k$ and $x_{k+1}, \ldots, x_n$, respectively where $2 \leqslant k \leqslant n-2$.

In a product of k or $n-k$ factors respectively one can insert parentheses in C_k or C_{n-k} ways respectively. This yields the following recurrence relation for the Catalan numbers:

$$C_n = \sum_{k=1}^{n-1} C_k C_{n-k},$$

where $C_1 = 1$.

As in the statement of the problem, let

$$f(x) = C_1 x + C_2 x^2 + \cdots + C_n x^n + \cdots.$$

It follows that

$$f^2(x) = C_1^2 x^2 + (C_1 C_2 + C_2 C_1) x^3 + \cdots + \left(\sum_{k=1}^{n-1} C_k C_{n-k}\right) x^n + \cdots = f(x) - x,$$

in view of the recurrence relation and the fact that $C_1 = C_2 = 1$.

The solution of the quadratic equation

$$f^2(x) - f(x) + x = 0$$

yields $f(x) = (1 \pm \sqrt{1-4x})/2$. In the sequel let $x < \tfrac{1}{4}$. Since $f(0) = 0$, one must take the minus sign in front of the radical. Thus

$$f(x) = \frac{1 - \sqrt{1-4x}}{2}.$$

Now expand the function $\sqrt{1-4x}$ in a power series in x, using the generalized Newton binomial formula for real exponents, which has the form

$$(x+a)^\alpha = a^\alpha + \alpha a^{\alpha-1} x + \frac{\alpha(\alpha-1)}{2!} a^{\alpha-2} x^2 + \cdots$$

$$+ \frac{\alpha(\alpha-1)\cdots(\alpha-k+1)}{k!} a^{\alpha-k} x^k + \cdots,$$

where $a > 0$.

This series converges for every x which satisfies $|x| < a$. If $\alpha \in \mathbb{R}$ is a positive integer, then only a finite number of terms of the series are different from zero and the resulting expansion is precisely the binomial formula named for Newton. (This formula was not actually discovered by Newton. Mathematicians from Central Asia such as Omar Khayyam knew it much earlier. In western Europe, Blaise Pascal also used the result before Newton. However, Newton proposed the generalization to noninteger exponents.)

In order to expand $(1-4x)^{1/2}$ in a series of powers of x, let $y = -4x$, $\alpha = \frac{1}{2}$, and expand the binomial $(y+1)^{1/2}$. It turns out that the coefficient of x^n is equal to

$$\frac{\frac{1}{2}(-\frac{1}{2})(-\frac{3}{2})\cdots(\frac{1}{2}-n+1)}{n!}(-4)^n = (-1)^{n-1}\frac{1}{2^n}\cdot\frac{1\times 3\times\cdots\times(2n-3)}{n!}(-1)^n 2^{2n}$$

$$= -\frac{(2n-2)!}{2^{n-1}(n-1)!n!}2^n$$

$$= -2\frac{(2n-2)!}{n(n-1)!(n-1)!} = -\frac{2}{n}\binom{2n-2}{n-1}.$$

However, C_n is the coefficient of x^n in the expansion of $f(x)$, and hence one obtains the coefficient of x^n from the expansion of $(1-4x)^{1/2}$ by multiplying by $-\frac{1}{2}$. Thus $C_n = (1/n)\binom{2n-2}{n-1}$.

3.24 Determine the number of sequences of letters which contain the letter a k times and the letter b m times, and which have property (P): For every i, $1 \leqslant i \leqslant m+k$ the number of letters a among the first i letters of the sequence is greater than or equal to the number of letters b. It is clear that the number of these sequences is nonzero if and only if the condition $k \geqslant m > 0$ is satisfied. The number of sequences of letters which contain the letter a k times and the letter b m times is equal to $P(m,k) = (m+k)!/m!k! = \binom{m+k}{m}$. If one determines the number of sequences which do not satisfy (P), then the desired number is obtained by subtracting this number from $\binom{m+k}{m}$. It will be shown that the number of sequences formed from m letters b and k letters a which do not satisfy (P) is equal to $P(m-1, k+1) = \binom{m+k}{m-1}$. It is equal to the number of sequences formed from $m-1$ letters b and $k+1$ letters a. The proof of this property follows.

Consider a sequence formed from m letters b and k letters a and which does not satisfy (P). There exists a position numbered $2s+1$, where $s \geqslant 0$, such that the sequence under consideration contains the letter b in position $2s+1$. In front of this position there are an equal number s of the letters a and b.

Now consider the smallest index s with this property, and add the letter a in front of the corresponding sequence to obtain a sequence of m letters b and $k+1$ letters a. The first letter of the sequence thus obtained is a, and among the first $2s+2$ letters there are an equal number of a's and b's. Interchange the letters a and b in the first $2s+2$ positions of the sequence. The total number of letters of each kind does not change, and one obtains a sequence with m b's and $k+1$ a's whose first letter is now equal to b.

In this manner one has associated with each sequence of m b's and k a's which does not satisfy (P) a sequence of m b's and $k+1$ a's which begins with the letter b. This mapping is injective, as follows immediately by considering two different sequences which do not satisfy (P) and which differ in the pth position, where $p \leqslant 2s+1$ or $p > 2s+1$.

It will be shown that in this manner it is possible to obtain every sequence formed from m b's and $k+1$ a's and which begins with b. Thus the mapping is also surjective.

Now consider a sequence which begins with b. Since $m \leqslant k$ and hence $m < k+1$, it follows that there exists a position numbered $2s$ such that the first $2s$ positions of the sequence contain an equal number s of the letters a and b. In front of the first position with this property interchange a and b and suppress the first letter a. One obtains a sequence of k a's and m b's which does not satisfy (P).

Now apply the indicated mapping to this sequence. The result is the original sequence. Thus in virtue of this bijection the number of sequences with m letters b and k letters a which do not satisfy (P) is equal to the number of sequences with m letters b and $k+1$ letters a which begin with b. If the first letter b is suppressed, one obtains all sequences consisting of $m-1$ letters b and $k+1$ letters a. The number of such sequences is equal to

$$P(m-1, k+1) = \binom{m+k}{m-1}.$$

It follows that the number of sequences which satisfy (P) is equal to

$$\binom{m+k}{m} - \binom{m+k}{m-1} = \frac{k-m+1}{k+1}\binom{m+k}{m}.$$

For $k = m = n-1$ one has

$$C_n = \frac{1}{n}\binom{2n-2}{n-1}.$$

It is clear that this number represents the solution to the problem, since if 1 is replaced by a and -1 by b, then condition (1) expresses the fact that the number of a's is at least equal to the number of b's in the first k positions for $1 \leqslant k \leqslant 2n-2$. Condition (2) expresses the fact that the number of a's is equal to the number of b's and both numbers are equal to $n-1$.

3.25 It will be shown that there exists a bijection from the triangulations of a convex polygon with $n+1$ vertices to the set of parenthesized products of n factors in the order $x_1, x_2, \ldots, x_n$.

Let $A_1 A_2 \cdots A_{n+1}$ be a polygon with $n+1$ vertices, and traverse the sides of the polygon from A_1 to A_2, and so on, until A_{n+1} is reached. One obtains a parenthesized product of n factors by using the following rules:

(1) When traversing a side, write a new factor x_i in the order $x_1, \ldots, x_n$.

(2) When one arrives at a vertex which is incident to some diagonals of the triangulation, write a number of closing parentheses equal to the number of diagonals which have an endpoint at this vertex and whose other endpoint has been traversed. Write a number of opening parentheses equal to the number of diagonals which are incident to this vertex and whose other endpoint has not been visited.

In this manner it turns out that for the triangulation of the convex polygon with eight vertices illustrated in Figure 3.1, one obtains the following product of seven factors:

$$(x_1((((x_2 x_3)x_4)x_5)x_6))x_7.$$

It is clear that this correspondence is injective. In order to show that it is surjective, consider a parenthesized product of n factors in the order $x_1, \ldots, x_n$. This product contains $n-2$ opening parentheses and $n-2$ closing parentheses. Each opening parenthesis corresponds to a unique closing parenthesis. For each pair of opening and closing parentheses consider the first letter x_i which occurs to the right of the opening parenthesis and the first letter x_j which occurs to the left of the closing parenthesis. Draw the diagonal $A_i A_{j+1}$.

Since each pair of parentheses contains two factors in its interior, and since the parentheses are correctly placed, it follows that the $n-2$ diagonals of the polygon constitute a triangulation for it. Now apply the indicated correspondence to this triangulation. The result is the original parenthesized product of n factors, which establishes that the transformation is a bijection.

Thus the number of triangulations is equal to the Catalan number C_n (Euler).

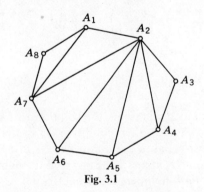

Fig. 3.1

Solutions

3.26 Consider the lines $x = k$, $y = l$ in a rectangular coordinate system where $0 \leq k, l \leq n$ are integers. One now studies the points of intersection of these lines in the first quadrant which occur on or beneath the line $y = x$.

For each increasing function $f : \{1, \ldots, n\} \to \{1, \ldots, n\}$ construct a path in this network as follows: Suppose that one is at the point $M(i, f(i))$. Go to the point $M_1(i+1, f(i))$ by a horizontal segment; then follow vertical segments until one arrives at the point $M_2(i+1, f(i+1))$. If $f(i+1) = f(i)$, it follows that $M_2 = M_1$. Otherwise the displacement involves an upward movement, since $f(i+1) > f(i)$.

By effecting this displacement for $i = 1, \ldots, n-1$ one obtains a rising path in the network with endpoints $(1, 0)$ and $(n, f(n))$. Also join the origin with the point $(1, 0)$ by a horizontal segment, and if $f(n) < n$, join the point $(n, f(n))$ to the point $A(n, n)$ by a sequence of vertical segments. This yields a path with endpoints $O(0, 0)$ and $A(n, n)$.

For the example illustrated in Figure 3.2, the increasing function f is defined as follows: $f(1) = f(2) = 1$, $f(3) = 2$, $f(4) = f(5) = 4$. The corresponding path is indicated by a heavy line.

It is clear that in general this path consists of n horizontal and n vertical segments. There are no descents from O to the point $A(n, n)$, and the path is situated beneath the line $y = x$. One has therefore defined a correspondence between the set of increasing functions $f : \{1, \ldots, n\} \to \{1, \ldots, n\}$ which satisfy the condition $f(x) \leq x$ for every $x = 1, \ldots, n$ and the set of paths with endpoints O and A with the given property. This mapping is a bijection. In fact, it is injective because distinct functions correspond to distinct paths.

In order to show that it is surjective, let d be a path with the given property and endpoints O and A. Define an increasing function as follows:

$$f_d(i) = \max \{j \mid (i, j) \in d\}$$

for every $i = 1, \ldots, n$. In this case the image of the function f_d under the correspondence is exactly the path d, which shows that the correspondence is a surjection.

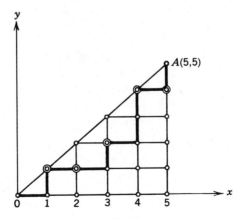

Fig. 3.2

To count the rising paths of length $2n$ with endpoints O and $A(n, n)$ which are situated beneath the line $y = x$, observe that there is a bijection from the set of these paths to the set of sequences $(x_1, x_2, \ldots, x_{2n})$ with $x_i = 1$ or $x_i = -1$ for $1 \leqslant i \leqslant 2n$ which satisfy the conditions

(1) $x_1 + \cdots + x_k \geqslant 0$ for every $k = 1, \ldots, 2n$ and
(2) $x_1 + \cdots + x_{2n} = 0$.

In order to define this correspondence one traverses a rising path d from O to A. The path d can be written as a sequence of segments of length 1, $d = (s_1, s_2, \ldots, s_{2n})$, where the order of the indices indicates the order of displacement of the segments from O to A. The sequence associated with the path is obtained from the sequence $(s_1, \ldots, s_{2n})$ by writing 1 in place of each horizontal segment and -1 in place of each vertical segment. In this way one obtains the following sequence for the path illustrated in Figure 3.2:

$$(1, -1, 1, 1, -1, 1, -1, -1, 1, -1).$$

Condition (1) expresses precisely the fact that the path d cannot pass through points which are located above the diagonal, and condition (2) implies that there are n horizontal and n vertical segments. The path must therefore terminate in the point A.

It follows from Problem 3.24 that the number of sequences of 1's and -1's with the given properties is equal to

$$C_{n+1} = \frac{1}{n+1}\binom{2n}{n}.$$

This observation ends the proof.

3.27 For $n = 4$ the number of solutions is equal to 2. Now let $n \geqslant 5$. A vertex v_1 of the polygon can be chosen in n ways. Join the two vertices which are adjacent to v_1 by a diagonal d_1. Consider the triangle which has side d_1 and a vertex in common with the polygon which is different from v_1. It must have a side in common with the convex polygon. Thus the third side must be one of the two diagonals which join a vertex located on d_1 with one of the neighbors of another vertex on d_1. Thus a second diagonal d_2 can be chosen in two ways.

Similarly, if $d_1, \ldots, d_i$ ($i < n-3$) have been chosen, then there are two possible choices for d_{i+1}. There are thus $n \, 2^{n-4}$ ways of selecting a vertex x_1 and a sequence $d_1, d_2, \ldots, d_{n-3}$ of diagonals with the desired property.

Each triangulation obtained in this manner has two triangles which contain two adjacent sides of the polygon, and hence each is counted twice. It follows that the desired number is $n \, 2^{n-5}$ for every $n \geqslant 4$.

3.28 Let $f_j(n+1)$ denote the number of sequences with $n+1$ terms and $a_1 = j$. It follows that $f_j(n+1) = f_{j-1}(n) + f_{j+1}(n)$ for $j \geqslant 1$ and $f_0(n+1) = f_1(n)$.

It will be shown by induction on $k \geqslant 1$ that for every $k < n$ the following relation holds:

Solutions

$$f_0(n) = \sum_{j \leqslant k/2} A_{k,j} f_{k-2j}(n-k), \tag{1}$$

where $A_{k,j} = \binom{k}{j} - \binom{k}{j-1}$ for $0 \leqslant 2j \leqslant k$. In fact, for $k=1$ relation (1) becomes $f_0(n) = f_1(n-1)$, which has been seen to be true. Now under the assumption that $k < n-1$ and (1) holds, replace $f_{k-2j}(n-k)$ by $f_{k-2j-1}(n-k-1) + f_{k-2j+1}(n-k-1)$ in (1). It follows that

$$f_0(n) = \sum_{j \leqslant (k+1)/2} A_{k+1,j} f_{k+1-2j}(n-(k+1)),$$

where

$$A_{k+1,j} = A_{k,j} + A_{k,j-1} = \binom{k}{j} - \binom{k}{j-1} + \binom{k}{j-1} - \binom{k}{j-2}$$

$$= \binom{k}{j} + \binom{k}{j-1} - \left[\binom{k}{j-1} + \binom{k}{j-2}\right]$$

$$= \binom{k+1}{j} - \binom{k+1}{j-1} \quad \text{for} \quad 0 \leqslant 2j \leqslant k+1,$$

and hence (1) is true for every $1 \leqslant k \leqslant n-1$.

If $n = k+1$ in (1), then

$$f_0(n) = \sum_{j \leqslant k/2} A_{k,j} = \binom{k}{[k/2]}.$$

[L. Carlitz, *Math. Nachr.*, **49** (1971), 125–147.]

3.29 Let $a_1 = j$ and $|a_i - a_{i+1}| \leqslant 1$, and denote the corresponding number of sequences by $g_j(n+1)$.

It is easy to show that $g_0(n+1) = g_0(n) + g_1(n)$ and $g_j(n+1) = g_{j-1}(n) + g_j(n) + g_{j+1}(n)$ for $j \geqslant 1$. Similarly, the fact that

$$(1+x+x^2)^{m+1} = \left[\sum_{k \geqslant 0} c(m,k) x^k\right](1+x+x^2),$$

yields

$$c(m+1, k) = c(m, k-2) + c(m, k-1) + c(m, k),$$

where $c(m, k) = 0$ for $k < 0$. One thus finds that

$$g_0(n) = g_0(n-1) + g_1(n-1) = 2g_0(n-2) + 2g_1(n-2) + g_2(n-2)$$

$$= 4g_0(n-3) + 5g_1(n-3) + 3g_2(n-3) + g_3(n-3)$$

$$= 9g_0(n-4) + 12g_1(n-4) + 9g_2(n-4) + 4g_3(n-4) + g_4(n-4)$$

$$= \cdots.$$

In general,
$$g_0(n) = \sum_{j=0}^{k} B_{k,j} g_{k-j}(n-k) \tag{1}$$

for every $0 \leq k \leq n-1$.

Relation (1) can be proved by induction on k. It has been seen that (1) is true for $k=0, 1, 2, 3, 4$. Supposing that (1) is true for all values $k \leq p \leq n-2$, it will be shown that

$$g_0(n) = \sum_{j=0}^{p+1} B_{p+1,j} g_{p+1-j}(n-p-1). \tag{2}$$

In fact, if one replaces k by p in (1) and replaces $g_{p-j}(n-p)$ by $g_{p-1-j}(n-p-1) + g_{p-j}(n-p-1) + g_{p-j+1}(n-p-1)$ for $j=0,\ldots,p$, then one obtains an expression of form (2) where the coefficients $B_{p+1,j}$ are given by the recurrence relation

$$B_{p+1,j} = B_{p,j-2} + B_{p,j-1} + B_{p,j}. \tag{3}$$

It will now be shown that $B_{k,j} = c(k,j) - c(k,j-2)$ for every $0 \leq j \leq k$. In fact, for $k=1$ the equation is satisfied, since $B_{1,0} = c(1,0) = 1$ and $B_{1,1} = c(1,1) = 1$. Suppose that the equation holds for every $k \leq p$ and every $0 \leq j \leq k$. It follows that

$$B_{p+1,j} = B_{p,j-2} + B_{p,j-1} + B_{p,j}$$
$$= c(p, j-2) - c(p, j-4) + c(p, j-1) - c(p, j-3) + c(p, j) - c(p, j-2)$$
$$= c(p+1, j) - c(p+1, j-2),$$

in view of the induction hypothesis and the recurrence relation satisfied by the number $c(m, k)$. One can thus write

$$g_0(n) = \sum_{j=0}^{k} \{c(k,j) - c(k,j-2)\} g_{k-j}(n-k)$$

for every $0 \leq k \leq n-1$. By taking $n = k+1$, one finds that

$$g_0(k+1) = \sum_{j=0}^{k} \{c(k,j) - c(k,j-2)\} g_{k-j}(1) = \sum_{j=0}^{k} \{c(k,j) - c(k,j-2)\}$$
$$= c(k, k) + c(k, k+1)$$

for every $k \geq 0$, since $g_{k-j}(1) = 1$. [L. Carlitz, *Math. Nachr.*, **49** (1971), 125–147.]

3.30 It was shown in Problem 3.24 that the number of sequences $(x_1, x_2, \ldots, x_{2n})$ with terms equal to ± 1 which satisfy the conditions $x_1 + \cdots + x_k \geq 0$ for every $1 \leq k \leq 2n$ and $x_1 + \cdots + x_{2n} = 0$ is equal to $\{1/(n+1)\}\binom{2n}{n}$. Observe that there exists a bijection from the set of sequences $x = (x_1, x_2, \ldots, x_{2n})$ to the set of sequences $a = (a_1, a_2, \ldots, a_{2n+1})$ which satisfy the given conditions. The mapping is defined by $f(x) = a$ if

$$a_1 = 0, \quad a_2 = x_1 = 1, \quad a_3 = x_1 + x_2, \ldots, \quad a_{k+1} = x_1 + x_2 + \cdots + x_k$$

for $1 \leq k \leq 2n$, and hence $a_{2n+1} = 0$.

Solutions

3.31 Consider a sequence which satisfies the condition of the problem, and suppress all terms equal to n. One obtains a sequence of k terms, where $0 \leq k \leq r$, which satisfies the given condition. For $i=n$ the condition implies that the number of such sequences is zero for $r=n$. Start with a sequence consisting of the numbers $1, \ldots, n-1$, of length k, with the property that at most $i-1$ terms are less than or equal to i for $i=1, \ldots, n-1$. One can insert $r-k$ new terms equal to n so as to obtain a sequence $(x_1, \ldots, x_r)$ with $1 \leq x_i \leq n$ which satisfies the same property. However, the $r-k$ terms which are equal to n can be inserted in $\binom{r}{k}$ ways, which corresponds to the $\binom{r}{r-k} = \binom{r}{k}$ possible ways of choosing the positions occupied by $r-k$ terms equal to n in a sequence of r terms.

Now start with all the sequences of length $k=0, \ldots, r$ composed of the numbers $1, \ldots, n-1$, which satisfy the given property. By inserting $r-k$ terms equal to n in all possible ways, one obtains, without repetitions, all the sequences $(x_1, \ldots, x_r)$ with $1 \leq x_i \leq n$ which satisfy the same property. This implies the recurrence relation

$$f(n, r) = \sum_{k=0}^{r} \binom{r}{k} f(n-1, k) \quad \text{for} \quad 1 \leq r \leq n-1, \tag{1}$$

where $f(n, r)$ represents the desired number of sequences.

It will be shown by induction on n that

$$f(n, r) = (n-r)n^{r-1}. \tag{2}$$

If $n=r$ it has been seen that $f(n, r)=0$, and this coincides with the value given by (2). Let $n>r$, and suppose that $f(n-1, k) = (n-1-k)(n-1)^{k-1}$ for every $k=0, \ldots, n-1$. In view of (1) it follows that

$$f(n, r) = \sum_{k=0}^{r} \binom{r}{k} f(n-1, k) = \sum_{k=0}^{r} \binom{r}{k} (n-1-k)(n-1)^{k-1}$$

$$= \sum_{k=0}^{r} \binom{r}{k} (n-1)^k - \sum_{k=0}^{r} \binom{r}{k} k(n-1)^{k-1}$$

$$= n^r - r \sum_{k=1}^{r} \binom{r-1}{k-1} (n-1)^{k-1}$$

$$= n^r - rn^{r-1} = (n-r)n^{r-1} \quad \text{for every } r=1, \ldots, n-1.$$

However, for $r=n$ it has been seen that formula (2) is also true. Thus (2) is true for every $n \geq r$. [H. E. Daniels, *Proc. Roy. Soc.*, **A183** (1945), 405–435.]

3.32 If $|f^{-1}(1)| = k \geq 1$, then these k elements can be chosen in $\binom{n}{k}$ ways and for the rest of the elements the function f can be defined in S_{n-k} ways. One thus obtains the recurrence relation

$$S_n = \sum_{k=1}^{n} \binom{n}{k} S_{n-k},$$

or
$$\frac{2S_n}{n!} = \sum_{k=0}^{n} \frac{1}{k!} \frac{S_{n-k}}{(n-k)!} \quad \text{for} \quad n \geq 1.$$

Thus the exponential generating function is

$$2s(x) = \sum_{n=0}^{\infty} \frac{2S_n}{n!} x^n = 1 + \sum_{n=0}^{\infty} \left(\sum_{k=0}^{n} \frac{1}{k!} \frac{S_{n-k}}{(n-k)!} \right) x^n$$

$$= 1 + \left(\sum_{m=0}^{\infty} \frac{S_m}{m!} x^m \right) \left(\sum_{k=0}^{\infty} \frac{x^k}{k!} \right)$$

$$= 1 + s(x) e^x,$$

and hence $s(x) = 1/(2 - e^x)$. Expand $s(x)$ in a power series:

$$s(x) = \frac{1}{2} \cdot \frac{1}{1 - e^x/2} = \frac{1}{2} \sum_{k=0}^{\infty} \left(\frac{e^x}{2} \right)^k = \sum_{k=0}^{\infty} \frac{1}{2^{k+1}} \sum_{n=0}^{\infty} \frac{k^n x^n}{n!},$$

from which the expression for S_n follows by identifying the coefficients of x^n in the two sides.

3.33 First determine the number $U_{n,k}$ of ordered systems consisting of k linearly independent vectors from V. As the first component of such a system one can choose any of the $q^n - 1$ vectors other than 0 in the space V. Each vector $v \neq 0$ generates a subspace of dimension one which contains q vectors. There thus exist $q^n - q$ vectors, each of which together with v forms a system of two linearly independent vectors and thus can be chosen as the second component of the system. Let w be one of them. The pair $\{v, w\}$ generates a two-dimensional subspace which contains q^2 vectors. Thus there exist $q^n - q^2$ vectors which are linearly independent of v and w, and any of them can be chosen as the third vector of the system. Continuing in this way, one has

$$U_{n,k} = (q^n - 1)(q^n - q) \cdots (q^n - q^{k-1}).$$

Each system consisting of k linearly independent vectors generates a k-dimensional subspace of V. Conversely every k-dimensional subspace has $U_{k,k}$ ordered bases. Thus

$$\begin{bmatrix} n \\ k \end{bmatrix}_q = \frac{U_{n,k}}{U_{k,k}},$$

and the formula follows by simplification.

3.34 The properties can be obtained by a direct calculation. By passing to the limit as $q \to 1$ in (b) and (c) the results are well-known properties of binomial coefficients.

3.35 The given equality can be proved by counting in two distinct ways the number of linear transformations of an n-dimensional vector space V_n over a finite field $GF(q)$ to a vector space Y over $GF(q)$ with y vectors.

Let $\{x_1, \ldots, x_n\}$ be a basis for V_n. The image of an arbitrary x_i can be any of the vectors in Y, and these n images uniquely determine a linear transformation. There are thus y^n such transformations.

It will be shown that the right-hand side of Cauchy's identity counts the number of linear transformations $f: V_n \to Y$ by taking into consideration the dimension of the subspace $\mathrm{Ker}(f)$ consisting of the vectors in V_n whose image is the zero vector in Y. By Problem 3.33 there exist $\begin{bmatrix}n\\k\end{bmatrix}_q$ subspaces V_k of V_n of dimension equal to k.

Let $z_1, \ldots, z_{n-k}, z_{n-k+1}, \ldots, z_n$ be a basis for V_n such that $z_{n-k+1}, \ldots, z_n$ generate the subspace V_k. A linear transformation $f: V_n \to Y$ has $\mathrm{Ker}(f) = V_k$ if and only if it maps the vectors $z_{n-k+1}, \ldots, z_n$ into the zero vector in Y and the $n-k$ remaining vectors $z_1, \ldots, z_{n-k}$ into a linearly independent set of vectors in Y. The vector z_1 can have as its image in Y any of these y vectors, other than the zero vector, the vector z_2 can have as its image any of the y vectors which do not belong to the linear subspace with dimension 1 generated by the image of z_1, and so on. One therefore obtains $(y-1)(y-q) \cdots (y-q^{n-k+1})$ linear mappings $f: V_n \to Y$ which have $\mathrm{Ker}(f) = V_k$ and $\dim(V_k) = k$. Thus $0 \leqslant k \leqslant n$. When $k = 0$ there are $(y-1)(y-q) \cdots (y-q^{n-1})$ such transformations, and when $k = n$ there is a unique transformation which maps all vectors in V_n onto the zero vector in Y.

For fixed y this establishes that the equality holds. Since y can take on an infinite number of values, it follows that the equation is a polynomial identity in the variable y. [J. Goldman, G.-C. Rota, *Studies in Applied Mathematics*, **XLIX**(3) (1970), 239–258.]

3.36 For $k = 1$ one has $f(n, 1) = 1$; for $k = 2$, $a_2 = n$ and $a_2 - a_1$ is an even number, and hence $f(n, 2) = [(n-1)/2]$. One can prove (a) by induction on n. This formula is valid for $n = 2$ and $n = 3$ for all $k \geqslant 1$, if by definition $\binom{0}{0} = 1$. Suppose (a) is true for $n \leqslant m - 1$. It is clear that $f(m, k) = N_1(m, k) + N_2(m, k)$, where $N_1(m, k)$ is the number of sequences, satisfying the same conditions, for which $a_1 = 1$ and $N_2(m, k)$ is the number of such sequences with $a_1 \geqslant 2$. It follows that $N_2(m, k) = f(m-1, k)$, since the sequence

$$1 \leqslant a'_1 < a'_2 < \cdots < a'_k = m - 1,$$

where $a'_j = a_j - 1$ for $1 \leqslant j \leqslant k$, satisfies the same conditions and this mapping is one-to-one. For $m - k \equiv 0 \pmod{2}$ it turns out that $N_1(m, k) = 0$, and for $m - k \equiv 1 \pmod{2}$ it can similarly be shown that $N_1(m, k) = f(m-1, k-1)$. Now consider two cases:

(a1) $m - k \equiv 0 \pmod{2}$. It follows that

$$f(m, k) = f(m-1, k) = \binom{[(m+k-4)/2]}{k-1}$$

by the induction hypothesis. If $m - k \equiv 0 \pmod{2}$ then $m + k \equiv 0 \pmod{2}$, and hence

$$\begin{bmatrix} m+k-4 \\ 2 \end{bmatrix} = \begin{bmatrix} m+k-3 \\ 2 \end{bmatrix} \quad \text{and} \quad f(m,k) = \binom{\left[\frac{m+k-3}{2}\right]}{k-1}.$$

(a2) $m-k \equiv 1 \pmod{2}$, whence $m+k \equiv 1 \pmod{2}$. In this case

$$\begin{aligned} f(m,k) &= f(m-1,k) + f(m-1,k-1) \\ &= \binom{[(m+k-4)/2]}{k-1} + \binom{[(m+k-5)/2]}{k-2} \\ &= \binom{[(m+k-3)/2]-1}{k-1} + \binom{[(m+k-3)/2]-1}{k-2} \\ &= \binom{[(m+k-3)/2]}{k-1} \end{aligned}$$

and hence (a) is also true for $n=m$. It is easy to see that the numbers $f(n,k)$ for $k \geq 1$ generate a zigzag line in the Pascal triangle of binomial coefficients. We denote their sum for $k \geq 1$ by G_n and prove that

$$G_{n+2} = G_{n+1} + G_n \tag{1}$$

for any $n \geq 2$. Suppose first that $n = 2p$, p integer. It follows that

$$G_n = \binom{p-1}{0} + \binom{p-1}{1} + \binom{p}{2}$$
$$+ \binom{p}{3} + \binom{p+1}{4} + \binom{p+1}{5} + \cdots + 1,$$

$$G_{n+1} = \binom{p-1}{0} + \binom{p}{1} + \binom{p}{2}$$
$$+ \binom{p+1}{3} + \binom{p+1}{4} + \cdots + 1,$$

$$G_{n+2} = \binom{p}{0} + \binom{p}{1} + \binom{p+1}{2}$$
$$+ \binom{p+1}{3} + \binom{p+2}{4} + \cdots + 1.$$

But

$$\binom{p}{0} = \binom{p-1}{0},$$
$$\binom{p}{1} = \binom{p-1}{1} + \binom{p-1}{0},$$
$$\binom{p+1}{2} = \binom{p}{2} + \binom{p}{1},$$

Solutions

$$\binom{p+1}{3} = \binom{p}{3} + \binom{p}{2}, \ldots,$$

$$\binom{q}{q} = \binom{q-1}{q-1},$$

where $\binom{q}{q} = 1$ is the last term of G_{n+2}, and so (1) is established in this case. For $n = 2p+1$ one has

$$G_n = \binom{p-1}{0} + \binom{p}{1} + \binom{p}{2} + \binom{p+1}{3} + \binom{p+1}{4} + \cdots + 1,$$

$$G_{n+1} = \binom{p}{0} + \binom{p}{1} + \binom{p+1}{2} + \binom{p+1}{3} + \cdots + 1,$$

$$G_{n+2} = \binom{p}{0} + \binom{p+1}{1} + \binom{p+1}{2} + \binom{p+2}{3} + \cdots + 1,$$

and (1) may be proved similarly.

Since $G_2 = 1 = F_1$ and $G_3 = 2 = F_2$, it follows that $G_n = F_{n-1}$ for any $n \geq 2$. Recurrence relation (1) may also be proved directly from Pascal's triangle in both of the cases when n is even or odd.

CHAPTER 4

4.1 Suppose that neither (1) nor (2) holds. Then X contains at most a pairwise distinct objects and there are at most a copies of each of these. Thus X has at most $a^2 \leq n-1$, objects which is a contradiction.

4.2 The k rows and k columns on which the rooks are found can be chosen in $\binom{m}{k}\binom{n}{k}$ distinct ways. The intersection of the k rows and k columns forms a table with k^2 squares, on which the k rooks can be situated in $k!$ ways in positions so that no rook can attack another.

In fact, the k rooks are found in k different columns, but the rook in the first column can be arranged in k different ways on the k rows; the rook on the second column can be arranged in $k-1$ different ways in $k-1$ rows other than the row on which the rook in the first column is stationed, and so on. The result is that there are $k!$ possible arrangements.

Thus the total number of possible arrangements is $k!\binom{m}{k}\binom{n}{k}$.

4.3 Each of the 19 numbers in A belongs to one of the following 18 pairwise disjoint sets:

$$\{1\}, \{52\}, \{4, 100\}, \{7, 97\}, \{10, 94\}, \ldots, \{49, 55\}.$$

Thus there exist two distinct integers in A which belong to one of the pairs $\{4, 100\}, \ldots, \{49, 55\}$ and which therefore have their sum equal to 104. [W. L. Putnam Competition of 1978, *American Math. Monthly* **86**(9) (1979), 752.]

4.4 Consider the set $\{k+1, k+2, \ldots, 2k+1\}$, which contains $k+1$ numbers. The sum of each pair of distinct numbers selected from this set lies between $2k+3$ and $4k+1$, and is therefore not divisible by $2k+1$. The difference of each two different numbers lies between 1 and k and is therefore not a multiple of $2k+1$. Let the number sought be denoted $n(k)$. It follows that $n(k) \geq k+2$.

Let A be a set which consists of $k+2$ integers. If there are two numbers in A which have the same residue modulo $2k+1$, then their difference is divisible by $2k+1$, and A satisfies the given condition.

In the opposite case, all the $k+2$ remainders of the numbers in A modulo $2k+1$, are pairwise distinct. Thus the set of remainders is a $(k+2)$-element subset of the set $\{0, 1, \ldots, 2k\}$. One can also consider the remainders as forming a $(k+2)$-element subset of the set $M = \{-k, -(k-1), \ldots, -1, 0, 1, \ldots, k\}$, since every integer $p = (2k+1)q + m$, where q is an integer and $m \in M$. However, for every choice of $k+2$ numbers from M, there will exist two among them whose sum is zero, since otherwise one would have $|M| \geq 2(k+1) + 1 = 2k+3$. Thus A contains two numbers whose sum is divisible by $2k+1$. But $|A| = k+2$, from which we deduce that $n(k) \leq k+2$. From the two opposite inequalities one can conclude that $n(k) = k+2$.

4.5 Sum the given inequality for $j = 1, \ldots, n$ to obtain
$$|A_i| + n|A \cap C_k| + |C_k| \geq n^2.$$
If one further sums over $i = 1, \ldots, n$, then
$$|M| + 2n|C_k| \geq n^3.$$
Finally by summing on k it is seen that $n|M| + 2n|M| \geq n^4$ and thus $|M| \geq n^3/3$. The lower bound is attained if $n \equiv 0 \pmod 3$.

Suppose now that $|M| = n^3/3$, and define a partition of M,
$$M = A_1^1 \cup A_2^1 \cup \cdots \cup A_n^1 \cup A_1^2 \cup \cdots \cup A_n^2 \cup \cdots \cup A_1^n \cup A_2^n \cup \cdots \cup A_n^n,$$
such that $|A_i^j| = n/3$ for every $1 \leq i, j \leq n$. Let $A_i = \bigcup_{j=1}^n A_j^i$, $B_i = \bigcup_{j=1}^n A_i^j$, and $C_i = \bigcup_{j=1}^n A_{i+j-1 \pmod n}^j$. The partitions (A_i), (B_i), and (C_i), $1 \leq i \leq n$, satisfy the identity $|A_i \cap B_j| + |A_i \cap C_k| + |B_j \cap C_k| = n$, since for every i, j, k the following relation holds:
$$|A_i \cap B_j| = |A_i \cap C_k| = |B_j \cap C_k| = \frac{n}{3}.$$
Thus in this case min $|M| = n^3/3$. [I. Tomescu, E 2582, *American Mathematical Monthly*, **83**(3) (1976), 197.]

4.6 First we show that f is idempotent if and only if the function $g : Y \to Y$ is the identity function, where $Y = f(X)$ and $g(x) = f(x)$ for every $x \in Y$. In fact, if f is idempotent, it follows that $g(x) = f(x)$ for $x \in Y$, and thus there exists $z \in X$ such that $x = f(z)$, and one can write
$$g(x) = f(x) = f(f(z)) = f(z) = x,$$

Solutions

and thus g is the identity function. But if g is the identity function, then

$$f(f(x)) = f(y) = g(y) = y = f(x)$$

for every $x \in X$, and hence f is idempotent.

Let $|Y| = k$. Then $1 \leq k \leq n$.

The set $Y \subset X$ can be chosen in $\binom{n}{k}$ ways; the restriction of the function f to Y is the identity function, and the number of functions $h: X \setminus Y \to Y$ is equal to k^{n-k}. Since f is the identity on Y, it follows that f is uniquely determined by its restriction h to $X \setminus Y$. This latter observation completes the proof of the formula for $i(n)$.

One can also show that the exponential generating function for the numbers $i(n)$ is $\exp(xe^x)$.

Observe that

$$\exp(xe^x) = \exp\left(x + \frac{x^2}{1!} + \frac{x^3}{2!} + \cdots + \frac{x^{n+1}}{n!} + \cdots\right)$$

$$= 1 + \left(x + \frac{x^2}{1!} + \frac{x^3}{2!} + \cdots\right) + \frac{1}{2!}\left(x + \frac{x^2}{1!} + \frac{x^3}{2!} + \cdots\right)^2 + \cdots$$

$$+ \frac{1}{n!}\left(x + \frac{x^2}{1!} + \frac{x^3}{2!} + \cdots\right)^n + \cdots.$$

Thus the coefficient of x^n in the expansion

$$\frac{1}{k!}\left(x + \frac{x^2}{1!} + \frac{x^3}{2!} + \cdots\right)^k = \frac{1}{k!} x^k e^{kx} = \frac{1}{k!} x^k \left(1 + \frac{kx}{1!} + \frac{k^2 x^2}{2!} + \cdots\right)$$

is precisely

$$\frac{1}{k!(n-k)!} k^{n-k} = \frac{1}{n!}\binom{n}{k} k^{n-k}.$$

This implies that the coefficient of x^n in the expansion of $\exp(xe^x)$ is $i(n)/n!$.

4.7 The proof is carried out by induction on m. For $m=1$ the statement is obviously true. Let $m \geq 2$, and suppose the property holds for $m-1$, with the antichains of the union being pairwise disjoint. Let P be a partially ordered set which does not contain a chain of cardinality $m+1$.

A chain will be said to be *maximal* if its element set is not a proper subset of the element set of another chain, and an element x of S will be said to be *maximal* if $y \leq x$ for every element y in S which is comparable with x.

The antichain M consisting of the maximal elements of P is nonempty, since the maximal element of an arbitrary maximal chain is contained in M. It follows that the partially ordered set $P \setminus M$ does not contain a chain of cardinality m. Suppose that $P \setminus M$ contains a chain $x_1 < x_2 < \cdots < x_m$ of cardinality m. Then there exists $z \in M$ such that $z > x_m$, since otherwise one has $x_m \in M$ and this contradicts the fact that $x_m \in P \setminus M$. Thus $x_1 < x_2 < \cdots < x_m < z$ is a chain of cardinality $m+1$ in P, which contradicts the hypothesis. Since $P \setminus M$ does not contain a chain of cardinality m, it follows from the induction hypothesis that $P \setminus M$ is the union of $m-1$ pairwise disjoint antichains. These, together with M, provide a partition of P into m antichains. [L. Mirsky, *American Math. Monthly*, **78**(8) (1971), 876–877.]

The dual of this proposition also holds for any partially ordered set P. It is called Dilworth's theorem: The smallest number of chains into which P can be partitioned is equal to the maximal number of elements in an antichain.

4.8 Each partition P_k with $1 \leq k \leq n-1$ has k classes and is obtained from P_{k+1} by combining two classes into a single class. Thus, if $P_n, P_{n-1}, \ldots, P_{k+1}$ are fixed, then the partition P_k can be chosen in $\binom{k+1}{2}$ ways. The number of chains of length n is equal to

$$\prod_{k=1}^{n-1} \binom{k+1}{2} = \frac{(n-1)!n!}{2^{n-1}}.$$

4.9 Suppose that $\bigcap_{i=1}^{s} E_i = \emptyset$ and let $E_1 = \{x_1, \ldots, x_r\}$. Since $x_i \notin \bigcap_{i=1}^{s} E_i$, there exists a set $F_i \in F$ such that $x_i \notin F_i$ for $i=1, \ldots, r$. It follows that $E_1 \cap F_1 \cap \cdots \cap F_r = \emptyset$, which contradicts the hypothesis. In fact, if, for example, $x_j \in E_1 \cap F_1 \cap \cdots \cap F_r$, then $x_j \in E_1$ and $x_j \in F_j$, but by construction $x_j \notin F_j$.

4.10 Let the family $S_1 = (X \setminus X_i)_{1 \leq i \leq r}$. It follows that all the sets of S_1 are pairwise distinct subsets which are distinct from subsets of S. In fact, if there exists an index i such that $X \setminus X_i \in S$ and $X_i \in S$, then $(X \setminus X_i) \cap X_i = \emptyset$, which contradicts the definition of the family S. One can thus conclude that $2r \leq |P(X)| = 2^n$, or $r \leq 2^{n-1}$, and thus $\max r \leq 2^{n-1}$. The upper bound is attained if $Y = X \setminus \{x\}$ for an element $x \in X$ and if S is the family of sets $\{Y_i \cup \{x\} \mid Y_i \subset Y\}$. In this case $|S| = |P(Y)| = 2^{n-1}$. If $A, B \in S$ then $A \cap B \neq \emptyset$, since $x \in A$ and $x \in B$ and hence $x \in A \cap B$. [G. Katona, *Acta Math. Acad. Sci. Hung.*, **15** (1964), 329–337.]

4.11 Fix a subset $C \in F$. The function $f : F \to P(X)$ defined by $f(A) = A \triangle C$ is injective. In fact, if $A_1 \triangle C = A_2 \triangle C$, then from the properties of $\triangle$ one can infer that $A_1 = (A_1 \triangle C) \triangle C = (A_2 \triangle C) \triangle C = A_2$ and thus $A_1 = A_2$. It follows that the number of sets of the form $A \triangle C$, where A runs through the family F, is at least equal to $|F| = m$.

4.12 Let $S = \{1, \ldots, n\}$, and denote by A_i the set of families of nonempty, pairwise distinct subsets of S which do not contain the element $i \in S$. Since the number of families of nonempty pairwise distinct subsets of S is equal to $2^{2^n - 1}$, it follows that

$$A(n) = 2^{2^n - 1} - |A_1 \cup \cdots \cup A_n|.$$

By applying the Principle of Inclusion and Exclusion one finds that

$$|A_1 \cup \cdots \cup A_n| = \sum_{i=1}^{n} |A_i| - \sum_{1 \leq i < j \leq n} |A_i \cap A_j| + \cdots.$$

Solutions

Since $|\bigcap_{i \in K} A_i| = 2^{2^{n-k}-1}$, where $|K| = k$, the formula for $A(n)$ follows immediately if one takes into account the fact that the set K can be selected in $\binom{n}{k}$ ways as a subset of S. [L. Comtet, C.R. Acad. Sci. Paris, **A262** (1966), 1091–1094.]

4.13 Let $A = (S_j)_{1 \leq j \leq k}$ be an irreducible covering of the set S. It follows from the definition that each subset S_j contains a nonempty subset T_j of S, consisting of elements which do not belong to the other subsets of the covering.

Let $T = \bigcup_{j=1}^{k} T_j$. It follows that the subsets $T_1, \ldots, T_k$ form a partition of the set T. If $|T| = i$, one sees that $k \leq i \leq n$ and all irreducible coverings of S can be obtained without repetitions in the following way (for every i): For each of the $\binom{n}{i}$ choices of T as a subset of S, consider the $S(i, k)$ partitions $T_1 \cup \cdots \cup T_k$ of T. Each of the $n - i$ remaining elements belong to a family of subsets chosen from among $S_1 \setminus T_1, \ldots, S_k \setminus T_k$ which contains at least two sets, and thus the number of possibilities is $2^k - \binom{k}{1} - \binom{k}{0} = 2^k - k - 1$. From this observation the formula for $I(n, k)$ follows.

The expression for $I(n, n-1)$ is obtained by using the fact that $S(n, n-1) = \binom{n}{2}$. In order to obtain the expression for $I(n, 2)$, suppose that $y \notin S$. Then there exists a bijection from the set of partitions with three classes of $S \cup \{y\}$ onto the set of irreducible 2-set coverings of S, defined as follows: The partition $S \cup \{y\} = S_1 \cup S_2 \cup S_3$ is associated with the irreducible covering of S whose sets are $S_1 \cup (S_3 \setminus \{y\})$ and $S_2 \cup (S_3 \setminus \{y\})$, where S_3 is the class of the partition which contains y. It follows that $I(n, 2) = S(n+1, 3)$. The same conclusion holds if one applies the recurrence relation (a) of Problem 3.5. [T. Hearne, C. Wagner, *Discrete Mathematics*, **5** (1973), 247–251.]

4.14 Let t be the smallest integer such that $[t^2/4] > n$. Consider the following $[t^2/4]$ sets of natural numbers:

$$A_{i,j} = \{x \mid i \leq x \leq j\}, \quad \text{where } 1 \leq i \leq \frac{t}{2} < j \leq t.$$

Let A_{i_k, j_k}, $1 \leq k \leq m$, denote a subfamily F of this family of sets, which is union-free. Label the number i_k (or j_k) if there does not exist another set A_{i_s, j_s} of this family with m sets F with the property that $i_k = i_s$ and $j_s < j_k$ (or $j_k = j_s$ and $i_s < i_k$).

At least one of the endpoints i_k and j_k of A_{i_k, j_k} must be labeled, since otherwise A_{i_k, j_k} would be the union of two sets in F, contrary to hypothesis. Since a labeled integer can be an endpoint for only one set A_{i_k, j_k}, it follows that $m \leq t$ and thus

$$f(n) \leq f\left(\left[\frac{t^2}{4}\right]\right) \leq t, \quad \text{or} \quad f(n) \leq 2\sqrt{n} + 1.$$

In order to obtain the lower bound, let $\{A_1, \ldots, A_n\}$ be an arbitrary family of n distinct sets. Construct a subfamily which is union-free in the following way: Let A_{i_1} denote a minimal A_i, that is, one which does not contain another set as a proper subset. If the sets $A_{i_1}, \ldots, A_{i_s}$ have been selected, then the set $A_{i_{s+1}}$ will be a minimal set in the family $\{A_1, \ldots, A_n\} \setminus \{A_{i_1}, \ldots, A_{i_s}\}$ which is not the union of two distinct sets of the family $\{A_{i_1}, \ldots, A_{i_s}\}$. One can select $A_{i_{s+1}}$

in this way if $n-s > s(s-1)/2$, since there are $\binom{s}{2}$ unions of two distinct sets in the family $\{A_{i_1}, \ldots, A_{i_s}\}$. This construction defines a subfamily $\{A_{i_1}, \ldots, A_{i_r}\}$ where $r + r(r-1)/2 \geq n$, that is, $r \geq \sqrt{2n} - 1$.

The equation $A_i \cup A_j = A_k$ (where $i, j, k \in \{i_1, \ldots, i_r\}$ are pairwise distinct) cannot hold, and thus the family $\{A_{i_1}, \ldots, A_{i_r}\}$ is union-free. In fact, if A_k were selected after A_i and A_j, this equality would also contradict the construction, for otherwise it would follow that A_k was not minimal when it was selected, which contradicts the definition of the family $\{A_{i_1}, \ldots, A_{i_r}\}$. [P. Erdös, S. Shelah, *Graph Theory and Applications*, Proceedings of the Conference at Western Michigan University, 1972, Lecture Notes in Mathematics, Springer-Verlag, 1972.]

4.15 For every $x \in S$ let $M(x) = \{i \mid x \in A_i\}$. It can be shown that for every subset $T \subset \{1, 2, \ldots, n\}$ with $|T| = n - k + 1$, there exists $x \in S$ such that $M(x) = T$. In fact, since $\bigcup_{i \notin T} A_i$ is a union of $k-1$ subsets of the given family, it follows that there is an element $x \in S$ which does not belong to this union. One can show that $M(x) = T$. By construction $M(x) \subset T$. If $M(x)$ is a proper subset of T, it follows that $|M(x)| \leq n - k$. Thus there exist k sets which do not contain the element x, which contradicts the hypothesis that every union of k sets of this family is equal to S. It follows that $M(x) = T$, and thus the function which associates with each element $x \in S$ the subset $M(x)$ has a surjective restriction defined on $U \subset S$, with values in the set $P_{n-k+1}(\{1, \ldots, n\})$. It follows that

$$|S| \geq |U| \geq \binom{n}{n-k+1} = \binom{n}{k-1}.$$

If $|S| = \binom{n}{k-1}$, then one can conclude from the hypothesis that

$$\sum_{i=1}^{n} |A_i| = n|A_i| = (n-k+1)|S| = (n-k+1)\binom{n}{k-1},$$

and it can be seen that

$$|A_i| = \frac{n-k+1}{n}\binom{n}{k-1} = \binom{n-1}{k-1}$$

for every $1 \leq i \leq n$. This deduction uses the fact that in this case $|M(x)| = n - k + 1$ for every $x \in S$.

4.16 Let m_0 be the smallest natural number with the property described in the statement of the problem, and set $\bigcup_{i=1}^{k} X_i = Y$. Since $|X_i| = h$ for $1 \leq i \leq k$ and $X_i \subset Y$, it follows that $k \leq \binom{|Y|}{h}$ and thus $m_0 \leq |Y|$, or $m_0 \leq \min |Y|$. In order to prove the opposite inequality, let $Y \subset X$ with $|Y| = m_0$. Since $\binom{m_0}{h} \geq k$, one can choose h-element subsets $X_1, \ldots, X_k$ such that $\bigcup_{i=1}^{k} X_i = Y$. In fact, this property follows from the inequality $m_0 \leq kh$. Suppose, however, that $m_0 > kh$. Since $\binom{m_0-1}{h} < k$ by the definition of m_0, and since $m_0 \geq kh + 1$, one can conclude that

$$\binom{m_0-1}{h} \geq \binom{kh}{h} \geq k,$$

Solutions

which is a contradiction. By construction $|\bigcup_{i=1}^{k} X_i| = |Y| = m_0$, and thus $\min |\bigcup_{i=1}^{k} X_i| \leq m_0$, which establishes the identity in question.

4.17 Let $Y \subset X$ with $|Y| = p$. The set Y may be written as the union of k sets, $Y = A_1 \cup \cdots \cup A_k$, in $(2^k - 1)^p$ different ways. In fact, each of the p elements of Y can belong to $2^k - 1$ nonempty families of subsets $A_1, \ldots, A_k$. It follows from the fact that Y can be chosen in $\binom{n}{p}$ different ways, that

$$\sum |A_1 \cup \cdots \cup A_k| = \sum_{p=1}^{n} p \binom{n}{p} (2^k - 1)^p = n(2^k - 1) \sum_{p=1}^{n} \binom{n-1}{p-1} (2^k - 1)^{p-1}$$

$$= n(2^k - 1) 2^{k(n-1)}.$$

4.18 From the previous problem one can also conclude that $\sum |A_1 \cap \cdots \cap A_k| = n 2^{k(n-1)}$ if $|X| = n$. Recall that the operation of complementation,

$$C(A_1 \cup \cdots \cup A_k) = CA_1 \cap \cdots \cap CA_k,$$

determines a bijection between the family of subsets of the form $A_1 \cup \cdots \cup A_k$ and the family of subsets $A_1 \cap \cdots \cap A_k$. Furthermore $|CA_1 \cap \cdots \cap CA_k| = n - |A_1 \cup \cdots \cup A_k|$. One can thus write

$$\sum |A_1 \cap \cdots \cap A_k| = \sum (n - |A_1 \cup \cdots \cup A_k|) = n 2^{nk} - n(2^k - 1) 2^{k(n-1)}$$

$$= n 2^{k(n-1)},$$

which is the desired result. One should also note that each of the indicated sums contains 2^{nk} terms and each of the subsets $A_1, \ldots, A_k$ can be selected from among the subsets of X in 2^n ways. [I. Tomescu, E 2764, *American Math. Monthly*, **86** (1979), 223.]

4.19 Suppose that a filter basis S contains the sets $A_1, A_2, \ldots A_p$. It follows that there exists $B_1 \in S$ such that $B_1 \subset A_1 \cap A_2$. Similarly, there exists $B_2 \in S$ such that $B_2 \subset B_1 \cap A_3$, and so forth. Finally there is a $B_{p-1} \in S$ such that $B_{p-1} \subset B_{p-2} \cap A_p$ and $B_i \subset A_{i+1}$ for $i = 1, \ldots, p-1$. It is also the case that $B_1 \subset A_1$. By construction,

$$B_{p-1} \subset B_{p-2} \subset \cdots \subset B_2 \subset B_1,$$

and hence there exists a set $B = B_{p-1} \in S$ such that $B \subset A_i$ for $i = 1, \ldots, p$.

Thus every filter basis S has the following form: There exists $B \subset X$, $B \neq \emptyset$, such that

$$S = \{B, B \cup C_1, \ldots, B \cup C_s\}, \tag{1}$$

where $C_1, \ldots, C_s$ are pairwise different nonempty subsets of $X \setminus B$. Let $|X \setminus B| = k$. Then $B \neq \emptyset$ implies that $0 \leq k \leq n-1$. Since $\{C_1, \ldots, C_s\}$ is a family of nonempty subsets of $X \setminus B$, it follows that $0 \leq s \leq 2^k - 1$. But $|B| = n - k$, and thus the set B can be chosen in $\binom{n}{n-k} = \binom{n}{k}$ different ways. For every choice of B the family $\{C_1, \ldots, C_s\}$ of subsets of $X \setminus B$ [which may be empty ($s = 0$)]

can be chosen in a number of ways which is equal to the number of subsets of a (2^k-1)-element set, namely, 2^{2^k-1}. (Note that $X\setminus B$ has 2^k-1 nonempty subsets.) If $|B|=n-k$, it follows that the number of ways in which a filter basis of form (1) can be selected is equal to

$$\binom{n}{k} 2^{2^k-1}$$

The number of filter bases for an arbitrary n-element set is obtained by summing these numbers over $k=0,\ldots,n-1$.

4.20 Let X be the union of all the elements of the sets A_i and B_j where $1 \leq i,j \leq m$. Suppose that $X=\{x_1,\ldots,x_n\}$, and let $(x_{p(1)},\ldots,x_{p(n)})$ be an arbitrary permutation of X.

There exists at most one index i for which each element of the set A_i has an index smaller than each element of the set B_i. In fact, suppose that this is not the case. Let $i \neq j$ be two indices such that each element of A_i has an index smaller than each element of B_i and each element of A_j has an index smaller than each element of B_j. By the hypothesis, there is an element $x_{p(k)} \in B_1 \cap A_2$ and an element $x_{p(r)} \in B_2 \cap A_1$. Since $x_{p(r)} \in A_1$ and $x_{p(k)} \in B_1$, it follows that $p(r)<p(k)$. But $x_{p(r)} \in B_2$ and $x_{p(k)} \in A_2$, and thus $p(r)>p(k)$, which establishes a contradiction.

Thus there exist at most $n!$ pairs (A_i, B_i) with the desired property. Let i be a fixed index, and consider permutations of the indices of the elements of X such that each element of A_i has an index which is smaller than the index of every element of B_i. The number of such permutations is equal to

$$\binom{n}{p+q} p!\, q!\,(n-p-q)! = \frac{n!}{\binom{p+q}{p}}.$$

To see this, note that one can choose $p+q$ elements from $A_i \cup B_i$ in $\binom{n}{p+q}$ ways. The first p elements in increasing order of indices are taken from A_i, and the remaining q elements (with indices larger than the elements of A_i) are associated with B_i. The indices of the elements from A_i can be permuted in $p!$ ways, the indices of the elements from B_i can be permuted in $q!$ ways, and the remaining elements' indices can be permuted in $(n-p-q)!$ ways. Thus in this way one obtains all permutations of $\{1,\ldots,n\}$ with the property that each element of A_i has an index which is smaller than each element of B_i.

Thus each pair (A_i, B_i) is calculated $n!/\binom{p+q}{p}$ times relative to all the permutations of indices for which each element of A_i has an index smaller than each element of B_i. Thus the number of pairs (A_i, B_i) satisfies the inequality

$$m \leq n! \Big/ \frac{n!}{\binom{p+q}{p}} = \binom{p+q}{p}.$$

4.21 A chain of length n formed from subsets of X is a sequence

$$M_1 \subset M_2 \subset \cdots \subset M_{n-1} \subset X$$

such that $|M_i| = i$ for $i = 1, \ldots, n-1$. The number of chains of length n which can be formed from elements of the set X is equal to $n!$. If $|A| = r$, then number of chains of length n which contain A of the form

$$M_1 \subset \cdots \subset M_{r-1} \subset A \subset M_{r+1} \subset \cdots \subset X$$

is equal to $r!(n-r)!$.

If A_i and A_j are noncomparable with respect to inclusion, then every chain which contains A_i is different from every chain which contains A_j. Thus one can write

$$\sum_{i=1}^{p} n_i!(n-n_i)! \leq n!,$$

where $|A_i| = n_i$. But this implies that

$$\frac{p}{\binom{n}{[n/2]}} \leq \sum_{i=1}^{p} \frac{1}{\binom{n}{n_i}} \leq 1,$$

since $\max_{n_i} \binom{n}{n_i} = \binom{n}{[n/2]}$. It follows that $\max p \leq \binom{n}{[n/2]}$. The opposite inequality is obtained by considering the family G of subsets of X which contain $m = [n/2]$ elements.

This proof is due to D. Lubell [*J. Comb. Theory*, **1** (1966), 299].

4.22 Let $\pi = (x_1, \ldots, x_n)$ be a cyclic permutation of the set X. Consider a cycle C with n edges, and associate the symbols $x_1, \ldots, x_n$ with the edges of C (in the usual cyclic order). If an r-element subset A of X contains consecutive elements in the permutation π, then it corresponds to a subwalk of the cycle C of length r.

By condition (2) the subwalks which correspond to the sets $A_1, \ldots, A_p$ have at least one edge pairwise in common. Let x be a vertex of C. Associate the subwalk having the same symbols with the corresponding r-element subset of X. It follows that x is a terminal vertex for at most one walk A_i. In fact, if A_i and A_j have a terminal vertex in common and $i \neq j$ (and thus A_i and A_j are distinct), then they must leave x in opposite directions on the cycle C. Since $n \geq 2r$, it follows that these walks are disjoint relative to edges, which contradicts the hypothesis. Now, since each walk A_j ($j = 2, \ldots, p$) has at least one edge in common with A_1, it follows that one of the endpoints of A_j is an interior point of A_1. But since two distinct walks cannot have a common endpoint, one can conclude that there exist at most $r - 1$ walks A_j with $2 \leq j \leq p$, and hence $p - 1 \leq r - 1$, or $p \leq r$.

Finally, for each cyclic permutation π of X, there exist at most r r-element subsets of X which consist of consecutive vertices relative to π and which satisfy (2). Since there exist $(n-1)!$ cyclic permutations of X, one can in this way obtain at most $r(n-1)!$ r-element subsets of X.

In order to determine a cyclic permutation for which an r-element set A consists of consecutive elements, one must first order the sets A and $X \setminus A$. Since $|A|=r$, it follows that each set A is counted $r!(n-r)!$ times. Thus the number of subsets A_i is $p \leqslant r(n-1)!/r!(n-r)! = \binom{n-1}{r-1}$. In order to show that $\max p \geqslant \binom{n-1}{r-1}$, consider the r-element subsets of X which contain a fixed element $x_i \in X$. There are $\binom{n-1}{r-1}$ such subsets.

This proof is due to G. O. H. Katona [*J. Comb. Theory*, **13** (1972), 183–184].

4.23 Let $X = \{x_1, \ldots, x_n\}$. Color these elements with one of the two colors a or b, so no set E_i is monochromatic. The element x_1 is assumed to be colored with a.

Suppose that one has colored the elements $x_1, \ldots, x_i$ with $1 < i < n$ with a or b so that no subset E_k is monochromatic, and consider the case which occurs when this process cannot be continued. Thus one cannot color the element x_{i+1} with the color a, since there exists a set $E \subset \{x_1, \ldots, x_{i+1}\}$ with $x_{i+1} \in E$ which has all the elements other than x_{i+1} colored with a. The element x_{i+1} can also not be colored with b, since there is a set $F \subset \{x_1, \ldots, x_{i+1}\}$ with $x_{i+1} \in F$ which has all elements distinct from x_{i+1} colored with b. It follows that E, F are distinct sets chosen from $E_1, \ldots, E_m$ with $E \cap F = \{x_{i+1}\}$, which contradicts the hypothesis. Thus one can color x_{i+1} with either a or b so that no monochromatic set E_k is produced.

It has thus been proved by induction that the coloring process can continue until one has colored X with two colors so that no set is monochromatic.

4.24 Color the elements of X randomly with red and blue. Assume that the colorings are independent and that each color has probability $\tfrac{1}{2}$. Let A_i be the event which consists in coloring all the elements of E_i with a single color. It follows that $P(A_i) = 1/2^{r-1}$ for every $i = 1, \ldots, n$, since there are 2^r ways of coloring E_i with two colors and among these only two lead to a monochromatic coloring. Thus the probability that a random coloring contains a monochromatic subset is equal to

$$P(A_1 \cup \cdots \cup A_n) < \sum_{i=1}^{n} P(A_i) = \frac{n}{2^{r-1}} \leqslant 1.$$

The first inequality is strict, since the events $A_1, \ldots, A_n$ are not independent; they occur simultaneously when all the elements of X have the same color.

Thus the probability of the complementary event is strictly positive, which shows that there exists a coloring with the desired property.

4.25 Assume that $n \geqslant 3$ and the family F of three-element subsets of M has the property that for each two distinct subsets $A, B \in F$ one has $|A \cap B| = 0$ or $|A \cap B| = 2$. In this case it will be shown that $|F| \leqslant n$, which establishes the desired property.

Thus suppose that F contains only three-element sets which either are

Solutions

pairwise disjoint or have exactly two elements in common. In particular, let $A, B \in F$ such that $A \cap B = \{a, b\}$, $A = \{a, b, c\}$, and $B = \{a, b, d\}$. Consider two cases:

(I) The element c belongs to another set $C \in F$. The set C cannot contain a unique element in common with A, and hence C contains one of the elements a or b. But C also cannot contain a unique element in common with B and hence $d \in C$. It follows that $C \subset A \cup B$. One can now show that if the elements a, b, or d also belong to a set of F other than A, B, or C, then this set must be contained in $A \cup B$. If the element d is selected instead of c the same conclusion can be obtained by an analogous argument. Suppose, for example, that $a \in C$. Let $a \in D$ and $D \in F$. From the fact that $|D \cap A| = |D \cap B| = |D \cap C| = 2$ (which holds because $\{a\} = A \cap B \cap C \cap D$) one can conclude that D is one of the sets A, B, or C. However, if $b \in D$, it follows from the fact that $b \in A \cap B$ that the pair D and A and the pair D and B each have two elements in common. One of these common elements also belongs to C, and hence D and C have two elements in common. It follows that if D is distinct from A, B, and C, then one must have $D = \{b, c, d\} \subset A \cup B$. On the other hand, consider a four-element set and its four three-element subsets. They satisfy the condition that the intersection of every two contains exactly two elements.

(II) The element a belongs to another set $C \in F$. If C also contains one of the elements c or d, it follows that $C \subset A \cup B$ and one again has case (I). Otherwise it turns out that $b \in C$ and C also contains an element $x_1 \neq c, d$. In this case the elements c and d do not belong to another set of the family F. Suppose, for example, that $c \in D$ and $D \neq A, B, C$. It follows that $a \in D$ and hence $d \in D$ and $x_1 \in D$, which establishes a contradiction, since $|D| = 3$.

If one of the elements a or b belongs to another set D in F, then D does not contain c or d. It follows that D contains both a and b and another element $x_2 \neq x_1$. From this observation one can deduce the structure of the family F of three-element subsets of M which pairwise have an intersection of cardinality zero or two.

The family F can contain three kinds of sets:

(a) three-element subsets which do not have an element in common with another subset in F;

(b) two subsets A and B such that $A \cap B = \{a, b\}$ and possibly subsets $C_1 = \{a, b, x_1\}$, $C_2 = \{a, b, x_2\}, \ldots$;

(c) two subsets A and B such that $|A \cap B| = 2$ and at most two other three-element subsets contained in $A \cup B$, but with every subset from (b) disjoint from the subsets from (c).

From this one can easily find the maximal cardinality of the family F as a function of n. Let max $|F| = f(n)$. Thus if $n = 4k$, it follows that $f(n) = n$, and this maximum is attained when F is obtained by starting from a partition of M into $n/4$ four-element classes and considering all four subsets with three elements of each class. For $n=4k+1$ one sees that $f(n)=n-1$, and this maximum is attained by applying the preceding construction to the set $M \setminus \{x\}$, where x is an arbitrary element of M. The maximum is attained only in this case. For $n=4k+2$ one has $f(n)=n-2$. The maximum is attained only in the following cases:

(1) The construction is applied for $n=4k$ to the set $M \setminus \{x, y\}$, where x, y are any two distinct elements of M.

(2) F contains $n-2$ subsets of type (b):
$$C_1=\{a, b, x_1\}, \quad C_2=\{a, b, x_2\}, \ldots, \quad C_{n-2}=\{a, b, x_{n-2}\}.$$

For $n=4k+3$ it turns out that $f(n)=n-2$ and the maximum occurs only in the following two cases:

(3) The construction is applied for $n=4k$ to the set $M \setminus \{x, y, z\}$, where x, y, z are distinct elements of M, and the set $\{x, y, z\}$ is then added to the family thus obtained.

(4) F contains $n-2$ subsets of type (b), as in case (2).

Since $f(n) \leq n$ for $n \geq 3$, the property in question has been established. One considers $n \geq 5$ in the statement of the problem, because then there are $n+1$ pairwise distinct three-element subsets. One can also observe that the limit $n+1$ can be reduced to n or $n-1$ when $n \equiv 1 \pmod 4$ or $n \equiv 2$ or $3 \pmod 4$ respectively.

4.26 One can assume that $a_i, b_i \geq 0$ for $i=1, \ldots, n$, since the problem remains the same if an arbitrary constant is added to all the numbers a_i and b_i. Define the following polynomials:
$$f(x) = \sum_{i=1}^n x^{a_i}, \quad g(x) = \sum_{i=1}^n x^{b_i}.$$

Then
$$f^2(x) = \sum_{i=1}^n x^{2a_i} + 2 \sum_{1 \leq i < j \leq n} x^{a_i+a_j} = f(x^2) + 2 \sum_{1 \leq i < j \leq n} x^{a_i+a_j},$$
$$g^2(x) = g(x^2) + 2 \sum_{1 \leq i < j \leq n} x^{b_i+b_j}.$$

Thus $f^2(x) - g^2(x) = f(x^2) - g(x^2)$, or
$$f(x) + g(x) = \frac{f(x^2) - g(x^2)}{f(x) - g(x)},$$

but the polynomial $f(x) - g(x)$ is not identically zero, since $\{a_1, \ldots, a_n\} \neq \{b_1, \ldots, b_n\}$.

Further, since $f(1) = g(1) = n$, one can write
$$f(x) - g(x) = (x-1)^k p(x)$$
where $k \geq 1$ and $p(1) \neq 0$. It follows that

Solutions

$$f(x)+g(x)=\frac{f(x^2)-g(x^2)}{f(x)-g(x)}=(x+1)^k\frac{p(x^2)}{p(x)}.$$

If one now sets $x=1$, it follows that

$$2n=f(1)+g(1)=(1+1)^k=2^k, \quad \text{or} \quad n=2^{k-1}.$$

For the case n a power of 2, one can construct an example of two collections which satisfy the given conditions as follows: $\{a_1, \ldots, a_n\}$ is formed from $\binom{k+1}{2p}$ copies of the integer $2p$ for $0 \leq p \leq (k+1)/2$. The set $\{b_1, \ldots, b_n\}$ is formed from $\binom{k+1}{2p+1}$ copies of the integer $2p+1$ for $0 \leq p \leq k/2$, where $n=2^k$. For $k=2$ these two collections are respectively $\{0, 2, 2, 2\}$ and $\{1, 1, 1, 3\}$. [J. Selfridge, E. G. Straus, *Pacific J. Math.*, **8** (1958), 847–856.]

4.27 If $\varnothing \in F$, it is clear that $F=\{\varnothing\}$, $m=1$, and thus $m \leq n$. If $X \in F$, then it is possible that $F=\{X\}$ and $m=1$ and hence $m \leq n$. Otherwise it follows that $m=2$, $A_1=X$, and $A_2=\{x\}$, where $x \in X$ and $n=|X| \geq 2$ because $A_1 \neq A_2$. Here too $m \leq n$.

Now consider the case in which neither $\varnothing$ nor X is in the family F. If all the sets $A_1, \ldots, A_m$ have an element x in common, then the sets $A_1 \setminus \{x\}, \ldots, A_m \setminus \{x\}$ are pairwise disjoint and they are subsets of the $(n-1)$-element set $X \setminus \{x\}$. It is easy to see that the maximum number of pairwise disjoint subsets of a set $Y=\{y_1, \ldots, y_{n-1}\}$ is equal to n, and this maximum is attained for the following sets: $\varnothing, \{y_1\}, \ldots, \{y_{n-1}\}$. Thus in this case one again has $m \leq n$.

Otherwise, for every $x \in X$ there is a set $A \in F$ which does not contain the element x. Let $d(x)$ be the number of sets of the family F which contain the element x.

For $x \in X$, let $A \in F$ be a set with the property that $x \notin A$, and let $A_1, \ldots, A_d$ [with $d=d(x)$] be the sets which contain x. By the hypothesis $A_1, \ldots, A_d$ each have an element in common with A. Let these elements be $x_1, \ldots, x_d$. If, for example, $x_i=x_j$, $1 \leq i, j \leq d$ and $i \neq j$, then it follows that $\{x, x_i\} \subset A_i \cap A_j$. Since $x \notin A$ and $x_i \in A$, one can conclude that $x \neq x_i$ and thus $|A_i \cap A_j| \geq 2$, which contradicts the hypothesis. The elements $x_1, \ldots, x_d$ are therefore pairwise distinct. From the fact that $\{x_1, \ldots, x_d\} \subset A$ it follows that $|A| \geq d = d(x)$.

Now suppose that $m>n$. Then $m-d(x)>n-d(x) \geq n-|A|$ for each pair (x, A) with $x \notin A$, and hence

$$\frac{d(x)}{m-d(x)} < \frac{|A|}{n-|A|}, \tag{1}$$

since $0<|A|<n$. Now take the sum of the inequalities of form (1) for every pair (x, A) with $x \notin A$.

It has been assumed that for every $x \in X$ there is a set $A \in F$ such that $x \notin A$, and thus every element $x \in X$ belongs to at least one pair (x, A). Similarly, for every $A \in F$ there exists an element $x \in X \setminus A \neq \varnothing$ with the property that $x \notin A$, and hence each set $A \in F$ belongs to at least one pair (x, A).

For a fixed element $x \in X$ there exist exactly $m-d(x)$ sets A with the property that $x \notin A$. For a fixed set A there are exactly $n-|A|$ elements x with the property that $x \notin A$. By summing the inequalities (1) and by grouping terms one obtains

$$\sum_{x \in X} (m - d(x)) \frac{d(x)}{m - d(x)} < \sum_{A \in F} (n - |A|) \frac{|A|}{n - |A|},$$

or

$$\sum_{x \in X} d(x) < \sum_{A \in F} |A|. \qquad (2)$$

It follows from the definition of $d(x)$ that the two sides of inequality (2) are equal, which establishes a contradiction. Thus the assumption that $m > n$ is false and hence $m \leq n$. [P. Crawley, R. P. Dilworth, *Algebraic Theory of Lattices*, Prentice-Hall, 1973.]

4.28 Let $\{p_1, \ldots, p_n\}$ denote a set of n points in the plane. Let a_i be the number of points at distance 1 from the point p_i for $1 \leq i \leq n$. The desired number of pairs will be equal to

$$m = \frac{a_1 + a_2 + \cdots + a_n}{2}.$$

Let C_i denote the unit circle with center at point p_i. Each two such circles have at most two points of intersection. Thus counting for each pair of circles either 0, 1, or 2 points of intersection and adding the numbers thus obtained, one finds that the sum is less than or equal to $2\binom{n}{2} = n(n-1)$.

It is clear that one needs to consider only the case in which each number a_i is at least equal to 1. Since each point p_i is an intersection point for exactly $\binom{a_i}{2}$ pairs of these circles, one can write

$$n(n-1) \geq \sum_{i=1}^{n} \binom{a_i}{2} = \frac{1}{2} \sum_{i=1}^{n} a_i^2 - m.$$

It follows from the Cauchy–Schwarz inequality that

$$\left(\sum_{i=1}^{n} a_i \right)^2 \leq n \sum_{i=1}^{n} a_i^2,$$

and thus from the preceding inequality one has

$$4m^2 = \left(\sum_{i=1}^{n} a_i \right)^2 \leq n(2n(n-1) + 2m),$$

or

$$4m^2 - 2nm - 2n^2(n-1) \leq 0.$$

Thus $m \leq (n + \sqrt{8n^3 - 7n^2})/4 < n\sqrt{n}$ for $n \geq 1$. This fact is immediate from the observation that the inequality is equivalent to $n > \sqrt{n} - 1$. [1978 W. L. Putnam Competition, *Amer. Math. Monthly*, 86 (1979), 752.]

4.29 Each line of intersection is determined by two planes, but each plane is determined by three points of the set M containing the n given points. The

Solutions

number of lines of intersection which contain no point of M is equal to $\frac{1}{2}\binom{n}{3}\binom{n-3}{3}$, since one can choose three points from n points in $\binom{n}{3}$ ways, and another plane in the remaining set in $\binom{n-3}{3}$ ways. The order of selection has no importance in determining the lines of intersection.

The number of lines which contain a unique point of M is equal to $\frac{3}{2}\binom{n}{3}\binom{n-3}{2}$, since one can choose three points in $\binom{n}{3}$ ways, and by considering each of the three points to be a common point of two planes, in turn, the second plane can be chosen in $\binom{n-3}{2}$ ways since one must select the two other points from the $n-3$ remaining points.

In a similar manner, one can see that the number of lines which contain two points of M is equal to $\binom{n}{2}$. Each line of intersection can contain at most two points of M, since no four points of M are coplanar.

The total number of lines of intersection is therefore equal to

$$\frac{1}{2}\binom{n}{3}\left\{\binom{n-3}{3}+3\binom{n-3}{2}\right\}+\binom{n}{2}.$$

4.30 One can choose a vertex of the triangle in n ways. The two other vertices must be chosen from among the $n-3$ vertices different from and not adjacent to the vertex already selected. Thus there are $\binom{n-3}{2}=(n-3)(n-4)/2$ possibilities, of which $n-4$ must be eliminated, since they correspond to the cases in which the two chosen vertices are joined by a side of the polygon. There therefore remain $(n-4)(n-5)/2$ ways of choosing the two other vertices. Since any one of the three vertices can be selected as the first vertex, it follows that there are $n(n-4)(n-5)/6$ ways of choosing a triangle which satisfies the given condition.

4.31 No three diagonals are concurrent in an interior point of the polygon. It follows that each interior point of intersection is uniquely determined by the two diagonals which meet on it and thus by the four vertices of the polygon which are the endpoints of these diagonals. Thus the number of points of intersection of the diagonals inside the polygon is equal to $\binom{n}{4}$.

Each diagonal AB intersects all diagonals which join pairs of vertices, other than A and B, in points other than the vertices of the polygon. It follows that the number of points of intersection other than vertices of the diagonal AB with other diagonals is equal to the number of diagonals which do not have an endpoint in common with AB. But this number is equal to

$$\frac{n(n-3)}{2}-2(n-3)+1=\frac{(n-3)(n-4)}{2}+1,$$

since in A and in B $n-3$ diagonals intersect, but AB has been counted twice, as a diagonal incident with A and with B.

Multiply this number by the total number of diagonals, $n(n-3)/2$. Each point of intersection will be counted twice, with respect to each diagonal on which it is found. Thus the total number of points of intersection other than vertices is equal to

$$\frac{n(n-3)\{(n-3)(n-4)+2\}}{8} = \frac{n(n-3)(n^2-7n+14)}{8}.$$

Subtract from this number the number of interior points of intersection. The number of exterior points of intersection is thus equal to

$$\frac{n(n-3)(n^2-7n+14)}{8} - \binom{n}{4} = \frac{n(n-3)(n-4)(n-5)}{12}$$

for every $n \geq 3$.

4.32 Let $A_1, \ldots, A_n$ be n points on a circle. The order is the order in which the points are encountered when the circle is traversed clockwise. It follows that $A_1 A_2 \cdots A_n$ is a convex polygon. If this polygon does not trace a diagonal, then the sides of the polygon, together with the circumference, determine $n+1$ regions. Draw the $n(n-3)/2$ diagonals, and number the new regions each time a new diagonal is drawn.

Each new diagonal produces exactly as many regions as the number of segments into which it is divided by already existing diagonals. But this number is one greater than the number of points of intersection (in the interior of the circle) of this diagonal with already existing diagonals. Observe that by this procedure each point of intersection of two diagonals located in the interior of the polygon is obtained once and only once. It follows that the total number of new regions is equal to the sum of the number of diagonals and the number of points of intersection located in the interior of the polygon. By hypothesis there do not exist three diagonals which are concurrent in an interior point of the polygon. Thus one can characterize as a bijection the mapping which associates each interior point of intersection of two diagonals to the four vertices of the polygon which are the endpoints of the two diagonals.

It follows that the total number of regions into which the interior of the circle is divided is equal to

$$n+1+\frac{n(n-3)}{2}+\binom{n}{4} = \binom{n}{4}+\binom{n}{2}+1,$$

since the number of interior points of intersection is equal to $\binom{n}{4}$. The formula is valid for every $n \geq 1$.

4.33 The property is established by induction on the number c of curves. If $c=1$, there are no points of intersection and hence the number of regions is equal to 1. Suppose that the property holds for every family of c curves which have n_p points of intersection having multiplicity p, for every $p \geq 2$, where the multiplicity of a point of intersection is defined as being the number of curves which intersect in it. By the induction hypothesis these c curves bound $1+n_2+\cdots+(p-1)n_p+\cdots$ closed regions. Consider a new curve, and suppose that it passes through k_p points of intersection of multiplicity p with the first c curves ($p \geq 2$). The total number of points of intersection of the new curve is equal to $k_2+k_3+\cdots+k_p+\cdots$, and hence the number of arcs induced on this curve is equal to $k_2+k_3+\cdots$. Each portion corresponds to a new region with respect to the family of c curves. Thus the total number of regions bounded by the family of $c+1$ curves is equal to

$$1+n_2+2n_3+\cdots+(p-1)n_p+\cdots+k_2+k_3+\cdots+k_p+\cdots. \quad (1)$$

Let m_p be the number of points of intersection of multiplicity p of the family of $c+1$ closed curves. This number can be expressed as a function of n_p and k_p in the form

$$m_p=n_p-k_{p+1}+k_p,$$

since the n_p points of intersection of multiplicity p of the c curves must be reduced by the number k_{p+1} of these points which become of multiplicity $p+1$ by being located on the new curve. One must also add the number k_p of points which have multiplicity $p-1$ and become of multiplicity p in the family of $c+1$ curves. It follows that

$$1+m_2+2m_3+\cdots+(p-1)m_p+\cdots$$
$$=1+(n_2-k_3+k_2)+2(n_3-k_4+k_3)+\cdots$$
$$+(p-1)(n_p-k_{p+1}+k_p)+\cdots$$
$$=1+n_2+2n_3+\cdots+(p-1)n_p+\cdots+k_2+k_3+\cdots+k_p+\cdots.$$

This is the number (1) of regions formed by the family of $c+1$ curves. The property is thus established by induction.

4.34 Suppose that $(S_1,\ldots,S_m)$ has an SDR, denoted $(a_1,\ldots,a_m)$. It follows that $S_{i_1}\cup\cdots\cup S_{i_k}\supset\{a_{i_1},\ldots,a_{i_k}\}$, and hence $|S_{i_1}\cup\cdots\cup S_{i_k}|\geq k$ for every k and every choice of pairwise distinct numbers $i_1,\ldots,i_k$.

The sufficiency will be established by induction on m. Suppose that the family $S_1,\ldots,S_m$ satisfies the condition for the existence of an SDR. For $m=1$ each element of S_1 forms an SDR.

Suppose further that the property is true for every $m'<m$, and let $M(S)=(S_1,\ldots,S_m)$. Two cases will be studied:

(1) $S_{i_1}\cup\cdots\cup S_{i_k}$ contains at least $k+1$ elements for $1\leq k\leq m-1$ and for each choice of distinct numbers $i_1,\ldots,i_k\in\{1,\ldots,m\}$. Let $a_1\in S_1$. Suppress a_1 each time it appears in the other sets of $M(S)$, and denote the sets thus obtained by $S'_2, S'_3,\ldots,S'_m$. Thus $M'(S)=(S'_2, S'_3,\ldots,S'_m)$ satisfies the necessary and sufficient conditions for the existence of an SDR, since the set $S_{i_1}\cup\cdots\cup S_{i_k}$ contains at least $k+1$ elements and a single element a_1 has been suppressed. By the induction hypothesis $M'(S)$ has an SDR. The element a_1 together with an SDR for $M'(S)$ forms an SDR for $M(S)$ in which a_1 is a representative of S_1 [a_1 no longer appears in the sets of $M'(S)$].

(2) Suppose that $1\leq k\leq m-1$, and let $\{i_1,\ldots,i_k\}\subset\{1,\ldots,m\}$ such that $|S_{i_1}\cup\cdots\cup S_{i_k}|=k$. Renumber the family $S_1,\ldots,S_m$ so that S_{i_1} becomes S_1,

S_{i_2} becomes $S_2, \ldots, S_{i_k}$ becomes S_k. By the induction hypothesis $(S_1, S_2, \ldots, S_k)$ has an SDR. Denote it by $D^* = (a_1, a_2, \ldots, a_k)$, and suppress the elements of D^* whenever they appear in the sets $S_{k+1}, S_{k+2}, \ldots, S_m$. Denote the sets thus obtained by $S^*_{k+1}, S^*_{k+2}, \ldots, S^*_m$. The system of $m-k$ sets $M^*(S) = (S^*_{k+1}, \ldots, S^*_m)$ satisfies necessary and sufficient conditions for the existence of an SDR. In fact, if $S^*_{i_1} \cup S^*_{i_2} \cup \cdots \cup S^*_{i_p}$, where $\{i_1, i_2, \ldots, i_p\} \subset \{k+1, \ldots, m\}$ contains fewer than p elements, then $S_1 \cup S_2 \cup \cdots \cup S_k \cup S^*_{i_1} \cup \cdots \cup S^*_{i_p}$ will contain fewer than $k+p$ elements, since $|S_1 \cup \cdots \cup S_k| = k$. By construction $S_1 \cup \cdots \cup S_k = \{a_1, \ldots, a_k\}$, and hence $|S_1 \cup \cdots \cup S_k \cup S_{i_1} \cup \cdots \cup S_{i_p}| < k+p$, which contradicts the hypothesis that $M(S)$ satisfies necessary and sufficient conditions for the existence of an SDR.

By the induction hypothesis $M^*(S)$ has an SDR, which together with D^* forms an SDR for $M(S)$, since $M^*(S)$ does not contain any element of D^*. [Phillip Hall, *J. London Math. Soc.*, **10** (1935), 26–30.]

4.35 It will first be shown that if F is a minimal family of h-element subsets of X, then there exists an element $x \in X$ such that

$$|\{E \mid E \in F, x \in E\}| \geq \frac{h}{n} M(n, k, h). \tag{1}$$

Suppose that the property is false, that is, for every element $x \in X$ the number of sets in the family F which contains it is smaller than $(h/n)M(n, k, h)$. In this case, the number of occurrences of the elements of X in the sets of F $[= hM(n, k, h)]$ is strictly smaller than the product of the number n of elements of X and the number $(h/n)M(n, k, h)$, which is an upper bound for the number of occurrences of each element in the sets of X. It has thus been shown that $hM(n, k, h) < hM(n, k, h)$, and this contradiction establishes (1). It follows that if $G = \{E \mid E \in F, x \notin E\}$, then

$$|G| = M(n, k, h) - |\{E \mid E \in F, x \in E\}| \leq \frac{n-h}{n} M(n, k, h).$$

Now it will be shown that $M(n-1, k, h) \leq |G|$, which will provide a proof of the first inequality. For this it is sufficient to show that the family G has the property that every k-element subset of the set $Y = X \setminus \{x\}$ contains at least one h-element subset of G.

Let $Z \subset Y$ with $|Z| = k$. Since Z is a k-element subset of X, it follows that there exists an h-element set T of F such that $T \subset Z$. From the fact that Z does not contain the element x, one can conclude that $x \notin T$ and hence the set T belongs to the family G. This observation establishes inequality (a). [G. Katona, T. Nemetz, M. Simonovits, *Mat. Lapok*, **15** (1964), 228–238.]

By iterating this inequality and using the fact that $M(k, k, h) = 1$ one can show that

$$M(n, k, h) \geq \left\{ \frac{n}{n-h} \left\{ \frac{n-1}{n-h-1} \cdots \left\{ \frac{k+1}{k-h+1} \right\} \cdots \right\} \right\},$$

where $\{x\}$ denotes the smallest integer greater than or equal to x. [J. Schönheim, *Pacific J. Math.*, **14** (1964), 1405–1411.]

In order to prove (b) use the following construction: Consider an element $x \in X$, and let G be an extremal family composed of $M(n-1, k, h)$ h-element subsets of $Y = X \setminus \{x\}$. Further let H be an extremal family made up of $M(n-1, k-1, h-1)$ $(h-1)$-element subsets of Y such that every set consisting of $k-1$ elements of Y contains at least one subset of H.

Let E be the family obtained from H by adding the element x to every subset of H. Then $G \cup E$ is such that $G \cap E = \emptyset$ and hence $|G \cup E| = M(n-1, k, h) + M(n-1, k-1, h-1)$. It can now be shown that every k-element subset Z of X contains an h-element subset of $G \cup E$, and this establishes (b).

In fact, if $x \in Z$, then $Z \setminus \{x\}$ is a $(k-1)$-element subset of Y, and hence it contains an $(h-1)$-element subset A in H, and thus $A \cup \{x\} \subset Z$ and $A \cup \{x\} \in E$. If $x \notin Z$, then Z is a k-element subset of Y and thus contains a subset $B \in G$. [J. Kalbfleisch, R. Stanton, International Conference on Combinatorial Mathematics, 1970; *Ann. New York Acad. Sci.*, **175** (1970), 366–369.]

In order to obtain the lower bound in (c), observe that an h-element subset A of X is contained in $\binom{n-h}{k-h}$ k-element subsets of X. For each of the $M(n, k, h)$ subsets of an extremal family F, construct all the $\binom{n-h}{k-h}$ k-element subsets which contain it. In this way, one obtains all the $\binom{n}{k}$ k-element subsets of X, since every k-element set contains an h-element set in F. It is thus possible to write $M(n, k, h) \times \binom{n-h}{k-h} \geq \binom{n}{k}$, from which one may deduce

$$M(n, k, h) \geq \frac{\binom{n}{k}}{\binom{n-h}{k-h}} = \frac{\binom{n}{h}}{\binom{k}{h}}.$$

The upper bound is proved by induction on the index n and by using inequality (b), the recurrence relation for binomial coefficients, and the fact that one obtains an equality for $n = k$ (and for $k = h$ or $h = 1$). In fact

$$M(n, k, h) \leq M(n-1, k-1, h-1) + M(n-1, k, h)$$

$$\leq \binom{n-k+h-1}{h-1} + \binom{n-k+h-1}{h} = \binom{n-k+h}{h}.$$

[I. Tomescu, *Cahiers du Centre d'Etudes de Recherche Opérationnelle*, **15**(3) (1973), 355–362.]

An expression for the number $M(n, k, h)$ is not known in the general case. It also represents the smallest number of tickets with h numbers needed to have at least one winning lottery ticket if a single drawing of k numbers is made from a total of n numbers. Turán's conjecture implies that

$$M(2n, 5, 3) = 2\binom{n}{3}.$$

4.36 (a) In order to prove (a) consider the two-element subsets to be the edges of a graph whose vertex set X contains n elements. One must find the minimum number of edges of this graph, such that each set of k vertices contains at least one edge. With respect to the complementary graph one must find the maximum number of edges in a graph G with n vertices which does not contain a complete subgraph with k vertices. The expression for $M(n, k, 2)$ is thus obtained in the form $\binom{n}{2} - M(n, k)$ where $M(n, k)$ is given by Turán's theorem (see Problem 9.9).

(b) If $k=1$, it can be shown that $M(n, n-h, 1) = n-(n-h)+1 = h+1$. Now let $k \geqslant 2$ and $S \subset X$, $|S|=h$. Suppose further that $P_{k-1}(X \setminus S)$ is the family of $(k-1)$-element subsets of $X \setminus S$. Since $n-h \geqslant h(k-1)$, one can define a function

$$f : S \to P_{k-1}(X \setminus S)$$

such that the sets $(f(x))_{x \in S}$ are pairwise disjoint. Let F be a family of h k-element subsets of X defined by

$$F = (\{x\} \cup f(x))_{x \in S}.$$

There exists an $(n-h)$-element subset of X, namely $X \setminus S$, which does not contain any set in F. An analogous result can be obtained for a family consisting of h k-element subsets of X which are not pairwise disjoint, and hence $M(n, n-h, k) \geqslant h+1$.

In order to prove the opposite inequality, let T be an $(h+1)$-element subset of X. There exists a function $g: T \to P_{k-1}(x \setminus T)$ such that the sets of the family $(g(x))_{x \in T}$ are pairwise disjoint. This follows from the fact that $n-h-1 \geqslant (h+1)(k-1)$. Let

$$G = (\{x\} \cup g(x))_{x \in T}.$$

One can at this point show that $|\{x\} \cup g(x)| = k$ for every $x \in T$ and also that G contains $h+1$ subsets. Let $U \subset X$, $|U|=n-h$, and suppose that there are indices $i_1, \ldots, i_r$ ($0 \leqslant r \leqslant h+1$) such that U does not contain n_p elements of the set $f(x_{i_p})$ for every $1 \leqslant p \leqslant r$. It follows that one can write

$$n-h = |U| = |U \cap T| + |U \cap (X \setminus T)|$$

$$\leqslant |U \cap T| + n-h-1 - \sum_{p=1}^{r} n_p,$$

and hence

$$|U \cap T| \geqslant \sum_{p=1}^{r} n_p + 1 \geqslant r+1.$$

Thus U contains at least one element $x \in T$ such that $x \notin \{x_{i_1}, \ldots, x_{i_r}\}$ and hence $U \supset \{x\} \cup g(x) \in G$. It has in fact been shown that U contains at least one k-element subset of G, that is, $M(n, n-h, k) \leqslant h+1$. One can finally conclude that $M(n, n-h, k) = h+1$ for every $k \geqslant 2$ and $n \geqslant k(h+1)$. [I. Tomescu, *Cahiers du Centre d'Etudes de Recherche Opérationnelle*, **15**(3) (1973), 355–362.]

It is also known that $M(n, n-h, k) = h+2$ for $k(h+1) - [k/2] \leq n < k(h+1)$ and $M(n, n-h, k) = h+3$ for $h \geq 2$ and $k(h+1) - [\frac{2}{3}k] \leq n < k(h+1) - [k/2]$. [N. N. Kuzyurin, *A Collection of Papers in Mathematical Cybernetics*, Akad. Nauk USSR, Moscow, 1981, 143–152.]

4.37 The first identity is verified by counting in two ways the number of occurrences of objects in the blocks. One can do this because each of the b blocks contains k objects, and each of the v objects belongs to exactly r blocks. With respect to the second identity one counts in two ways the occurrences in the blocks of pairs which contain a fixed object a_1.

The object a_1 occurs in r blocks, and in each of them it forms $k-1$ pairs with the other $k-1$ objects. On the other hand the object a_1 forms pairs with each of the $v-1$ remaining objects, and each pair is contained in λ blocks.

4.38 The matrices $(k-\lambda)I + \lambda J$ are of the form

$$\begin{pmatrix} k & \lambda & \cdots & \lambda \\ \lambda & k & \cdots & \lambda \\ \vdots & & & \vdots \\ \lambda & \cdots & \lambda & k \end{pmatrix},$$

that is, the elements on the main diagonal are equal to k and all the other elements are equal to λ. If A is the incidence matrix of a (v, k, λ)-configuration, it follows that each column of the matrix contains k elements equal to 1 and the remaining $v-k$ elements are equal to zero.

Let $B = A^T A$. The element b_{ij} of the matrix B is the scalar product of columns i and j of the matrix A. Thus $b_{ii} = k$, that is, the number of 1's in column i of the matrix A. If $i \neq j$, then columns i and j of A have a 1 in row t if and only if the element t of X belongs to both of the sets X_i and X_j. Thus the off-diagonal element b_{ij} is equal to $|X_i \cap X_j| = \lambda$. Analogously, if a v-by-v binary matrix A satisfies $A^T A = (k-\lambda)I + \lambda J$ where $0 < \lambda < k < v-1$, then define

$$X_j = \{i \mid a_{ij} = 1\}$$

for every $1 \leq j \leq v$. It follows that (1) and (2) are verified.

4.39 One first evaluates the determinant of the matrix $A^T A$ where A is the incidence matrix of a (v, k, λ)-configuration. Recall the form of the matrix $A^T A$ obtained in the preceding problem; subtract the first row from rows 2 through v, and then add columns 2 through v to the first column. In this way one obtains an upper triangular matrix with elements on the main diagonal equal to $k + \lambda(v-1) = v\lambda - \lambda + k$ and the remaining $v-1$ elements equal to $k-\lambda$. Thus $\det(A^T A) = (k-\lambda)^{v-1}(v\lambda - \lambda + k) > 0$, that is, the matrix A is nonsingular, since $\det(A^T A) = (\det A)^2$.

In order to prove that the (v, k, λ)-configuration of the matrix A is a BIBD with parameters (v, v, k, k, λ) it is necessary to show that each element belongs to exactly k blocks and each pair of distinct objects is contained in exactly λ blocks. These conditions hold if and only if the matrix A satisfies the equation

$$AA^T = (k-\lambda)I + \lambda J \qquad (1)$$

which is similar to the equation satisfied by the incidence matrix of a (v, k, λ)-configuration. The proof of relation (1) is analogous to that given in the preceding problem, since if $C = AA^T$, the element c_{ij} (of the matrix C) is the scalar product of rows i and j of the matrix A. Thus $c_{ii} = k$ if the number of 1's in row i of the matrix A is equal to k, and hence each object belongs to exactly k blocks. If $i \neq j$, then rows i and j of the matrix A have a 1 in column t if and only if the elements i and j both belong to block t. It follows that the number of blocks which contain the pair $\{i, j\}$ is equal to λ.

One must finally show that if the matrix A satisfies the equation $A^T A = (k-\lambda)I + \lambda J$, then $AA^T = (k-\lambda)I + \lambda J$. Since each column of the matrix A contains k elements equal to 1 and $v - k$ 0's, one can show that $JA = kJ$. A further implication is that

$$(JA)A^{-1} = (kJ)A^{-1}.$$

Multiply both sides of the equation on the right by the matrix AA^{-1}. It follows that

$$J = kJA^{-1}AA^{-1} = kJA^{-1},$$

and thus

$$k^{-1}J = JA^{-1}. \qquad (2)$$

But it follows from (2) that

$$A^T = (A^T A)A^{-1} = ((k-\lambda)I + \lambda J)A^{-1} = (k-\lambda)A^{-1} + \lambda JA^{-1}$$
$$= (k-\lambda)A^{-1} + \lambda k^{-1}J.$$

Thus

$$A^T J = (k-\lambda)A^{-1}J + \lambda k^{-1}J^2 = (k-\lambda)A^{-1}J + \lambda k^{-1}vJ,$$

since $J^2 = vJ$ and the matrix J has order v.

On the other hand $A^T J = (JA)^T = (kJ)^T = kJ$.

By identifying the two expressions for $A^T J$, one sees that

$$kJ = (k-\lambda)A^{-1}J + \lambda k^{-1}vJ,$$

and hence,

$$A^{-1}J = mJ, \quad \text{or} \quad J = mAJ,$$

where $m = (k - \lambda k^{-1}v)/(k - \lambda)$. But $vJ = J^2 = J(mAJ) = m(JA)J = mkJ^2 = mkvJ$, and hence $v = mkv$, or $mk = 1$. Thus $k - \lambda = k^2 - \lambda v$, so that $k^2 - k = \lambda(v - 1)$; and $A^{-1}J = mJ = k^{-1}J$, that is, $J = k^{-1}AJ$, or

$$AJ = kJ.$$

It has thus been shown that $JA = AJ$; this property is essential to the verification of identity (1). The proof proceeds as follows:

$$AA^T = A(A^T A)A^{-1} = A((k-\lambda)I + \lambda J)A^{-1}$$
$$= (k-\lambda)I + \lambda(AJ)A^{-1} = (k-\lambda)I + \lambda J(AA^{-1}) = (k-\lambda)I + \lambda J.$$

Solutions

4.40 Observe that a Steiner triple system is a BIBD for which $k=3$ and $\lambda=1$. By using the relations of Problem 4.37 one can show that $r=(v-1)/2$ and $b=v(v-1)/6$. Thus $v\equiv 1$ or $3\pmod 6$ because b must be an integer. Kirkman showed in 1847 that this condition is also sufficient for the existence of a Steiner triple system.

A simple proof was recently given by A. J. W. Hilton [*J. Combinatorial Theory*, **A13** (1972), 422–425.]

4.41 In the solution to Problem 4.39 it was shown that $\det(A^T A)=(\det A)^2 = (k-\lambda)^{v-1}(v\lambda-\lambda+k)$, where A is the incidence matrix of a (v, k, λ)-configuration or a BIBD with parameters (v, v, k, k, λ). Since $r=k$ and $v=b$, the relations which the parameters of a BIBD must satisfy imply that $k(k-1)=\lambda(v-1)$ and hence $v\lambda-\lambda+k=k^2$. Thus $(k-\lambda)^{v-1}k^2$ is the square of an integer. Since $v-1$ is odd, it follows that $k-\lambda$ is a perfect square.

4.42 Let p be the maximum number of subsets with the desired property. Then by considering the family of sets

$$\{A\cup B\mid A\subset Y, |A|=r;\ B\subset X\setminus Y, |B|=[(n-k)/2]\},$$

one can show that

$$p\geqslant \binom{k}{r}\binom{n-k}{[(n-k)/2]}.$$

In order to obtain the opposite inequality, let F be a maximum family of subsets with the desired property, and suppose that $A\in F$. Let

$$F_1=\{B\mid B\in F, B\cap Y=A\cap Y\},$$

and let F_2 be the family obtained from F_1 by suppressing all elements of $A\cap Y$ in the sets B of F_1. It turns out that F_2 is a family of subsets of $X\setminus Y$ which are noncomparable with respect to inclusion. It then follows from Sperner's theorem that

$$|F_2|\leqslant \binom{n-k}{[(n-k)/2]}$$

which implies that

$$p\leqslant \binom{k}{r}\binom{n-k}{[(n-k)/2]},$$

since $|A\cap Y|=r$.

4.43 Let $K_a(f)=\{x\in X\mid f(x)=a\}$. It will be shown that $f(f(x))=a$ for every $x\in X$ if and only if $f(X)\subset K_a(f)$. In fact, if $f(X)\subset K_a(f)$ it follows that $f(f(x))=a$, since $f(x)\in f(X)$, and hence $f(x)\in K_a(f)$. Suppose that $f(f(x))=a$ for every $x\in X$, and let $y\in f(X)$; it follows that there exists $x_y\in X$ such that $f(x_y)=y$. This implies that $f(y)=f(f(x_y))=a$, and hence $y\in K_a(f)$ or $f(X)\subset K_a(f)$.

Since $b=f(a)\in X$ and $f(f(a))=a$, it follows that $f(f(f(a)))=f(a)$. But $f(f(f(a)))=f(f(b))=a$, or $f(a)=a$, which implies that $a\in K_a(f)$.

Suppose that $|K_a(f)|=1$, so that $K_a(f)=\{a\}$. It follows that $f(x)=a$ for every $x\in X$, or $K_a(f)=X$, which is a contradiction, since $1=|K_a(f)|=|X|=n\geqslant 2$. Hence $|K_a(f)|\geqslant 2$.

Let $K_a(f)=\{a, x_{i_1},\ldots, x_{i_p}\}$ for $1\leqslant p\leqslant n-1$. This implies that $f(a)=f(x_{i_1})=\cdots=f(x_{i_p})=a$, and hence for any $x\in X\setminus K_a(f)$ one has $f(x)\in\{x_{i_1},\ldots,x_{i_p}\}$ because $f(X)\subset K_a(f)$.

Thus the number of functions $f:X\to X$ such that $f(f(x))=a$ for any $x\in X$ and $|K_a(f)|=p+1\geqslant 2$ is equal to $\binom{n-1}{p}p^{n-p-1}$, since elements $x_{i_1},\ldots, x_{i_p}$ can be chosen from $X\setminus\{a\}$ in $\binom{n-1}{p}$ ways. The number of functions $f:X\setminus K_a(f)\to\{x_{i_1},\ldots,x_{i_p}\}$ is equal to p^{n-p-1}. By summing these numbers for $p=1,\ldots,n-1$ one obtains the expression in the statement of the problem.

4.44 There exist $\binom{n+1}{r+1}$ subsets with $r+1$ elements taken from the set $\{0, 1, \ldots, n\}$. Delete the smallest element of each subset X to obtain a subset Y with r elements of the set $\{1,\ldots, n\}$. In this manner, each subset Y is obtained with multiplicity equal to its minimum element.

For example, if $Y=\{y_1,\ldots, y_r\}$ and $1\leqslant y_1<y_2<\cdots<y_r$, then Y is the image of y_1 subsets X, namely $\{0, y_1,\ldots, y_r\}$, $\{1, y_1,\ldots, y_r\},\ldots,\{y_1-1, y_1,\ldots, y_r\}$. It follows that the sum of the smallest elements is equal to $\binom{n+1}{r+1}$, and their arithmetical mean is $\binom{n+1}{r+1}\binom{n}{r}^{-1}=(n+1)/(r+1)$.

(Problem proposed at the 22nd International Mathematical Olympiad, Washington, 1981.)

4.45 The number of nonempty subsets of X is equal to $2^{15}-1=32{,}767$. The largest 15 elements of M have their sum equal to $2048+2047+\cdots+2034=30{,}615$. Hence for any $X\subset M$, $|X|=15$, it follows that $1\leqslant\sum_{i\in Y}i\leqslant 30{,}615$ for any $Y\subset X$, $Y\neq\varnothing$. Using Dirichlet's principle it can be seen that there exist two subsets U, $V\subset X$ such that $\sum_{i\in U}i=\sum_{j\in V}j$. Now let $A=U\setminus(U\cap V)$ and $B=V\setminus(U\cap V)$. This property does not hold for 12-element subsets of M. To see this let $Y=\{1, 2, 2^2,\ldots, 2^{11}=2048\}\subset M$. The existence of A, $B\subset Y$, $A\cap B=\varnothing$ such that $\sum_{i\in A}i=\sum_{j\in B}j=n$ would imply that n has two distinct representations in base 2, which is false.

4.46 The p vectors $(1, 1), (2, 2),\ldots, (p, p)$ constitute a covering set H, and hence $\sigma(2, p)\leqslant p$. Since (a, b) is not covered by any vector in the set $(a_1, b_1),\ldots, (a_{p-1}, b_{p-1})$ where $a\neq a_i$ and $b\neq b_i$ for $i=1,\ldots, p-1$, it follows that $\sigma(2, p)\geqslant p$, and hence $\sigma(2, p)=p$. The number of vectors which differ from a given vector in at most one component is $n(p-1)+1$. Thus each vector of H covers $n(p-1)+1$ vectors of F, and H contains at least $p^n/\{n(p-1)+1\}$ vectors. [O. Taussky, J. Todd, *Ann. Soc. Polonaise Math.*, **21** (1948), 303–305.]

J. G. Kalbfleisch and R. G. Stanton [*J. London Math. Soc.*, **44** (1969), 60–64] showed that $\sigma(3, p)=[(p^2+1)/2]$, and S. Zaremba [*ibid.*, **26** (1950), 71–72] proved that if p is a prime or a prime power and $n(p-1)+1$ is a power of p, then $\sigma(n, p)=p^n/\{n(p-1)+1\}$.

4.47 Every positive integer can be written uniquely in the form $2^p q$ where p is a non-negative integer and q is a positive odd integer, called the odd part of n. Therefore the odd part of an integer from the set $M = \{1, 2, \ldots, 2n\}$ must be one of the n integers $1, 3, 5, \ldots, 2n-1$. Given $n+1$ integers in M, at least two must have the same odd part; that is, they must be of the form $2^{p_1} q$ and $2^{p_2} q$, where $p_1 \neq p_2$. If, for example, $p_1 < p_2$ then $2^{p_1} q$ divides $2^{p_2} q$. (W. L. Putnam Math. Competition, 1958.)

4.48 Let $X = \{x_1, \ldots, x_m\}$, and denote by $n_1, \ldots, n_m$ the numbers of sets among $A_1, \ldots, A_{100}$ which contain $x_1, \ldots, x_m$ respectively. Then

$$n_1 + \cdots + n_m = \sum_{i=1}^{100} |A_i| > \tfrac{3}{4} m \times 100 = 75m,$$

which implies that there is an index i such that $n_i \geq 76$. Suppose for example that $i=1$ or $n_i \geq 76$. If x_1 is deleted from X, one obtains a subset X_1 with $m-1$ elements. Let $B_1, \ldots, B_s$ be the sets among $A_1, \ldots, A_{100}$ not containing x_1. If $s=0$ let $Y = \{x_1\}$. Otherwise, it follows that $s \leq 100 - 76 = 24$.

Since $|B_i| > \tfrac{3}{4}|X| > \tfrac{3}{4}|X_1|$, it can similarly be shown that there exists an element x_2 which belongs to more than $\tfrac{3}{4} s$ of the sets B_i. Let $C_1, \ldots, C_p$ be the sets which do not contain x_2 (and hence not x_1, by construction), where $p < s - 3s/4 = s/4 \leq 6$, or $p \leq 5$. If $p = 0$ let $Y = \{x_1, x_2\}$. Otherwise, it follows that there exists x_3 which belongs to more than $\tfrac{3}{4} p$ sets from $C_1, \ldots, C_p$. Because $p \leq 5$, it follows that $p/4 \leq \tfrac{5}{4}$, and hence at most one set from $C_1, \ldots, C_p$ does not include x_3. If all the sets $C_1, \ldots, C_p$ contain x_3, the set $Y = \{x_1, x_2, x_3\}$. Otherwise, let x_4 be an element of the unique set C_i which does not contain x_3 (and hence not x_1 or x_2). In this case one can choose $Y = \{x_1, x_2, x_3, x_4\}$; Y has at least one element in common with every set A_i for $1 \leq i \leq 100$. A similar problem was proposed at the W. L. Putnam Math. Competition in 1980 (Problem B-4).

4.49 Suppose that the digital plane has a finite metric basis B. Then there exists a sufficiently large rectangle which contains B. Examination of Figure 4.1 shows that the digital points x and y have equal d_4-distances from the rectangle.

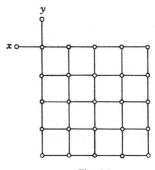

Fig. 4.1

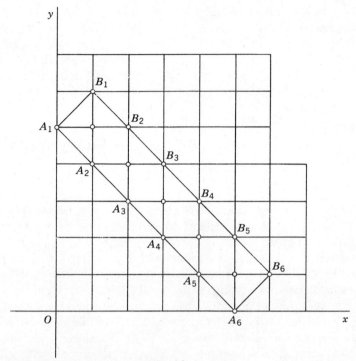

Fig. 4.2

This contradicts the assertion that B is a metric basis, since any set which contains a metric basis is clearly also a metric basis.

In order to show (b) consider the region in the plane bounded by the lines $x=0$, $y=0$, $x+y=n-1$ and $x+y=n+1$. We shall choose E_n to be the rectangle R, composed of all digital points in the region with the exception of the four points having coordinates $(0, n)$, $(0, n+1)$, $(n, 0)$, $(n+1, 0)$ (see Figure 4.2). It follows that R contains $3n-1$ digital points. Denote the (digital) intersection points of the opposite sides of R with lines of slope 1 by $A_1, B_1, A_2, B_2, \ldots, A_n, B_n$. It is clear that for $1 \leqslant i \leqslant n$, $d_4(A_i, M) = d_4(B_i, M)$ for any digital point M in R such that $M \neq A_i, B_i$. It follows that any 4-metric basis for R must contain at least one point from every pair $\{A_i, B_i\}$ for $1 \leqslant i \leqslant n$. This implies that $\dim_4(R) \geqslant n$, where $\dim_4(R)$ denotes the number of elements in a minimal metric basis for R (i.e., the metric dimension of R).

It will in fact be shown that $B = \{C_1, \ldots, C_n\}$, where $C_i = A_i$ or $C_i = B_i$ for $1 \leqslant i \leqslant n$, is a 4-metric basis for R.

One observes that the pairs of digital points of R which have equal city-block distances to A_i and A_{i+1} are $\{B_i, B_{i+1}\}$ and the pairs of the form $\{A_1, B_1\}, \ldots, \{A_{i-1}, B_{i-1}\}, \{A_{i+2}, B_{i+2}\}, \ldots, \{A_n, B_n\}$. Suppose now that B is not a 4-metric basis for R. Then there exist two distinct digital points M, N such that $M, N \notin B$ and M and N have equal distances to all points of B. If B contains A_i and A_{i+1}, it follows that $\{M, N\} = \{B_i, B_{i+1}\}$, since M and N do not belong to B. In a similar manner one can show that if A_i, $B_{i+1} \in B$ then $\{M, N\} = \{B_i, A_{i+1}\}$; if B_i, $A_{i+1} \in B$ then $\{M, N\} = \{A_i, B_{i+1}\}$; and if B_i, $B_{i+1} \in B$ then $\{M, N\} = \{A_i, A_{i+1}\}$. Since $n \geqslant 3$, one can apply this argument to the pairs $\{C_1, C_2\}$ and $\{C_2, C_3\}$. It turns out that $\{M, N\} = \{A_1, A_2, B_1, B_2\} \cap \{A_2, A_3, B_2, B_3\} = \{A_2, B_2\}$. This establishes a contradiction, since M and N do not belong to B. It follows that $\dim_4(R) = n$ and R has exactly 2^n minimal 4-metric bases. [R. A. Melter and I. Tomescu, *Computer Vision, Graphics and Image Processing*, (**25**) (1984), 113–121].

4.50 Suppose first that Δ is a finite projective plane. Note that the condition $v \geqslant 4$ implies that the order n of Δ is at least 2 and that the block size k is at least 3. Condition (1) is satisfied, since $\lambda = 1$. It will be shown that every two lines of Δ intersect. Suppose that there exist two lines L_1 and L_2 such that $L_1 \cap L_2 = \varnothing$. If $L_1 = \{x_1, \ldots, x_k\}$, then by hypothesis x_i belongs to exactly k lines $L_1, L_i^2, \ldots, L_i^k$ for every $i = 1, \ldots, k$. Condition (1) implies that $L_i^m \neq L_j^p$ for every $1 \leqslant i, j \leqslant k$, $i \neq j$, and $2 \leqslant m, p \leqslant k$. In fact, if for example $L_i^m = L_j^p$, then the pair $\{x_i, x_j\}$ is contained in two lines L_1 and L_i^m, which contradicts (1). In this case the number of lines of Δ is at least

$$v \geqslant k(k-1) + 2 > k^2 - k + 1,$$

which contradicts the hypothesis.

Since every two lines L_1 and L_2 intersect, it follows from (1) that $|L_1 \cap L_2| = 1$ and (2) holds. Since $r = k \geqslant 3$, one may choose a point x_0 and two lines L_1 and L_2 containing it. Now choose x_1 and x_3 on L_1 distinct from x_0, and choose x_2 and x_4 on L_2 distinct from x_0. Since (1) holds, one easily verifies that $\{x_1, x_2, x_3, x_4\}$ satisfies condition (3).

Conversely, suppose Δ satisfies (1), (2), and (3). First it will be shown that every two lines contain the same number of points; this involves choosing two lines and establishing a bijection between them.

Given any two lines L and L', there exists a point $x_0 \notin L \cup L'$. To see this let $\{x_1, x_2, x_3, x_4\}$ be a set whose existence is implied by (3), and let $L \cap L' = \{y\}$. If $\{x_1, \ldots, x_4\} \not\subset L \cup L'$, then clearly one may choose $x_0 \in \{x_1, \ldots, x_4\}$. If $\{x_1, \ldots, x_4\} \subset L \cup L'$, then condition (3) implies that $y \notin \{x_1, \ldots, x_4\}$, and hence one may assume that $x_1, x_2 \in L \setminus L'$ and $x_3, x_4 \in L' \setminus L$. Let L_1 be the line through x_1 and x_3, and let L_2 be the line through x_2 and x_4 which exists by (1). Let $\{x_0\} = L_1 \cap L_2$. By (2), $x_0 \notin L \cup L'$.

Now define the function $\gamma : L \to L'$ as follows. If $x \in L$, let L_x denote the unique line [in virtue of (2)] which contains x and x_0, and let $\gamma(x)$ be the unique point [in virtue of (1)] which is contained in $L_x \cap L'$. By (2), γ is an injection, since exactly one line contains both x_0 and $\gamma(x)$. Thus $|L| \leqslant |L'|$, and equality holds by symmetry.

Thus every line contains exactly $k \geqslant 2$ points. Now suppose that $k = 2$. This leads to a contradiction since $x_1, x_3 \in L_1$, $x_2, x_4 \in L_2$, $L_1 \cap L_2 = \{x_0\}$, and (3) implies that $x_0 \notin \{x_1, \ldots, x_4\}$. It follows that $k \geqslant 3$ and the number of points $v \geqslant 4$. Since $x_2, x_4 \notin L_1$, one can also see that $k \leqslant v - 2$ or $k < v - 1$. This implies that Δ is a $(v, k, 1)$-configuration. By Problem 4.39, Δ is a finite projective plane of order $k-1$.

It is not difficult to see that there exists only one projective plane of order 2 (up to isomorphism). It has point set $V=\{1,\ldots,7\}$ and line set $E=(\{1, 2, 3\}, \{1, 4, 5\}, \{1, 6, 7\}, \{2, 4, 6\}, \{2, 5, 7\}, \{3, 4, 7\}, \{3, 5, 6\})$. This plane is also the unique Steiner triple system on seven vertices (up to isomorphism).

CHAPTER 5

5.1 Let
$$A=\{m\,|\,1\leqslant m\leqslant 3n,\ m\equiv 0\ (\mathrm{mod}\ 3)\},$$
$$B=\{m\,|\,1\leqslant m\leqslant 3n,\ m\equiv 1\ (\mathrm{mod}\ 3)\},$$
$$C=\{m\,|\,1\leqslant m\leqslant 3n,\ m\equiv 2\ (\mathrm{mod}\ 3)\}.$$

It follows that $|A|=|B|=|C|=n$. The sum $x+y+z\equiv 0\ (\mathrm{mod}\ 3)$ if and only if $x, y, z \in A$ or $x, y, z \in B$ or $x, y, z \in C$ or x, y, z each belong to a different set among A, B, C. Thus the number of solutions is equal to
$$3\binom{n}{3}+n^3=\frac{n(3n^2-3n+2)}{2}.$$

5.2 The partitions of a number n into at most k parts form a set with $P(n, 1)+P(n, 2)+\cdots+P(n, k)$ elements. Each partition of n into at most k parts can be expressed in the form $n=a_1+a_2+\cdots+a_m+0+\cdots+0$, where the sum contains k terms and $a_1\geqslant a_2\geqslant\cdots\geqslant a_m\geqslant 1$ $(1\leqslant m\leqslant k)$. From this expression for n one can obtain a partition of $n+k$ into k parts in the following manner:
$$n+k=(a_1+1)+(a_2+1)+\cdots+(a_m+1)+1+\cdots+1$$
where the sum contains k terms and $a_1+1\geqslant a_2+1\geqslant\cdots\geqslant a_m+1\geqslant 1$.

The mapping thus defined is an injection, since different partitions of n into at most k parts correspond to different partitions of $n+k$ into k parts. The mapping is also surjective, since every partition of $n+k$ into k parts results from the partition of n with $m\leqslant k$ parts obtained by subtracting 1 from each term of the partition of $n+k$ and retaining the first nonzero terms. The existence of a bijection between the set of partitions of n into at most k parts and the set of partitions of $n+k$ into k parts implies the validity of the given recurrence relation. This permits the computation, by iteration, of all values of $P(n, k)$. One starts with $P(n, 1)=P(n, n)=1$ for all n and $P(n, k)=0$ for $n<k$.

5.3 A bijection will be defined between the set of partitions of n into odd parts and the set of partitions of n into pairwise distinct parts. Thus suppose that in a partition of n into odd parts the number $2k+1$ appears p times. Write p as a sum of powers of 2:
$$p=2^{i_1}+2^{i_2}+\cdots+2^{i_s}, \tag{1}$$
where $i_1>i_2>\cdots>i_s\geqslant 0$.

Solutions

The $(2k+1)p$ entries in the Ferrers diagram associated with the partition of n can be arranged as follows: Put in different rows respectively $(2k+1)2^{i_1}$ entries, $(2k+1)2^{i_2}$ entries, ..., $(2k+1)2^{i_s}$ entries. Maintain the decreasing order of the number of entries from top to bottom. For example, the partition $7+5+5+3+3+3+1+1+1+1$ is associated in this way with the partition $10+7+6+4+3$ into distinct parts. The partitions thus obtained have distinct parts because every integer can be uniquely expressed as a product of an odd number and a power of 2.

The injectivity of the mapping follows from the uniqueness of the representation (1). In order to show that this correspondence is a surjection, consider a partition P of n into distinct parts. Each part can be uniquely written as a product of an odd number and a power of 2. By regrouping the terms of the sum and combining common factors one obtains only terms of the form $(2k+1)p$ where $k \geqslant 0$. This will generate p terms equal to $2k+1$.

By arranging these terms in decreasing order one obtains a partition Q of n into odd parts. Apply the previously defined mapping to Q. The result is the partition P and this establishes the surjectivity of this mapping.

5.4 Let $a_1 + \cdots + a_m = n$ be a partition of n into m pairwise distinct parts. It follows that $a_1 > a_2 > \cdots > a_m \geqslant 1$, which implies that

$$a_1 - (m-1) \geqslant a_2 - (m-2) \geqslant \cdots \geqslant a_{m-1} - 1 \geqslant a_m \geqslant 1.$$

As a result one can write

$$\{a_1 - (m-1)\} + \{a_2 - (m-2)\} + \cdots + a_m = n - \binom{m}{2},$$

which is a partition of $n - \binom{m}{2}$ into m parts.

Thus each partition of n into m pairwise distinct parts is associated with a partition of $n - \binom{m}{2}$ into m parts. This correspondence is injective. In order to show that this mapping is a surjection, consider a partition of $n - \binom{m}{2}$ into m parts. Add the numbers $m-1, m-2, \ldots, 1, 0$, to each of the m parts respectively in decreasing order. One thus obtains a partition of n into m pairwise distinct parts. If one applies the transformation previously defined to this partition, one obtains the partition of $n - \binom{m}{2}$ into m parts with which the process originated. Thus the given correspondence between the partitions of n into m pairwise distinct parts and the set of the $P(n - \binom{m}{2}, m)$ partitions of $n - \binom{m}{2}$ into m parts is a bijection. This completes the proof of the property in question.

5.5 Suppose that the partition of n contains k_1 terms equal to 1. The partial sums formed by the elements equal to 1 uniquely represent every number between 1 and k_1. Thus the numbers $2, \ldots, k_1$ cannot appear in this partition of n because of the uniqueness of the representation. Since the number $k_1 + 1$ must be represented as a partial sum, it must also be a term of the partition with multiplicity $k_2 > 0$. It follows that all numbers between 1 and $k_1 + k_2(k_1+1)$ have a unique representation as a partial sum.

Using this argument, one finds that the numbers which occur in the partition of n are

$$1, k_1+1, (k_1+1)(k_2+1), \ldots, (k_1+1)(k_2+1)\cdots(k_m+1),$$

where $(k_1+1)\cdots(k_i+1)$ has multiplicity equal to k_{i+1} for $i=1,\ldots,m$. Since the sum of these terms is equal to n, one can write

$$n = k_1 + k_2(k_1+1) + \cdots + k_{m+1}(k_1+1)\cdots(k_m+1)$$

and hence $n+1 = (k_1+1)\cdots(k_{m+1}+1)$.

If $n+1$ is prime, then $m=0$ and the only partition of n which satisfies the condition of the problem contains only terms equal to 1. If $n+1$ is not prime, then there is another solution to the problem. In fact it follows that

$$n+1 = (k_1+1)(k_2+1)$$

with $k_1, k_2 \geq 1$, and the partition

$$n = \underbrace{1 + \cdots + 1}_{k_1} + \underbrace{(k_1+1) + \cdots + (k_1+1)}_{k_2}$$

has the property that every number between 1 and n can be uniquely represented as a partial sum of this partition.

5.6 It follows from the rules for removing parentheses that $\psi(n) = Q_e(n) - Q_o(n)$ where $Q_e(n)$ represents the number of partitions of n into an even number of distinct parts and $Q_o(n)$ represents the number of partitions of n into an odd number of distinct parts.

In order to prove Euler's identity, one defines a transformation of a Ferrers diagram with an even number of rows into a diagram with the same number of cells and an odd number of rows and vice versa. Since one considers only partitions into pairwise distinct parts, the diagram of this kind of partition is formed of several trapezoids placed next to each other as in Figure 5.1. Let the number of cells in the last row of the diagram be equal to m, and let the number of rows in the upper trapezoid be equal to k.

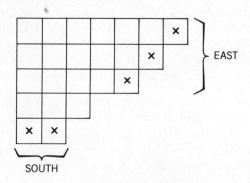

Fig. 5.1

Solutions

If $m \leq k$, one suppresses the last row of the diagram (labeled SOUTH) and adds one cell to each of the first m rows of the upper trapezoid (in a line inclined at $45°$ to the east in the diagram). This transformation does not change the total number of cells. One obtains in this way a new diagram in which each row has a different length. The parity of the number of rows is changed with respect to the initial diagram. The diagram of Figure 5.1 corresponds to the partition

$$23 = 7 + 6 + 5 + 3 + 2.$$

After performing this transformation, one finds the row SOUTH laid against the diagonal EAST. Since $m=2$ and $k=3$, the partition $23 = 8 + 7 + 5 + 3$ is thereby obtained.

If the diagram contains at least two trapezoids and if $m > k$, then take one cell from EAST in each row of the upper trapezoid, and with these cells make a new row SOUTH in the new diagram. This construction is possible because $m > k$ and thus the row SOUTH is shorter than the old row SOUTH in the diagram. The length of each row in the upper trapezoid has been shortened by one. It follows that all the rows of the new diagram are pairwise different in length. The new diagram contains the same number of cells as the old diagram, but the parity of the number of rows has changed. The new diagram contains one more or one less row than the original diagram.

This operation can be carried out when the diagram consists of a single trapezoid (when the diagonal EAST contains k cells and k is equal to the number of parts of n) if $m \neq k$ and $m \neq k+1$.

The transformation just described is an involution on the set of partitions of n into pairwise distinct parts. (This means that if this transformation is applied twice, one obtains the original diagram.) It follows that the transformation is bijective. Thus the Ferrers diagrams for partitions of n which admit this transformation can be divided into an equal number of diagrams with an odd and even number of rows, respectively. One can now find the diagrams which do not admit this transformation. They consist of a single trapezoid for which $m = k$ or $m = k+1$ (Figure 5.2).

In the first case

$$n = k^2 + 1 + 2 + \cdots + (k-1) = k^2 + \frac{k^2 - k}{2} = \frac{3k^2 - k}{2},$$

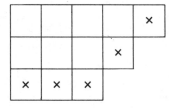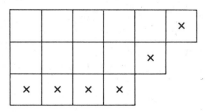

Fig. 5.2

and in the second case

$$n = \frac{3k^2 - k}{2} + k = \frac{3k^2 + k}{2}.$$

Thus if n is not a number of the form $(3k^2 \pm k)/2$, it has the same number of partitions into an odd and even number of pairwise distinct terms. If $n = (3k^2 \pm k)/2$, then $Q_e(n) = Q_o(n) + (-1)^k$, since there remains a unique diagram with k rows outside this bijection. It follows that $\psi(n) = 0$ for $n \neq (3k^2 \pm k)/2$ and $\psi(n) = (-1)^k$ if $n = (3k^2 \pm k)/2$.

5.7 The proof will be given for case (a). The other cases can be established analogously.

Consider the expansion

$$\frac{1}{(1-a_1 x)(1-a_2 x^2) \cdots (1-a_k x^k) \cdots}$$
$$= (1 + a_1 x + a_1^2 x^2 + \cdots)(1 + a_2 x^2 + a_2^2 x^4 + \cdots) \cdots (1 + a_k x^k + a_k^2 x^{2k} + \cdots) \cdots$$
$$= 1 + a_1 x + (a_1^2 + a_2) x^2 + \cdots + (a_1^{\lambda_1} a_2^{\lambda_2} \cdots a_k^{\lambda_k} + \cdots) x^n + \cdots.$$

Observe that the term $a_1^{\lambda_1} a_2^{\lambda_2} \cdots a_k^{\lambda_k}$ which appears in the coefficient of x^n is such that $\lambda_1 + 2\lambda_2 + \cdots + k\lambda_k = n$, and thus it determines the following partition of n:

$$n = \underbrace{k + k + \cdots + k}_{\lambda_k} + \cdots + \underbrace{2 + 2 + \cdots + 2}_{\lambda_2} + \underbrace{1 + 1 + \cdots + 1}_{\lambda_1}.$$

The rules for removing parentheses imply that in this case the exponents of the symbols which appear in the coefficient of x^n generate, without repetitions, all the partitions of n. If one sets $a_1 = a_2 = \cdots = 1$, then the coefficient of x^n will be exactly equal to $P(n)$, the number of partitions of n. Part (c) of the problem now follows immediately, since $\lambda_2 = \lambda_4 = \cdots = 0$.

The proof of part (e) is contained in the solution of Problem 5.13; the proof of part (d) is analogous to that of part (e).

The property expressed in Problem 5.3 follows from an algebraic calculation due to Euler and based on the use of generating functions:

$$\frac{1}{(1-x)(1-x^3)(1-x^5) \cdots} = \frac{1-x^2}{1-x} \cdot \frac{1-x^4}{1-x^2} \cdot \frac{1-x^6}{1-x^3} \cdot \frac{1-x^8}{1-x^4} \cdots$$
$$= (1+x)(1+x^2)(1+x^3)(1+x^4) \cdots.$$

5.8 By using Euler's identity and the expression for the generating function of the numbers $P(n)$, one can write

$$\left(1 + \sum_{i=1}^{\infty} P(i) x^i\right)\left(1 + \sum_{j=1}^{\infty} \psi(j) x^j\right) = 1.$$

Solutions

After equating to zero the coefficient of x^n on the left-hand side of this equation, it turns out that

$$\sum_{j \geq 0} P(n-j)\psi(j) = 0,$$

or in other words

$$P(n) + \sum_{k \geq 1} (-1)^k \left\{ P\left(n - \frac{3k^2-k}{2}\right) + P\left(n - \frac{3k^2+k}{2}\right) \right\} = 0.$$

This is in fact Euler's Pentagonal Theorem.

5.9 Consider the expansion

$$(1 + a_1 x + a_1^2 x^2)(1 + a_2 x^2 + a_2^2 x^4)(1 + a_3 x^3 + a_3^2 x^6) \cdots,$$

and observe that each monomial $a_1^{\lambda_1} a_2^{\lambda_2} \cdots a_k^{\lambda_k}$ which occurs in the coefficient of x^n has the property that $0 \leq \lambda_i \leq 2$ for $1 \leq i \leq k$ and $\lambda_1 + 2\lambda_2 + \cdots + k\lambda_k = n$, and thus determines a partition of n of the form

$$n = (k + \cdots + k) + \cdots + (2 + \cdots + 2) + (1 + \cdots + 1).$$

The numbers of repetitions are equal respectively to $\lambda_k, \ldots, \lambda_2, \lambda_1$. The rules for removing parentheses imply that the exponents of the symbols which make up the coefficient of x^n generate, without repetitions, all partitions of n in which no integer occurs more than twice.

Taking $a_1 = a_2 = \cdots = 1$, one finds that the generating function of these partitions of n is given by

$$G_1(x) = (1 + x + x^2)(1 + x^2 + x^4) \cdots (1 + x^p + x^{2p}) \cdots$$

Using the same reasoning for the generating function of the numbers $P(n)$, it is seen that the generating function for the number of partitions of n into parts which are not divisible by 3 is

$$G_2(x) = (1-x)^{-1}(1-x^2)^{-1}(1-x^4)^{-1}(1-x^5)^{-1}(1-x^7)^{-1} \cdots$$

$$= \prod_{\substack{p=1 \\ p \not\equiv 0 \,(\text{mod } 3)}}^{\infty} (1-x^p)^{-1}.$$

It remains to show that $G_1(x) = G_2(x)$ by using the identity

$$(1 + x^p + x^{2p})(1 - x^p) = 1 - x^{3p}.$$

In fact

$$G_1(x) = \frac{1-x^3}{1-x} \cdot \frac{1-x^6}{1-x^2} \cdot \frac{1-x^9}{1-x^3} \cdots$$

$$= (1-x)^{-1}(1-x^2)^{-1}(1-x^4)^{-1}(1-x^5)^{-1}(1-x^7)^{-1} \cdots$$

$$= G_2(x).$$

5.10 It is known (from Problem 5.7) that the generating function of the numbers $P(n)$ is

$$P(0)+P(1)x+\cdots+P(n)x^n+\cdots=\frac{1}{(1-x)(1-x^2)(1-x^3)\cdots}, \quad (1)$$

while the generating function of the numbers $Q(n)$ can be expressed as

$$Q(0)+Q(1)x+\cdots+Q(n)x^n+\cdots=\frac{1}{(1-x)(1-x^3)(1-x^5)\cdots}, \quad (2)$$

where $P(0)=Q(0)=1$. The substitution of $-x$ for x in (2) yields

$$\sum_{i\geqslant 0}(-1)^i Q(i)x^i=\frac{1}{(1+x)(1+x^3)(1+x^5)\cdots}; \quad (3)$$

after substituting x^2 for x in (2) the result is

$$\sum_{i\geqslant 0}Q(i)x^{2i}=\frac{1}{(1-x^2)(1-x^6)(1-x^{10})\cdots}. \quad (4)$$

The identity

$$\frac{1}{(1-x)(1-x^3)(1-x^5)\cdots}\cdot\frac{1}{(1+x)(1+x^3)(1+x^5)\cdots}=\frac{1}{(1-x^2)(1-x^6)(1-x^{10})\cdots}$$

then leads to the following relation between the associated formal series:

$$\sum_{i\geqslant 0}Q(i)x^i\sum_{j\geqslant 0}(-1)^j Q(j)x^j=\sum_{i\geqslant 0}Q(i)x^{2i}.$$

The proof of part (a) is completed by equating the coefficients of x^{2n} on the two sides of this identity.

The proof of part (b) starts with the identity

$$\frac{1}{(1-x)(1-x^2)(1-x^3)\cdots}=\frac{1}{(1-x)(1-x^3)\cdots}\cdot\frac{1}{(1-x^2)(1-x^4)\cdots}.$$

This implies that $\sum_{i\geqslant 0}P(i)x^i=\sum_{i\geqslant 0}Q(i)x^i\sum_{j\geqslant 0}P(j)x^{2j}$, and the proof is finished by again equating the coefficients of x^n. [R. Karpe, *Casopis Pest. Mat.*, **94** (1969), 108–114.]

5.11 Each partition of n into m parts of the form $n=n_1+\cdots+n_m$ with $n_1\geqslant\cdots\geqslant n_m\geqslant 1$ corresponds to a partition of $n-m$ obtained by writing

$$n-m=(n_1-1)+(n_2-1)+\cdots+(n_m-1),$$

and the possible elimination of zero terms.

The mapping thus defined is injective. One can also show that it is surjective for $m\geqslant n/2$. In fact, starting from the partition

$$n-m=r_1+\cdots+r_k \quad (1)$$

of $n-m$, it follows that $k\leqslant m$, since otherwise one would have $k\geqslant m+1$ and

Solutions

hence $n-m \geq k \geq m+1$. This implies that $m \leq (n-1)/2$. The latter inequality contradicts the hypothesis $m \geq n/2$. Now add one to each term of (1) and $m-k \geq 0$ terms equal to 1 to obtain a partition of n into m parts: $n=(r_1+1)+ \cdots + (r_k+1)+1+ \cdots +1$. Its image under the given mapping is precisely the partition (1).

5.12 Suppose that $n=x+y+z$ with $x>y>z \geq 1$ is a partition of n into three distinct parts. It follows that

$$(x+y)+(y+z)+(x+z)=2n$$

and

$$(y+z)+(x+z)=x+y+2z>x+y.$$

Thus $x+y$, $y+z$, $x+z$ are the lengths of the sides of a triangle with perimeter $2n$. It is also the case that no two sides are equal, since

$$x+y>x+z>y+z.$$

Conversely, suppose that $a>b>c$ are the integral lengths of the sides of a triangle with perimeter equal to $2n$. Let

$$x=n-a, \qquad y=n-b, \qquad z=n-c.$$

It follows that

$$x=\frac{b+c-a}{2}>0, \qquad y=\frac{a-b+c}{2}>0, \quad \text{and} \quad z=\frac{a+b-c}{2}>0.$$

Further, $x<y<z$ and $x+y+z=n$, while $x+y=c$, $x+z=b$, and $y+z=a$.

There is thus defined a mapping of the set of partitions of n into three distinct parts onto the set of triangles of perimeter $2n$ with integral sides, no two of which are equal. This mapping is injective. The construction shows that it is also surjective. This follows from the fact that, starting with a triangle of perimeter $2n$ with sides $a>b>c$, one obtains a partition of n of the form $n=x+y+z$ where $x+y=c$, $x+z=b$, and $y+z=a$. Thus the image of this partition under the mapping being considered is the original triangle. Since the mapping is bijective, one can conclude that the number of triangles which satisfy the given condition is equal to $Q(n, 3)$.

The number n can be represented as a sum of three positive integers (two representations which differ only in the order of the terms are to be considered distinct) in

$$\binom{n-1}{2} = \frac{n^2-3n+2}{2}$$

different ways (Problem 1.19). It follows from this that two terms are equal in each of the following representations of n:

$$n = 1+1+(n-2) = 1+(n-2)+1 = (n-2)+1+1$$
$$= 2+2+(n-4) = 2+(n-4)+2 = (n-4)+2+2$$
$$= 3+3+(n-6) = 3+(n-6)+3 = (n-6)+3+3 = \cdots.$$

There are thus $3(n-2)/2$ representations if n is even and $3(n-1)/2$ representations if n is odd and $n \geqslant 4$. But for $n=1, 2, 3$ one has $Q(n, 3)=0$, and hence the given formula is verified.

If n is a multiple of 3, one must also subtract 2 from the numbers obtained, since there is a representation of n with three equal terms which is counted three times instead of being counted only once. For example, for $n=6$ one obtains a unique partition with two terms equal to 2, namely $6=2+2+2$.

In order to obtain the number of representations with three pairwise distinct terms, one must subtract from the number of all representations of n as the sum of three integers, the number of representations which contain equal terms. The result is that this number is equal to

$$\frac{n^2-3n+2}{2} - \tfrac{3}{2}(n-2)+2 = \frac{n^2-6n+12}{2} \qquad \text{if} \quad n=6k,$$

$$\frac{n^2-3n+2}{2} - \tfrac{3}{2}(n-1) = \frac{n^2-6n+5}{2} \qquad \text{if} \quad n=6k+1,$$

$$\frac{n^2-3n+2}{2} - \tfrac{3}{2}(n-2) = \frac{n^2-6n+8}{2} \qquad \text{if} \quad n=6k+2,$$

$$\frac{n^2-3n+2}{2} - \tfrac{3}{2}(n-1)+2 = \frac{n^2-6n+9}{2} \qquad \text{if} \quad n=6k+3,$$

$$\frac{n^2-3n+2}{2} - \tfrac{3}{2}(n-2) = \frac{n^2-6n+8}{2} \qquad \text{if} \quad n=6k+4,$$

$$\frac{n^2-3n+2}{2} - \tfrac{3}{2}(n-1) = \frac{n^2-6n+5}{2} \qquad \text{if} \quad n=6k+5.$$

Since these representations of n as the sum of three terms contain only pairwise distinct terms, in order to find the partitions of n into three pairwise distinct parts, one must divide the number obtained in each case by $3!=6$, since the order of the terms is no longer of any importance. It can immediately be seen that all the expressions obtained have the form

$$\left[\frac{n^2-6n+12}{12}\right] = \left[\frac{n(n-6)}{12}\right]+1 \qquad \text{for every } n.$$

For example, for $n=6k+1$ one has

$$\frac{n^2-6n}{12} = \frac{36k^2-24k-5}{12} = 3k^2-2k-\tfrac{5}{12}$$

and thus

$$\left[\frac{n^2-6n}{12}\right]+1=3k^2-2k=\frac{n^2-6n+5}{12}.$$

It follows from the inequalities

$$\frac{n^2-6n+5}{12}\leqslant Q(n,3)\leqslant\frac{n^2-6n+12}{12}$$

that

$$Q(n,3)=\left[\frac{n^2-6n+12}{12}\right].$$

5.13 One first shows that the number of partitions of n into pairwise distinct odd terms is equal to the number of partitions of n with a symmetric Ferrers diagram. (The axial symmetry of a Ferrers diagram is with respect to a line drawn from the top left of the diagram at an angle of 45° with the horizontal.)

Define a bijection from the set of partitions of n with symmetric Ferrers diagram onto the set of partitions of n into pairwise distinct odd terms as follows: Suppose that a symmetric Ferrers diagram has k cells on the diagonal, and let a_1 be the total number of cells which are found in the first row and first column of the diagram. Since the diagram is symmetric, it follows that a_1 is odd. Now let a_2 denote the number of cells which are found in the first row and first column of the diagram obtained by suppressing the a_1 cells. Similarly let a_3 be the number of cells which are found in the first row and first column of the diagram obtained when the a_2 cells are suppressed, and so on. One thus obtains k odd numbers $a_1>a_2>\cdots>a_k$ which define a partition of n: $n=a_1+a_2+\cdots+a_k$.

For example, consider the partition of $n=30$ with the symmetric Ferrers diagram of Figure 5.3:

$$30=7+7+5+4+3+2+2$$

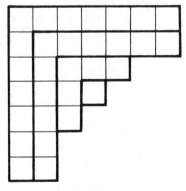

Fig. 5.3

In this case $a_1=13$, $a_2=11$, $a_3=5$, $a_4=1$, which corresponds to the partition of 30 into four odd parts: $30=13+11+5+1$.

It is clear that this correspondence is injective. In order to establish the surjectivity let $n=a_1+\cdots+a_k$ be a partition of n into k pairwise distinct odd parts. A symmetric Ferrers diagram will now be constructed having a_r cells in the rth row and column for $1 \leqslant r \leqslant k$. It will turn out that for the described correspondence the image of this partition with symmetric Ferrers diagram is precisely the original partition of n into odd parts, and this establishes the surjectivity and hence the bijectivity of the correspondence.

The number of partitions of n with pairwise distinct odd parts is equal to the coefficient of x^n in the product

$$(1+x)(1+x^3)(1+x^5)\cdots$$

In fact, in order to obtain a term equal to x^n in this expansion one must multiply terms of the form $x^{a_1}, x^{a_2}, \ldots, x^{a_k}$, where $a_1+a_2+\cdots+a_k=n$, $a_1,\ldots,a_k$ are pairwise distinct odd numbers, and we set $a_1 \geqslant a_2 \geqslant \cdots \geqslant a_k$ in order not to count these monomials twice. Since the coefficient of each product of the form $x^{a_1}\cdots x^{a_k}$ is equal to 1, it turns out that the coefficient of x^n is equal to the number of partitions of n into pairwise distinct odd parts.

To obtain the generating function for the number of symmetric Ferrers diagrams with k cells on the diagonal, one suppresses the square with k cells on a side which lies in the upper left-hand portion of the diagram.

Now consider the number of cells in row i plus the number of cells in column i ($i \geqslant k+1$) as a new term in a partition of n with a term equal to k^2 and a sequence of other terms less than or equal to $2k$. For the diagram of Figure 5.3 this new partition can be written

$$30 = 4^2 + 6 + 4 + 4.$$

One thus has shown that the number of partitions of n with symmetric Ferrers diagrams having k cells on the diagonal is equal to the number of partitions of n of the form

$$n = k^2 + 2a_1 + 2a_2 + \cdots + 2a_r,$$

where $k \geqslant a_1 \geqslant a_2 \geqslant \cdots \geqslant a_r$. Similarly one finds that the number of these partitions of n is equal to the coefficient of x^n in the expansion of the product

$$x^{k^2}(1+x^2+x^4+\cdots)(1+x^4+x^8+\cdots)\cdots(1+x^{2k}+x^{4k}+\cdots)$$
$$= \frac{x^{k^2}}{(1-x^2)(1-x^4)\cdots(1-x^{2k})}. \quad (1)$$

In fact the number of partitions of $m=n-k^2$ into even parts which are less than or equal to $2k$ is the coefficient of x^m in the expansion of the product

$$(1+x^2+x^4+\cdots)(1+x^4+x^8+\cdots)\cdots(1+x^{2k}+x^{4k}+\cdots).$$

Solutions

In order to show this, start with the expansion

$$\frac{1}{(1-a_1 x^2)(1-a_2 x^4)\cdots(1-a_k x^{2k})}$$
$$=(1+a_1 x^2+a_1^2 x^4+\cdots)(1+a_2 x^4+a_2^2 x^8+\cdots)\cdots(1+a_k x^{2k}+a_k^2 x^{4k}+\cdots)$$
$$=1+a_1 x^2+(a_1^2+a_2)x^4+\cdots+(a_1^{\lambda_1} a_2^{\lambda_2}\cdots a_k^{\lambda_k}+\cdots)x^m+\cdots.$$

The term $a_1^{\lambda_1} a_2^{\lambda_2}\cdots a_k^{\lambda_k}$ which appears in the coefficient of x^m satisfies the relation

$$2\lambda_1+4\lambda_2+\cdots+2k\lambda_k=m,$$

and thus it defines a partition of m of the form

$$\underbrace{2k+\cdots+2k+\cdots+}_{\lambda_k}\underbrace{4+\cdots+4}_{\lambda_2}+\underbrace{2+\cdots+2}_{\lambda_1}.$$

It follows from the rules for removing parentheses that the exponents of the symbols which appear in the coefficient of x^m generate, without repetition, all the partitions of m into even parts. Thus if $a_1=a_2=\cdots=a_k=1$, then the coefficient of x^m will be equal to the number of partitions of m into even parts which are less than or equal to $2k$. This justifies the expression which has been obtained for the generating function of the number of symmetric Ferrers diagrams with k cells on the diagonal. Thus

$$\sum_{k=0}^{\infty}\frac{x^{k^2}}{(1-x^2)(1-x^4)\cdots(1-x^{2k})}$$

represents the generating function for the number of symmetric Ferrers diagrams with n cells; this completes the proof of the first identity.

In order to prove Euler's identity, observe that (as in the discussion of the generating function of the number of partitions of n into distinct odd parts) the number of partitions of n into k distinct odd parts is the coefficient of $x^n y^k$ in the expansion of the product

$$(1+xy)(1+x^3 y)(1+x^5 y)\cdots.$$

On the other hand, it has been seen that there exists a bijection of the set of partitions of n with symmetric Ferrers diagrams with k cells on the diagonal onto the set of partitions of n into k distinct odd parts. Thus the number of partitions of n into k distinct odd parts is equal to the coefficient of x^n in (1) or the coefficient of $x^n y^k$ in

$$\frac{x^{k^2} y^k}{(1-x^2)(1-x^4)\cdots(1-x^{2k})}.$$

One thus obtains the identity

$$(1+xy)(1+x^3 y)(1+x^5 y)\cdots=\sum_{k=0}^{\infty}\frac{x^{k^2} y^k}{(1-x^2)(1-x^4)\cdots(1-x^{2k})}.$$

Also, if the substitution $y=x$, is made then the resulting identity is

$$(1+x^2)(1+x^4)(1+x^6)\cdots = \sum_{k=0}^{\infty} \frac{x^{k^2+k}}{(1-x^2)(1-x^4)\cdots(1-x^{2k})}.$$

5.14 Let $2n+1 = 2^{i_1} + 2^{i_2} + \cdots + 2^{i_r} + 1$ be a partition of $2n+1$ into powers of two, where $i_1 \geq i_2 \geq \cdots \geq i_r \geq 0$. It follows that $2n = 2^{i_1} + \cdots + 2^{i_r}$; this correspondence is a bijection, and hence $B(2n+1) = B(2n)$. In order to prove (b) note that the set of partitions of $2n$ whose parts are powers of two can be written as $A_n \cup B_n$, where A_n is the set of partitions of the form

$$2n = 2^{i_1} + 2^{i_2} + \cdots + 2^{i_r} + 1,$$

where $i_1 \geq i_2 \geq \cdots \geq i_r \geq 0$ and B_n contains all the partitions of the form

$$2n = 2^{i_1} + 2^{i_2} + \cdots + 2^{i_r},$$

where $i_1 \geq i_2 \geq \cdots \geq i_r \geq 1$, and hence $A_n \cap B_n = \varnothing$.

Since in the first case one has $2n-1 = 2^{i_1} + \cdots + 2^{i_r}$, and in the second $n = 2^{i_1-1} + \cdots + 2^{i_r-1}$, where $i_1 - 1 \geq \cdots \geq i_r - 1 \geq 0$, and these correspondences between sets of partitions are bijections, one can conclude that (b) follows.

(c) may be proved by induction on n, since $B(2) = B(3) = 2$, $B(4) = B(5) = 4$. Suppose (c) to be true for all $m \leq n-1$. If $n = 2m$, then $B(2m) = B(2m-1) + B(m)$ is even, and if $n = 2m+1$, then $B(2m+1) = B(2m)$ is also even by the induction hypothesis.

5.15 Let $r = [\sqrt{n}]$, and let K be a k-element subset $\{a_1, \ldots, a_k\}$ of $\{1, \ldots, r\}$ for $0 \leq k \leq r$. It follows that K generates a partition of n whose parts are $a_1, \ldots, a_k$, $n - (a_1 + \cdots + a_k)$, in view of the fact that $a_1 + \cdots + a_k < kr \leq r^2 \leq n$. Thus $P(n) \geq 2^r$, since in this case different subsets of $\{1, \ldots, r\}$ induce distinct partitions of n.

CHAPTER 6

6.1 There is a unique walk which joins each pair of vertices of a tree. The subgraph induced by B does not contain cycles, since A itself does not contain cycles. If $x, y \in B$ and $x \neq y$, then x, y are vertices of each subtree $A_1, \ldots, A_p$ and thus each of these subtrees contains the unique walk $[x, z_1, \ldots, z_k, y]$ which joins x and y in A. Hence $z_1, \ldots, z_k \in X_1, \ldots, X_p$ or $z_1, \ldots, z_k \in B$, and thus the subgraph induced by the set of vertices B is connected and is in fact a tree.

6.2 The proof uses induction on the number of vertices in the tree G.

If G has two vertices the property is immediate. Suppose that the property holds for all trees with at most n vertices, and let G be a tree with $n+1$ vertices. G contains a vertex x of degree 1 which is adjacent to a vertex y. By the induction hypothesis the property is valid for the subtree G_x obtained from G by suppressing the vertex x and the edge $[x, y]$. If no subtree $G_1, \ldots, G_k$ is the graph consisting of only the vertex x, then all the subtrees $G_1^x, \ldots, G_k^x$ obtained from

$G_1, \ldots, G_k$ by suppressing the vertex x have at least one vertex pairwise in common. In fact, if x is a common vertex for G_i and G_j, then y is also a vertex common to G_i and G_j, and thus also for G_i^x and G_j^x. By the induction hypothesis $G_1^x, \ldots, G_k^x$ have at least one vertex in common, and thus the subtrees $G_1, \ldots, G_k$ have this property. If, for example, G_1 contains only the vertex x, then x is common to all the subtrees $G_1, \ldots, G_k$ and the property is again verified.

6.3 The proof will show by induction on the number n that a tree with n vertices has $n-1$ edges.

If $n=1$, the tree has a single vertex and no edges. Suppose that the property is true for all trees with n vertices, and let A be a tree with $n+1$ vertices. The tree A contains at least one vertex of degree 1, since otherwise A would contain a cycle, which contradicts the definition of a tree. In fact, let

$$L = [x_1, \ldots, x_k]$$

be an elementary walk of maximal length in A. Since $d(x_k) \geq 2$, it follows that x_k is adjacent to at least one of the vertices $x_1, \ldots, x_{k-2}$. Otherwise one would obtain a walk longer than L, which would contradict the maximality of L. One thus obtains a cycle which passes through the vertex x_k.

It has been shown that A contains a vertex of degree 1. Let $d(x) = 1$, and let y be the vertex adjacent to x. If one suppresses the vertex x and the edge $[x, y]$, then a subgraph A_1 is obtained which is connected and without cycles, and hence is a tree. By the induction hypothesis A_1 has n vertices and $n-1$ edges, and thus A has $n+1$ vertices and n edges, and the property is established.

The necessity of the condition in the problem is now immediate, since $d_1 + \cdots + d_n = 2m = 2n - 2$, where m denotes the number of edges of a tree with n vertices.

Suppose now that $d_1 + \cdots + d_n = 2n - 2$. At this point we use induction on n. For $n=1$ it follows that $d_1 = 0$, and for $n=2$ one has $d_1 = d_2 = 1$ and hence the trees are K_1 and K_2 respectively. Assume that the property is true for $n-1$ integers; we establish it for integers $d_1, \ldots, d_n$ whose sum is equal to $2n-2$ with $n \geq 3$. It follows that $d_1 = 1$, since $d_1 \geq 2$ would imply that $d_1 + \cdots + d_n \geq 2n > 2n - 2$. Similarly $d_n > 1$, since otherwise $d_1 + \cdots + d_n = n < 2n - 2$. Thus $d_2 + \cdots + d_{n-1} + (d_n - 1) = 2n - 4 = 2(n-1) - 2$, and by the induction hypothesis there is a tree A having $n-1$ vertices of degrees $d_2, \ldots, d_{n-1}, d_n - 1$. One can add to the tree A a new vertex which is connected to the vertex of A of degree $d_n - 1$. In this way a tree is constructed with degrees $d_1, \ldots, d_n$.

6.4 Suppose U_1 is the set of edges of the tree A_1, and U_2 is the set of edges of the tree A_2; let $u \in U_1 \setminus U_2$. If the edge u is suppressed from the tree A_1, then a graph G_1 is obtained which contains two connected components C_1 and C_2. In fact, if the graph obtained had contained at least three connected components, then by adding the edge u between two vertices located in different components the resulting graph would not have been connected and hence A_1 would not be connected. Thus the definition of a tree would be contradicted.

On the other hand, it has been seen that every tree with n vertices has $n-1$ edges (Problem 6.3), and thus by suppressing edge u the resulting graph G_1 is no longer connected, since it has $n-2$ edges.

There is an edge $v \in U_2$ which joins two vertices located in components C_1 and C_2. For otherwise it would follow that the edges of the tree A_2 join only pairs of vertices located either in C_1 or C_2, which would imply that A_2 is not connected. But A_2, being a tree, is connected, and hence there exists an edge $v \in U_2$ which joins vertices located in components C_1 and C_2. Since G_1 does not contain a cycle, it follows that the graph obtained from G_1 by adding the edge v also does not contain cycles and is connected. It is thus a tree, which will be denoted B_2.

But B_2 is a spanning tree of G which has more edges in common with A_2 than A_1. Repeat this transformation, replacing all the edges $u \in U_1 \setminus U_2$ by edges of U_2, to obtain finally a tree $B_r = A_2$ where $r \geqslant 2$.

6.5 (a) We show that if all the vertices of degree 1 of the tree G are suppressed then $e(x)$ is decreased by 1 for every vertex of the resulting subgraph. All the vertices at a distance $e(x)$ from x have degree 1, and thus by suppressing them $e(x)$ decreases for all of the remaining vertices. One can also observe that by this operation $e(x)$ decreases by exactly one, since the longest walk which leaves x ends in a vertex of degree 1 in G, which is then suppressed. The property is true for a graph with a single vertex. Assume therefore that it is true for all trees with at most $n-1$ vertices. Let G be a tree with $n \geqslant 2$ vertices. Denote by C the set of vertices x in the center of G, that is, those for which $e(x)$ is a minimum. Suppose that C does not contain a vertex of degree 1, and suppress all vertices of degree 1 in the graph G. For all the remaining vertices the value of $e(x)$ is reduced by one, and hence by this operation a new tree G' is obtained with the same center C. Since G' has at most $n-2$ vertices, it follows from the induction hypothesis that C consists of one vertex or two adjacent vertices, and property (a) is therefore established. If C contains a vertex of degree $d(x)=1$, then x will be adjacent to a unique vertex y. It is clear that y is strictly nearer than x to every other vertex of G. Thus $e(x)$ can be a minimum only if $e(x)=1$ and G is a tree consisting of x and y joined by an edge. In this case $C=\{x, y\}$ and the property is established.

By Problem 8.4, the walks of maximum length in a tree have a nonempty intersection. It can be shown that this intersection contains the center of the tree.

(b) Let L be a walk of length $e(x)$ which starts at x. If L does not contain either y or z, then

$$e(y)=e(x)+1, \qquad e(z)=e(x)+1,$$

and hence $2e(x)<e(y)+e(z)$. The walk L cannot contain both y and z, since both vertices are adjacent to x. For example, if L contains y, then

$$e(y) \geqslant e(x)-1, \qquad e(z)=e(x)+1,$$

and thus $2e(x) \leqslant e(y)+e(z)$.

Solutions

(c) It is easily shown that the distance thus defined on the set of vertices of a connected graph is a metric for this set and thus satisfies the triangle inequality. Let x, y be vertices such that $d(x, y) = d(G)$, and let z be a vertex of minimal eccentricity $[= \rho(G)]$. It can be shown that

$$d(G) = d(x, y) \leqslant d(x, z) + d(z, y) \leqslant \rho(G) + \rho(G) = 2\rho(G).$$

6.6 Let $L = [x_1, \ldots, x_{2k-2}]$ be a walk of length $2k-3$ of a tree A. To every vertex y which is not on L associate a walk, L_y, in the following manner: Let M be the walk which joins y and a vertex z of L and which has only the endpoint z in common with L. If the length of M is greater than $k-1$, let L_y be the subwalk of M of length k, and with endpoint y. Otherwise L_y consists of M and a subwalk of L. The construction is always possible, since L has length $2k-3$ and thus there is a subwalk of L of length $k-1$ incident with an endpoint of M. In fact, in the opposite case L would have length less than or equal to $2(k-2)$, which would contradict the hypothesis.

It is also the case that L contains the subwalks $[x_1, \ldots, x_{k+1}], \ldots, [x_{k-2}, \ldots, x_{2k-2}]$ of length k. One thus obtains $n - (2k-2) + k - 2 = n - k$ pairwise distinct walks of length k.

6.7 (a) Let G_x denote the subgraph obtained from G by suppressing the vertex x and the edges incident with x. Since G is a tree, it follows that G_x is not connected and y and z are found in different connected components of G_x which contain k_1 and k_2 vertices respectively. This implies that $k_1 + k_2 \leqslant n - 1$. It follows from the definition of the function $s(x)$ that in going from x to y at each step one moves closer by a distance 1 to k_1 vertices, but further away by 1 from $n - k_1$ vertices. It is hence the case that

$$s(n) = s(y) + k_1 - (n - k_1) = s(y) + 2k_1 - n.$$

One can show analogously that

$$s(x) = s(z) + 2k_2 - n.$$

By adding these two identities it turns out that

$$2s(x) = s(y) + s(z) + 2(k_1 + k_2 - n) \leqslant s(y) + s(z) - 2.$$

(b) Suppose that there are two nonadjacent vertices x and y such that $s(x) = s(y) = $ minimum. Let $[x, x_1, x_2, \ldots, x_p, y]$ be the unique walk which joins x and y in the tree ($p \geqslant 1$). By hypothesis $s(x_1) \geqslant s(x)$. The inequality proved in (a) shows that

$$s(x) + s(x_2) > 2s(x_1) \geqslant s(x_1) + s(x),$$

and thus

$$s(x_2) > s(x_1) \geqslant s(x).$$

It also follows that $s(x_1) + s(x_3) > 2s(x_2) > s(x_1) + s(x_2)$, and thus $s(x_3) > s(x_2) > s(x_1) \geqslant s(x)$, and so forth. Finally one finds that $s(y) > s(x_p) > \cdots > s(x_1) \geqslant s(x)$, which contradicts the equation $s(x) = s(y)$, since $p \geqslant 1$.

The points in which $s(x)$ attains its minimum form the barycenter of the tree G. By considering a walk with an odd (or respectively even) number of vertices one observes that the barycenter is formed of a single vertex (or of two adjacent vertices). If $x_1, \ldots, x_p$ are the vertices of a walk and $y_1, \ldots, y_q$ are other vertices joined by an edge to x_1, then if p is even, it follows that the center of the tree thus obtained is the vertex $x_{p/2}$. If q is sufficiently large, for example $q = \binom{p}{2}$, then the barycenter of the tree is x_1, while the distance between the center and the barycenter is $p/2 - 1$. This number can be arbitrarily large for values of p which are sufficiently large.

6.8 If the tree G has vertex set X of cardinality n, it follows that it also has $n-1$ edges. Then in the sum $\sum_{x,y \in X} d(x, y)$ there are exactly $2(n-1)$ terms equal to 1, and the other nonzero terms are greater than or equal to 2. If G is a star (the graph $K_{1,n-1}$), all the terms different from 0 (for $x = y$) and 1 are equal to 2. If G is not a star, there is at least one term equal to 3, and thus the desired minimum is attained only for the tree $K_{1,n-1}$. In order to find the maximum of this sum one can show that $s(x)$ is a maximum in the set of terminal vertices of a tree only when the tree is a walk and x is one of its endpoints. If L is a walk with n vertices and x is one of its endpoints, then

$$s(x) = 1 + 2 + \cdots + (n-1).$$

If G is a tree and x is a terminal vertex with $e(x) = d$, then there will exist at least one vertex at a distance $1, 2, \ldots, d$ from x. It follows from the definition of $s(x)$ that the sum which enters into $s(x)$ is of the form

$$s(x) = 1 + 2 + \cdots + d + d_1 + \cdots + d_{n-1-d},$$

where $d_1, \ldots, d_{n-1-d} \leq d$. By comparing this with the expression for $s(x)$ in the case of the walk L, one finds that the maximum of $s(x)$ is attained only when $d = n-1$, that is, when G is a walk and x is one of its endpoints. One now uses induction on n to prove that $\sum_{x,y \in X} d(x, y)$ is a maximum in the set of trees G with n vertices from the set X when G is a walk. For $1 \leq n \leq 3$ the property is immediate, since in this case every tree with n vertices is a walk.

Suppose now that the property is true for all trees with at most $n-1$ vertices, and let G be a tree with n vertices which consist of the set X. If a is a vertex of degree 1 in the tree G, then

$$\sum_{x,y \in X} d(x, y) = 2s(a) + \sum_{x, y \in Y} d(x, y),$$

where $Y = X \setminus \{a\}$.

The first term $s(a)$ is a maximum only if G is a walk and a is one of its endpoints. In this case the second term can also be shown to be maximal by applying the induction hypothesis to the subtree whose $n-1$ vertices make up the set Y.

6.9 Consider three pairwise different indices $i, j, k \in \{1, \ldots, r\}$. [If, for example, $i = j$ then $d_{ij} + d_{jk} - d_{ik} = 0$ and (a) is satisfied.] Since A is a tree, there exists a unique walk $L_{ij} = [x_i, \ldots, x_j]$ between the terminal vertices x_i and x_j.

Solutions

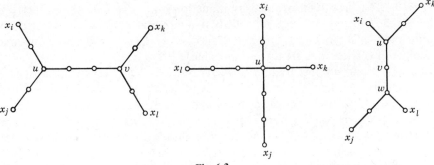

Fig. 6.3

There is also a vertex $v \in L_{ij}$ such that the unique walk from v to x_k has no vertices in common with L_{ij} other than the endpoint v. It follows that

$$d(x_i, x_j) + d(x_j, x_k) - d(x_i, x_k) = 2d(x_j, v),$$

which implies (a).

Consider $i, j, k, l \in \{1, \ldots, r\}$ which are pairwise different. If, for example, $i = j$ the three numbers become $d_{kl}, d_{ki} + d_{il}, d_{ki} + d_{il}$, and thus two are equal and the third satisfies the inequality

$$d_{kl} \leq d_{ki} + d_{il}.$$

In fact, the distance defined between the vertices of a graph satisfies the conditions for a metric, including the triangle inequality.

Let L_{ij} and L_{kl} be walks which join vertices x_i to x_j and x_k to x_l, respectively. These walks can have no, one, or two or more vertices in common.

If the walks have no vertices in common, let u be a vertex of the walk L_{ij} and v a vertex of the walk L_{kl} such that the unique walk with endpoints u and v has in common with the walks L_{ij} and L_{kl} only the endpoints u and v (Figure 6.3). In this case, it follows that

$$d_{ik} + d_{jl} = d_{il} + d_{jk} = d_{ij} + d_{kl} + 2d(u, v),$$

which establishes (b).

If L_{ij} and L_{kl} have exactly one vertex u in common, then $d_{ij} + d_{kl} = d_{ik} + d_{jl} = d_{il} + d_{jk}$.

If the walks L_{ij} and L_{kl} have at least two vertices u, v in common (Figure 6.3) then (b) is also true. If u and w denote the endpoints of the subwalk common to the two walks, then it follows that

$$d_{ij} + d_{kl} = d_{il} + d_{jk} = d_{ik} + d_{jl} + 2d(u, w).$$

Let $r \geq 2$ and $(d_{ij})_{i,j=1,\ldots,r}$ be a symmetric matrix with non-negative integer entries, such that $d_{ij} = 0$ if and only if $i = j$, which satisfies (a) and (b). It is possible to show (by induction on r) that there exists a tree A with r terminal vertices $x_1, \ldots, x_r$ such that $d(x_i, x_j) = d_{ij}$ for $i, j = 1, \ldots, r$. [K. A. Zaretskii, Uspehi Mat. Nauk, **20** (6) (1965), 94–96.]

6.10 The proof will proceed by induction on r. For $r=2$ the property is immediate, since every tree with exactly two terminal vertices is an elementary walk and the fact that $d_A(1, 2) = d_B(1, 2)$ shows that the two walks have the same number of vertices and are thus isomorphic.

Let $r > 2$, and suppose that the property is true for all trees with at most $r-1$ terminal vertices. Let v_A denote the vertex of degree greater than or equal to 3 of the tree A, which is closest to the terminal vertex labeled r. Similarly let v_B denote the corresponding vertex of the tree B. It follows that there exists an elementary walk $[v_A, \ldots, r]$ in the tree A, such that all vertices located between v_A and r have degree 2 in the graph A (their set may possibly be empty). There is a similar elementary walk $[v_B, \ldots, r]$ in B. Suppress the terminal vertices with label r as well as all internal vertices on the walks $[v_A, \ldots, r]$ and $[v_B, \ldots, r]$ respectively. In this manner one obtains trees A_1 and B_1 which each have $r-1$ terminal vertices with labels selected from the set $\{1, \ldots, r-1\}$. The distances between the terminal vertices i and j are the same in A and B for $1 \leq i \leq j \leq r-1$ and thus will remain unchanged for the trees A_1 and B_1. By the induction hypothesis A_1 and B_1 are isomorphic trees, and thus there exists a bijection f from the set of vertices of A_1 onto the set of vertices of B_1 which preserves the adjacency of vertices.

One can assume that $f(i) = i$ for $i = 1, \ldots, r-1$, because it is possible to relabel the terminal vertices of B_1 so that this condition is satisfied and the distances between the terminal vertices i and j are the same in A_1 and B_1. It is now possible to show that $f(v_A) = v_B$. Suppose that $f(v_A) \neq v_B$. There exists a unique walk in B_1 which joins $f(v_A)$ and v_B which can be extended in an arbitrary fashion to a walk which joins the terminal vertices i and j in B_1. It follows that $d_B(i, j) + d_B(i, r) - d_B(j, r) = 2d_B(i, v_B)$. Since $f(v_A)$ is found on the walk which joins i and j in B_1 (which is isomorphic to A_1), it follows that v_A is found on the walk which joins i and j in A_1. Thus, one can conclude similarly that

$$d_A(i, j) + d_A(i, r) - d_A(j, r) = 2d_A(i, v_A).$$

Since the distances between terminal vertices are the same in the trees A and B, it also follows that $d_B(i, v_B) = d_A(i, v_A)$. But $d_B(i, v_B) = d_{B_1}(i, v_B)$ and $d_A(i, v_A) = d_{A_1}(i, v_A)$. Since A_1 and B_1 are isomorphic under f, it follows that $d_{A_1}(i, v_A) = d_{B_1}(i, f(v_A)) = d_{B_1}(i, v_B)$, which contradicts the fact that $f(v_A) \neq v_B$ and the vertices $f(v_A)$, v_B, and i are found on the same walk. Finally, $f(v_A) = v_B$. Let i, j with $1 \leq i < j \leq r-1$ be labels for two terminal vertices such that v_A and $f(v_A) = v_B$ are found on the walk with endpoints i and j in A_1 and B_1, respectively, and hence in A and B.

One can thus write

$$d_A(r, i) + d_A(r, j) - d_A(i, j) = 2d_A(r, v_A),$$

$$d_B(r, i) + d_B(r, j) - d_B(i, j) = 2d_B(r, v_B).$$

Since the left-hand sides are equal by the hypothesis, it follows that $d_A(r, v_A) = d_B(r, v_B)$, and hence the walks $[v_A, \ldots, r]$ and $[v_B, \ldots, r]$ have the same length.

Now let $[v_A, \ldots, r] = [v_A, x_1, \ldots, x_k, r]$ and $[v_B, \ldots, r] = [v_B, y_1, \ldots, y_k, r]$, and define an isomorphism g between A and B as follows: $g(x) = f(x)$ for every vertex x of A_1, and

$$g(x_1) = y_1, \ldots, g(x_k) = y_k; \qquad g(r) = r.$$

[E. A. Smolenskii, *Jurnal Vicisl. Mat. i. Matem. Fiz.* **2**(2) (1962), 371–372.]

6.11 The proof will use induction on the number of vertices. If $|X| = 1$ or 2 the property is immediate.

Suppose that the property is true for every tree with at most $n-1$ vertices. It will be shown that it is true for a tree G with $n \geq 3$ vertices. If f is a bijection, then $f(x) \neq f(y)$ for $x \neq y$. Also $[x, y] \in U$ implies that $[f(x), f(y)] \in U$ and thus f is an automorphism of G. Every terminal vertex (of degree 1) is thus mapped by f into a terminal vertex.

Let G' denote the subtree of G obtained by suppressing all terminal vertices. It follows that G' is nonempty, since $n \geq 3$. If X' denotes the vertex set of G', then $f(X') = X'$ and the restriction f' of f to X' has the same property as f. Thus by the induction hypothesis f' (and consequently f) has a fixed point or a fixed edge ($|X'| \leq n-2$). If f is not a bijection, then $f(X)$ is a proper subset of the vertex set of G. It follows from the conditions on f that these vertices induce a connected subgraph of G, and hence $f(X)$ is the vertex set of a tree and $|f(X)| \leq n-1$.

Since $f(f(X)) \subset f(X)$, one can consider the restriction of f to the subtree generated by the vertex set $f(X)$, which has the same properties as f. By the induction hypothesis this restriction, and hence also f, has a fixed point or a fixed edge.

The property is no longer valid if G contains cycles. For example, suppose $G = K_3$ contains vertices x, y, z and let $f(x) = y$, $f(y) = z$, $f(z) = x$. In this case the mapping f has neither fixed points nor fixed edges.

6.12 It has been seen (Problem 6.5) that the center of a tree always consists of a single vertex u or two adjacent vertices u and v. The proof uses induction on the number m of vertices of X to show that if the tree A has a single vertex u as its center then $f(u) = u$. If the center is $\{u, v\}$, then either $f(u) = u$ and $f(v) = v$ or $f(u) = v$ and $f(v) = u$. Since the bijection f preserves adjacency of vertices, it follows that x and $f(x)$ have the same degree in the tree A.

For $m = 1$ the tree is equal to its center and hence $f(u) = u$. For $m = 2$ the tree is identical with its center and the property is again satisfied. Suppose that the property is true for all trees with at most $m-1$ vertices ($m \geq 3$), and let A be a tree with m vertices. If $x_1, \ldots, x_r$ is the set of terminal vertices of A, then it follows that $f(x_1), \ldots, f(x_r)$ are vertices of degree 1 and thus constitute a permutation of the set of terminal vertices. Consider the restriction of the function f to the set of vertices of degree at least 2 in A:

$$g : X_r \to X_r, \qquad \text{where} \quad X_r = X \setminus \{x_1, \ldots, x_r\},$$

and $g(x) = f(x)$ for every $x \in X_r$. It follows that g is an automorphism of the subtree A_r of A with vertex set X_r. In the solution of Problem 6.5 it was shown that A and A_r have the same center. It therefore follows from the induction hypothesis that either the center of A is equal to u, in which case $g(u) = f(u) = u$, or the center of A consists of the adjacent vertices u and v, in which case either $g(u) = f(u) = u$ and $g(v) = f(v) = v$, or $g(u) = f(u) = v$ and $g(v) = f(v) = u$.

The stated property is thus valid for every m. Since it must be shown that f has a fixed point, the only case which must be investigated is that in which A has center $\{u, v\}$ with $f(u) = v$ and $f(v) = u$. Let A_u and A_v denote the subtrees of A obtained by suppressing the edge $[u, v]$ and which contain the vertices u and v respectively. Let $u_1, \ldots, u_s$ denote the vertices in A_u which are adjacent to u. Since f preserves adjacency in A, it follows that $f(u_1), \ldots, f(u_s)$ are adjacent to v in the subtree A_v, and so on.

Let X_u and X_v denote the vertex sets of the trees A_u and A_v respectively. It follows that $f(X_u) = X_v$. Consider the restriction of the function f to the set X_u; denote it by $h: X_u \to X_v$. The function h is a bijection, and $[x, y]$ is an edge in the tree A_u if and only if $[h(x), h(y)]$ is an edge in the tree A_v. Thus h is an isomorphism of the trees A_u and A_v. It follows that A_u and A_v have the same number of vertices, which implies that $|X| = |X_u| + |X_v| = 2|X_u|$ and is thus an even number.

It was assumed that $|X| = 2n + 1$, and thus the case in which $f(u) = v$ and $f(v) = u$ does not occur. It follows that f has at least one fixed point.

6.13 Let $A_1 - x$ and $A_2 - x$ denote the graphs obtained from A_1 and A_2 respectively by suppressing x and the edges incident with x. Since a tree with n vertices has $n - 1$ edges (Problem 6.3), the degree of the vertex x in the tree A_1 is equal to

$$d_{A_1}(x) = |X| - 1 - m(A_1 - x), \tag{1}$$

where $m(A_1 - x)$ denotes the number of edges in the graph $A_1 - x$. It also follows that

$$d_{A_2}(x) = |X| - 1 - m(A_2 - x). \tag{2}$$

The fact that the graphs $A_1 - x$ and $A_2 - x$ are isomorphic implies that $m(A_1 - x) = m(A_2 - x)$.

By using (1) and (2) one can show that $d_{A_1}(x) = d_{A_2}(x)$ for every vertex $x \in X$ and thus the trees A_1 and A_2 have the same terminal vertices (of degree 1). Let T be the set of terminal vertices for the trees A_1 and A_2. If $|T| = 2$, it follows that A_1 and A_2 are walks of length $|X| - 1$, and thus A_1 and A_2 have the same diameter. Suppose that $|T| \geq 3$, and let L be an elementary walk of maximal length in the tree A_1. The length of L is by definition equal to $d(A_1)$, the diameter of A_1.

The endpoints of this walk are two terminal vertices in the set T. The set T also contains at least one other terminal vertex x which does not belong to the walk L. By hypothesis $A_2 - x$ is isomorphic to $A_1 - x$. Since the walk L is contained in the graph $A_1 - x$, it follows that $A_2 - x$ contains an elementary walk

Solutions

L_1 which has the same length as the walk L. Thus the graph A_2 contains an elementary walk L_1 of the same length as L. It follows that $d(A_2) \geqslant d(A_1)$.

By interchanging the roles of A_1 and A_2 in the preceding argument one can conclude that $d(A_1) \geqslant d(A_2)$ and hence A_1 and A_2 have the same diameter.

It can also be shown that under the given conditions the trees A_1 and A_2 are isomorphic. [P. J. Kelly, *Pacific J. Math.*, **7** (1957), 961–968.]

6.14 One can obtain an arborescence from the tree G by considering the vertex x_1 to be the root and directing all the edges of the tree so that for every vertex $y \neq x_1$ there is a unique path which originates in x_1 and terminates in y. In this way a partial order is defined on the set X: Let $x_i \leqslant x_j$ if the unique path from x_1 to x_j contains the vertex x_i.

Let the matrix $Z = (z_{ij})_{i,j=1,\ldots,n}$ be defined as follows: $z_{ij} = 1$ if $x_i \leqslant x_j$, and $z_{ij} = 0$ if x_i is not less than or equal to x_j. In the solution to Problem 2.19 it was shown that one can renumber the elements of X so that Z is upper triangular with $z_{ii} = 1$ for $i = 1, \ldots, n$. It follows that $\det Z = 1$.

Let

$$A = \begin{pmatrix} 0 & 1 & 1 & \cdots & 1 \\ 1 & -2 & 0 & \cdots & 0 \\ 1 & 0 & -2 & \cdots & 0 \\ \vdots & \vdots & & & \vdots \\ 1 & 0 & \cdots & 0 & -2 \end{pmatrix}.$$

It will be shown that $Z^T A Z = D$. In fact, the element in row i and column j of the matrix $Z^T A Z$ is

$$c_{ij} = \sum_{k=1}^{n} \sum_{l=1}^{n} z_{ki} a_{kl} z_{lj} = \sum_{x_k \leqslant x_i} \sum_{x_l \leqslant x_j} a_{kl}.$$

But $a_{kl} \neq 0$ only if $k = l$ or $k = 1$ or $l = 1$, and hence

$$c_{ij} = \sum_{x_k \leqslant x_i, x_j} (-2) + \sum_{x_k \leqslant x_i} 1 + \sum_{x_l \leqslant x_j} 1,$$

since $x_1 \leqslant x_i$ for every $i = 1, \ldots, n$. Let x_r denote the last common vertex of the paths from x_1 to x_i and from x_1 to x_j, with x_1 being considered as the initial point. It follows that

$$c_{ij} = -2(d(x_1, x_r) + 1) + (d(x_1, x_i) + 1) + (d(x_1, x_j) + 1)$$
$$= d(x_1, x_i) + d(x_1, x_j) - 2d(x_1, x_r) = d(x_i, x_j).$$

Thus it has been shown that $Z^T A Z = D$ and finally that $\det D = \det A$. By adding the other columns to the first column of the matrix A and expanding the resulting determinant on the first column one sees that

$$\det A = (n-1)(-2)^{n-1} - \det A, \quad \text{and hence} \quad \det A = -(n-1)(-2)^{n-2}.$$

[R. L. Graham, H. C. Pollak, *Bell Syst. Techn. J.*, **50** (1971), 2495–2519.]

6.15 (a) Since every tree has at least two terminal vertices, it follows that by using the procedure previously described one obtains a tree consisting only of the vertex x_n and another vertex adjacent to x_n. The Prüfer code of A is completed by a new position $a_{n-1} = n$.

Let $b_1, \ldots, b_{n-1}$ be the indices of the terminal vertices which are suppressed when the algorithm determining the Prüfer code is applied to the tree A. It will now be shown how to determine the numbers b_i from the Prüfer code $(a_1, \ldots, a_{n-1})$.

It is clear that b_i is different from $b_1, \ldots, b_{i-1}$ and also from a_i, since $[x_{a_i}, x_{b_i}]$ is an edge in the tree A. Since the vertex x_{b_i} has been suppressed, it cannot be adjacent to a terminal vertex at a later step. Thus $b_i \neq a_j$ for $j > i$.

Conversely, if $k \notin \{b_1, \ldots, b_{i-1}, a_i, \ldots, a_{n-1}\}$, then the vertex x_k is a terminal vertex of the tree A-$\{x_{b_1}, \ldots, x_{b_{i-1}}\}$ obtained from A by suppressing the vertices with indices $b_1, \ldots, b_{i-1}$. For otherwise it would have to be adjacent to a vertex which will be suppressed at a later step. It follows that $k \in \{a_i, \ldots, a_{n-1}\}$, which contradicts the hypothesis. Hence

$$b_i = \min \{k \mid k \notin \{b_1, \ldots, b_{i-1}, a_i, \ldots, a_{n-1}\}\}. \tag{1}$$

Thus the Prüfer code uniquely determines the numbers b_i, and hence the tree A which consists of the edges $[x_{a_i}, x_{b_i}]$. It follows that the correspondence which associates to every tree its Prüfer code is injective.

Now let $(a_1, \ldots, a_{n-1})$ be an arbitrary sequence of integers such that $1 \leq a_i \leq n$ and $a_{n-1} = n$. Define the numbers b_i recursively by identity (1). Join the vertices x_{a_i} and x_{b_i} by an edge for $i = 1, \ldots, n-1$.

One can now show that the graph A obtained in this manner is a tree whose Prüfer code is exactly $(a_1, \ldots, a_{n-1})$; thus the correspondence under consideration is surjective, and hence bijective. In order to establish this property it is sufficient to show that x_{b_i} is a terminal vertex with minimal index of the graph

$$A_i = A\text{-}\{x_{b_1}, \ldots, x_{b_{i-1}}\}.$$

It follows from (1) that $b_i \neq a_j$ for $j > i$, and thus $a_i \neq b_1, \ldots, b_{i-1}$, which implies that x_{a_i} is a vertex in A_i. By construction the vertices x_{a_i} and x_{b_i} are adjacent, and thus x_{b_i} is adjacent with a vertex in the graph A_i. The vertex x_{b_i} cannot be adjacent to a vertex of A_i other than x_{a_i}, since if $[x_{a_j}, x_{b_j}]$ were another edge of A_i incident with x_{b_i}, it would follow that $j > i$, since x_{b_j} is a vertex of A_i.

But x_{b_i} is one of the vertices x_{a_j} or x_{b_j}, and thus $b_i = b_j$ or $b_i = a_j$, which contradicts identity (1), since $j > i$. It follows that x_{b_i} is a terminal vertex of the graph A_i, and hence A and all of the graphs A_i are trees. This property follows by induction on i. For $i = n$ the graph A_n is composed of the vertex x_{b_n} and is thus a tree. The fact that A_i is a tree is a consequence of the fact that A_{i+1} is a tree.

Suppose now that A_i has a terminal vertex x_k with $k < b_i$. It follows from (1) that either $k = b_s$ with $s < i$ or $k = a_j$ with $j \geq i$. The first alternative is impossible, since x_{b_s} is a vertex in the tree A_i, which does not contain the vertices $x_{b_1}, \ldots, x_{b_{i-1}}$. If $k = a_j$ with $j \geq i$, then $j \leq n-2$, since $a_{n-1} = n \geq b_i > k$. But it has been shown that x_{b_j} is a terminal vertex of A_j which is adjacent only to $x_{a_j} = x_k$.

Solutions

The vertex x_k, being a terminal in A_i, is also a terminal vertex in A_j, since $j \geqslant i$. It follows that x_{a_j} and x_{b_j} are both terminal vertices of A_j, and thus A_j reduces to the edge $[x_{a_j}, x_{b_j}]$. This is a contradiction, since A_j has $n-j+1 \geqslant 3$ vertices, since $j \leqslant n-2$.

(b) The number of trees with n vertices $x_1, \ldots, x_n$ is thus equal to the number of sequences $(a_1, \ldots, a_{n-1})$ with $1 \leqslant a_i \leqslant n$ and $a_{n-1} = n$. Hence it is equal to n^{n-2}.

(c) Observe that each vertex x_{a_i} occurs in the sequence $(a_1, \ldots, a_{n-2})$ exactly $d_{a_i} - 1$ times. In fact, the indices of the terminal vertices do not appear in the sequence, but a vertex x_i of degree $d_i \geqslant 2$ will become a terminal vertex after exactly $d_i - 1$ of its neighbors have been eliminated. It follows that the number of trees with degrees $d(x_1) = d_1, \ldots, d(x_n) = d_n$, whose sum is equal to twice the number of edges (i.e., $2n - 2$) is equal to the number of sequences $(a_1, \ldots, a_{n-2})$ which contain the number k $d_k - 1$ times for $1 \leqslant k \leqslant n$. But this number is equal to the number of arrangements of $n-2$ objects in n cells such that the kth cell contains $d_k - 1$ objects for $1 \leqslant k \leqslant n$. The objects in cell i represent the number orderings of positions in which the number i is found in the sequence $(a_1, \ldots, a_{n-2})$. By Problem 1.15 this number is equal to

$$\frac{(n-2)!}{(d_1-1)! \cdots (d_n-1)!}.$$

6.16 Let A be a tree with vertices $x_1, \ldots, x_n$. If an arbitrary edge is suppressed, the result is two disjoint trees which together contain all the vertices of A. Label the endpoints of the suppressed edge. Since A has $n-1$ edges, starting from the t_n trees with n vertices, one obtains $(n-1)t_n$ pairs of trees of this kind with one vertex labeled in each tree. Suppose that A_1 and A_2 are two disjoint trees with k and $n-k$ vertices respectively and which together contain the vertices $x_1, \ldots, x_n$. One can label a vertex of A_1 and a vertex of A_2 in $k(n-k)$ ways for $1 \leqslant k \leqslant n-1$. The vertex sets of A_1 and A_2 can be chosen in $\binom{n-1}{k-1}$ ways under the condition that a fixed vertex x_1 belongs to the tree A_1 in order to eliminate repetition.

One can find t_k and t_{n-k} trees with vertex sets A_1 and A_2, respectively, and thus, by counting (in two ways) the pairs of disjoint trees which together contain the vertices $x_1, \ldots, x_n$ and have a labeled vertex in each tree, one finds that

$$\sum_{k=1}^{n-1} \binom{n-1}{k-1} t_k t_{n-k} k(n-k) = (n-1) t_n. \tag{1}$$

Since

$$\binom{n-1}{k-1} = \frac{n-1}{n-k}\binom{n-2}{k-1},$$

the desired identity is obtained, after dividing both sides of (1) by $n-1$.

Recall that $\binom{n-1}{k-1} = \binom{n-1}{n-k}$. As a result of interchanging the indices k and $n-k$, (1) becomes

$$\sum_{k=1}^{n-1} \binom{n-1}{k} t_k t_{n-k} k(n-k) = (n-1) t_n. \tag{2}$$

Addition of (1) and (2) yields

$$\sum_{k=1}^{n-1} \binom{n}{k} t_k t_{n-k} k(n-k) = 2(n-1)t_n. \tag{3}$$

This identity can also be found directly if x_1 is not a fixed vertex in A_1.

It will now be shown by induction on n that $t_n = n^{n-2}$. For $n = 1$ there is a unique tree with one vertex and the formula is satisfied. Suppose that $t_m = m^{m-2}$ for every $1 \leq m \leq n-1$. It will be shown that $t_n = n^{n-2}$. Since (3) holds, one must show that

$$\sum_{k=1}^{n-1} \binom{n}{k} k^{k-1}(n-k)^{n-k-1} = 2(n-1)n^{n-2}.$$

This equation is in fact identity (c) of Problem 1.29 and implies that $t_n = n^{n-2}$.
[O. Dziobek, *Sitzungsber. Berl. Math. G.*, **17** (1917), 64–67.]

6.17 Suppose that the terminal vertices are fixed and in fact that they are $x_1, \ldots, x_p$. Since $d_1 = \cdots = d_p = 1$, it follows from the preceding problem that the desired number is equal to

$$\sum_{(d_{p+1},\ldots,d_n)} \frac{(n-2)!}{(d_{p+1}-1)! \cdots (d_n-1)!} = \sum_{(k_{p+1},\ldots,k_n)} \frac{(n-2)!}{k_{p+1}! \cdots k_n!}, \tag{1}$$

where the first summation is taken over all values

$$d_{p+1}, \ldots, d_n \geq 2 \quad \text{and} \quad d_{p+1} + \cdots + d_n = 2n - 2 - p.$$

The second summation is obtained by substituting the variables $k_{p+1} = d_{p+1} - 1, \ldots, k_n = d_n - 1$ and hence $k_{p+1}, \ldots, k_n \geq 1$ and $k_{p+1} + \cdots + k_n = n - 2$. The number

$$\frac{(n-2)!}{k_{p+1}! \cdots k_n!}$$

represents the number of ways of arranging $n-2$ objects in $n-p$ cells so that the first cell contains k_{p+1} objects, $\ldots$, the $(n-p)$th cell contains k_n objects. Since $k_i \geq 1$ for $i = p+1, \ldots, n$, this number also represents the number of surjective functions

$$f : X \to Y$$

where $|X| = n-2$, $|Y| = n-p$, and if $Y = \{y_{p+1}, \ldots, y_n\}$ then $|f^{-1}(y_i)| = k_i \geq 1$ for $p+1 \leq i \leq n$. It follows that the sum (1) is equal to the number of surjective functions $s_{n-2, n-p} = (n-p)! \, S(n-2, n-p)$, where $S(n-2, n-p)$ is the Stirling number of the second kind (Problem 3.4).

Since the p terminal vertices are not specified, they can be chosen from the set of n vertices in $\binom{n}{p}$ ways. Thus the number of trees with n vertices p of which have degree 1 is equal to

$$\binom{n}{p}(n-p)! \, S(n-2, n-p) = \frac{n!}{p!} S(n-2, n-p).$$

[A. Rényi, *Mat. Kut. Int. Közl.*, **4** (1959), 73–85.]

Solutions

6.18 (a) Let $f(n)$ be the desired number of possible ways of selection. It follows that $f(1)=1$ and $f(2)=2$. This corresponds to the choice of edge $[x_1, y_1]$ and respectively to $[x_1, x_2]$ and $[y_1, y_2]$ or $[x_1, y_1]$ and $[x_2, y_2]$. If the ladder graph has $2n$ vertices, then one can select $[x_1, y_1]$ and for the remaining edges there are $f(n-1)$ possible choices. One may also select $[x_1, x_2]$ and $[y_1, y_2]$, leaving $f(n-2)$ possible further choices. It then follows that

$$f(n) = f(n-1) + f(n-2),$$

and this relation together with the initial values $f(1)=1$ and $f(2)=2$ implies that $f(n) = F_n$, the Fibonacci number.

(b) Let $g(n)$ denote the number of spanning trees of the ladder graph. It follows that $g(1)=1$ when the spanning tree consists of $[x_1, y_1]$. Also $g(2)=4$ for the trees which are obtained from a cycle with four vertices if one suppresses, in turn, an edge of the cycle.

In order to prove the recurrence relation for $g(n)$, consider the graph of Figure 6.1 with $2n+2$ vertices:

$$x_1, y_1, \ldots, x_{n+1}, y_{n+1}.$$

The set of its spanning trees can be written in the form

$$A_1 \cup A_2 \cup A_3 \cup A_4,$$

where A_1 is the set of spanning trees which do not contain the edge $[x_1, x_2]$, A_2 is the set of spanning trees which do not contain the edge $[x_1, y_1]$, A_3 is the set of spanning trees which do not contain $[y_1, y_2]$, and A_4 is the set of spanning trees which contain $[x_1, x_2], [x_1, y_1]$, and $[y_1, y_2]$ and do not contain $[x_2, y_2]$. It is clear that these sets are pairwise disjoint. If, for example, there were a spanning tree in $A_1 \cap A_2$, then it would not contain the edges $[x_1, x_2]$ and $[x_1, y_1]$. The vertex x_1 would thus be isolated, which contradicts the definition of a spanning tree. It follows that $|A_1|=|A_2|=|A_3|=g(n)$. [Using the vertices $x_2, y_2, \ldots, x_{n+1}, y_{n+1}$ there are $g(n)$ possible spanning trees.] Let $|A_4|=h(n+1)$. Then

$$g(n+1) = 3g(n) + h(n+1). \tag{1}$$

Now consider the graph of Figure 6.1 with $2n$ vertices denoted $x_2, y_2, \ldots, x_{n+1}, y_{n+1}$. Its set of spanning trees can be written in the form $B_1 \cup B_2$, where B_1 is the set of spanning trees which do not contain the edge $[x_2, y_2]$ and B_2 is the set of spanning trees which do contain $[x_2, y_2]$. Since $B_1 \cap B_2 = \emptyset$, it follows that $g(n) = |B_1| + |B_2|$. But $|B_1| = g(n-1)$ and $|B_2| = |A_4|$. There thus exists a bijection between these two sets defined as follows: In each spanning tree of B_2 replace the edge $[x_2, y_2]$ by a walk of length 3 having the same endpoints: $[x_2, x_1, y_1, y_2]$. One thus obtains a spanning tree in A_4. It is obvious that this correspondence is a bijection and hence

$$g(n) = g(n-1) + h(n+1),$$

from which it follows by using (1) that

$$g(n+1) = 4g(n) - g(n-1). \tag{2}$$

The characteristic equation of relation (2) is

$$r^2 - 4r + 1 = 0,$$

which has the solutions $r_1 = 2 + \sqrt{3}$ and $r_2 = 2 - \sqrt{3}$. Thus the general solution of the recurrence relation (2) is of the form

$$g(n) = C_1 r_1^n + C_2 r_2^n,$$

where C_1 and C_2 are determined by the system

$$C_1(2 + \sqrt{3}) + C_2(2 - \sqrt{3}) = 1,$$
$$C_1(7 + 4\sqrt{3}) + C_2(7 - 4\sqrt{3}) = 4,$$

which has the solution $C_1 = 1/2\sqrt{3}$, $C_2 = -1/2\sqrt{3}$. Thus

$$g(n) = \frac{1}{2\sqrt{3}} \{(2 + \sqrt{3})^n - (2 - \sqrt{3})^n\}.$$

[J. Sedláček, *Časopis pro pěstování matem.*, **94**(2) (1969), 217–221.]

6.19 Let $D = (d_{ij})$ be the distance matrix of a connected graph G. Properties (1)–(4) are precisely the expression of the fact that the distance function is a metric on the set of vertices $\{1, \ldots, p\}$. In order to prove (5), let $1 \leq i < j \leq p$, so that $d_{ij} = d(i, j) > 1$, and let L be a shortest walk from i to j, which contains at least two edges. Let k be a vertex of L, different from the endpoints i and j. It follows that both subwalks of L from i to k and from k to j are subwalks of minimal length, that is, $d_{ij} = d_{ik} + d_{kj}$.

It will be shown, conversely, that a square matrix D of order p which satisfies properties (1)–(5) is the distance matrix of a given graph. Let G be a graph with vertex set $X = \{1, \ldots, p\}$ and with edge set $U = \{[i, j] \mid d_{ij} = 1\}$. It remains to show that for each two vertices i and j it is the case that $d(i, j) = d_{ij}$.

If $i = j$ then $d(i, i) = 0 = d_{ii}$. If $[i, j] \in U$, one has $d(i, j) = 1 = d_{ij}$ in view of the definition of the graph G. Thus suppose that $i \neq j$, $[i, j] \notin U$, and hence $d_{ij} \geq 2$. By repeated use of property (5) one finds integers $i_1, i_2, \ldots, i_k$ such that

$$d_{ij} = d_{ii_1} + d_{i_1 i_2} + \cdots + d_{i_k j};$$

each term on the right-hand side of this equation is equal to 1. Thus $[i, i_1]$, $[i_1, i_2], \ldots, [i_k, j] \in U$, which implies the existence of a walk of length d_{ij} in G with endpoints i and j. In fact G also contains an elementary walk with endpoints i and j of length at most equal to d_{ij}. The graph G is connected, and $d(i, j) \leq d_{ij}$. If $d(i, j) < d_{ij}$, there is an elementary walk $[i, j_1, j_2, \ldots, j_m, j]$ of length less than d_{ij}. The existence of this walk implies that $d_{ij_1} = d_{j_1 j_2} = \cdots = d_{j_m j} = 1$. In this case (4) implies that $d_{ij} \leq d_{ij_1} + \cdots + d_{j_m j} < d_{ij}$, which is a contradiction. It has thus been shown that $d(i, j) = d_{ij}$ for every i and j.

Solutions

6.20 It will be shown that (1) is equivalent to (2) and that (2) is equivalent to (3).

Suppose that G is a tree, that is, G is connected and without cycles. If the graph G_1 obtained from G by deleting edge $[x, y]$ is connected, then there is a walk between x and y in G_1. It follows that in G_1 there exists an elementary walk between x and y, which together with $[x, y]$ generates a cycle in G, and this is a contradiction. Hence (1) implies (2). Suppose G is a graph satisfying (2) and G has a cycle $[x, z_1, \ldots, z_k, y, x]$. By deleting one edge $[x, y]$ in this cycle one obtains a new graph G_1 which is connected; this is a contradiction. It follows that G is connected and without cycles, and hence is a tree.

(1) implies (3): If G is a tree, it follows that G does not contain a cycle. For any two nonadjacent vertices x, y of G there is an elementary walk between x and y. This walk together with the edge $[x, y]$ generates a cycle in G_1, and hence (3) holds.

It remains to show that (3) implies (1). To show this let G be a graph satisfying (3). One must prove that G is connected. Suppose that G is not connected. There are therefore two vertices x and y belonging to different connected components of G. By inserting the edge $[x, y]$ one obtains a graph G_1 which does not contain a cycle, which is a contradiction.

6.21 If T_n is a tree with n labeled vertices, one may choose a root a in n ways. For any such selection a unique arborescence is obtained directing each edge so that any vertex $x \neq a$ to be reached from a by a unique path. Hence the number of arborescences with n labeled vertices is equal to $n \cdot n^{n-2} = n^{n-1}$ by Cayley's formula.

6.22 Choose a vertex of a labeled tree T_n with vertex set $\{1, \ldots, n\}$ (say the vertex n), and call it the root. There exists a unique walk from any other vertex $i < n$ to the root. If $[i, j]$ is the first edge in this walk, let $f(i) = j$. The function f is called the tree function of T_n. Assign the label i to the edge $[i, j]$, where $j = f(i)$, for $i = 1, \ldots, n-1$. This defines a mapping of the set of n^{n-2} vertex-labeled trees T_n onto the set of edge-labeled trees.

When $n \geq 3$, each edge-labeled tree is the image of n vertex-labeled trees, since the vertex n may be chosen in n ways and the labels of the other vertices are uniquely determined by the labels of the edges. It follows that the number of trees with n unlabeled vertices and $n-1$ labeled edges is equal to $n^{n-2}/n = n^{n-3}$ by Cayley's formula. [E. M. Palmer, *J. Combinatorial Theory*, 6 (1969), 206–207.]

6.23 Each column of an incidence matrix A contains one $+1$, one -1, and $n-2$ zeros, and hence the sum of all n rows vanishes. The sum of any r rows of A must contain at least one nonzero entry if $r < n$, for otherwise G would not be connected. This implies that no r rows are linearly dependent if $r < n$. In fact, if there exist r rows $a_{i_1}, \ldots, a_{i_r}$ of A ($r < n$) whose sum equals the null vector with m components, then it follows that there is no edge which is directed away from or towards the vertex set $\{i_1, \ldots, i_r\}$ of G; this contradicts the hypothesis that G is connected.

By applying this result to the submatrices corresponding to the connected components of G it follows that if G has s connected components then the rank of A is $n-s$. [G. Kirchhoff, *Annalen der Physik und Chemie*, **72** (1847), 497–508.]

6.24 If B is nonsingular, then each column of B must contain at least one nonzero entry, but not all columns can contain two nonzero entries. Hence some column of B must contain exactly one nonzero entry. The desired result now follows by induction on the order $r \geqslant 1$ of B by expanding the determinant of B along this column. [H. Poincaré, *Proc. London Math. Soc.*, **32** (1901), 277–308.]

6.25 Let F denote the spanning graph of G whose $n-1$ edges correspond to the columns of B. It follows that B is the reduced incidence matrix of F, but B is nonsingular if and only if rank$(B) = n-1$. It has been seen that rank$(B) = n-1$ if and only if F is connected (Problem 6.23). In fact, if C is the incidence matrix of F, then rank$(C) =$ rank(B), and if F is connected, then rank$(C) = n-1$. If F is not connected, then it has $s \geqslant 2$ connected components and rank$(C) =$ rank$(B) = n-s < n-1$. It remains to prove that an F which has n vertices and $n-1$ edges is connected if and only if it is a tree. If F is a tree with n vertices, it is connected and has $n-1$ edges by Problem 6.3. Conversely, since F is connected and has n vertices and $n-1$ edges, it will be shown that F is a tree.

Indeed, if F is connected, it contains a spanning tree T. To see this, suppose that for any edge $[x, y]$ of F the graph F_1 obtained from F by deleting edge $[x, y]$ is not connected. In this case, by Problem 6.20, F is itself a tree and we define $T = F$. Otherwise, the same argument can be applied to F_1 and so on by obtaining a spanning tree T of F.

The tree T has n vertices and $n-1$ edges; hence $T = F$, or F is a tree. [J. Chuard, *Rend. Circolo Mat. Palermo*, **46** (1922), 185–224.]

6.26 The Binet–Cauchy theorem states that if P and Q are matrices of size p by q and q by p where $p \leqslant q$, then det $PQ = \sum$ det B det C, where the sum taken is over the square submatrices B and C of P and Q of order p such that the columns of P in B are numbered the same as the rows of Q in C. Applying this to A and A^T, assuming that $m \geqslant n-1$, and using Problem 6.24, one can show that

$$\det A_r A_r^T = \sum \det B \det B^T = \sum (\det B)^2 = \sum 1,$$

where the last sum is taken over all nonsingular $(n-1)$-by-$(n-1)$ submatrices of A_r. The desired result now follows from Problem 6.25. [R. L. Brooks, C. A. B. Smith, A. H. Stone, W. T. Tutte, *Duke Math. J.*, **7** (1940), 312–340.]

6.27 Let $B = A_r A_r^T$, and let a_i denote the ith row of the incidence matrix A of G. It is clear that b_{ij} is equal to the scalar product $a_i a_j$ for $1 \leqslant i, j \leqslant n-1$. It follows that b_{ii} is equal to the number of nonzero entries of a_i, that is, to the number of vertices adjacent to i in G. If $i \neq j$, then b_{ij} is equal to -1 if $[i, j]$ is an edge of G and $b_{ij} = 0$ otherwise. It follows that

Solutions

$$t_n = \begin{vmatrix} n-1 & -1 & \cdots & -1 \\ -1 & n-1 & \cdots & -1 \\ \vdots & \vdots & & \vdots \\ -1 & -1 & \cdots & n-1 \end{vmatrix} = \begin{vmatrix} 1 & 1 & \cdots & 1 \\ -1 & n-1 & \cdots & -1 \\ \vdots & \vdots & & \vdots \\ -1 & -1 & \cdots & n-1 \end{vmatrix} = \begin{vmatrix} 1 & 1 & \cdots & 1 \\ 0 & n & \cdots & 0 \\ \vdots & \vdots & & \vdots \\ 0 & 0 & \cdots & n \end{vmatrix} = n^{n-2},$$

where all matrices are of order $n-1$.

This method for obtaining t_n was first pointed out by L. Weinberg [*Proc. IRE*, **46** (1958), 1954–1955].

6.28 For $n=1$ one obtains three independent sets, namely $\varnothing$, $\{1\}$, $\{2\}$, and hence $I_1 = 3$. In a similar manner one can show that $I_2 = 8$. For the graph R_n let M_n denote the family of all independent sets, and let A_n denote the family of independent sets containing vertex $n+1$. Let B_n be the family of independent sets containing 1, and let C_n be the family of independent sets of R_n which contain neither $n+1$ nor 1. It follows that $M_n = A_n \cup B_n \cup C_n$, where A_n, B_n, C_n are pairwise disjoint sets. It is clear that $|A_n| = |C_n| = I_{n-1}$ and $|B_n| = 2I_{n-2}$, since the independent sets containing 1 do not contain $n+1$ and 2, but they may or may not contain the vertex $n+2$. Thus the numbers I_n satisfy the recurrence relation

$$I_n = 2I_{n-1} + 2I_{n-2}$$

with initial conditions $I_1 = 3$ and $I_2 = 8$. The characteristic equation is $r^2 - 2r - 2 = 0$ with roots $r_{1,2} = 1 \pm \sqrt{3}$, which implies that

$$I_n = C_1(1+\sqrt{3})^n + C_2(1-\sqrt{3})^n.$$

From the initial conditions one can conclude that $C_1 = (3+2\sqrt{3})/6$ and $C_2 = (3-2\sqrt{3})/6$. [H. Prodinger, R. F. Tichy, *Fibonacci Quarterly*, **20**(1) (1982), 16–21.]

6.29 The proof is by induction on n. For $n=2$ this inequality becomes an equality and coincides with the Principle of Inclusion and Exclusion.

Suppose that the inequality holds for any $n-1$ subsets of X and any choice of a tree on vertices $1, \ldots, n-1$. Without loss of generality one can suppose that n is a terminal vertex of the tree T. Denote by T_1 the tree obtained from T by suppressing n and its incident edge $[k, n]$. It follows from the induction hypothesis that

$$|(A_1 \cup \cdots \cup A_{n-1}) \cup A_n| = |A_1 \cup \cdots \cup A_{n-1}| + |A_n| - |(A_1 \cup \cdots \cup A_{n-1}) \cap A_n|$$

$$\leq \sum_{i=1}^{n} |A_i| - \sum_{[i,j] \in E(T_1)} |A_i \cap A_j| - |A_k \cap A_n|$$

$$= \sum_{i=1}^{n} |A_i| - \sum_{[i,j] \in E(T)} |A_i \cap A_j|,$$

since $(A_1 \cup \cdots \cup A_{n-1}) \cap A_n \supset A_k \cap A_n$. [K. J. Worsley, *Biometrika* **69**(2) (1982), 297–302.]

6.30 It is clear that (1) is satisfied. In order to show (2) suppose that the edges of I induce a spanning graph (X, I) with p components $C_1, \ldots, C_p$ and that these components contain respectively $n_1, \ldots, n_p$ vertices $(n_1 + \cdots + n_p = n)$. Since (X, I) does not contain a cycle, it follows that $C_1, \ldots, C_p$ are trees having respectively $n_1 - 1, \ldots, n_p - 1$ edges. By hypothesis (X, J) also has no cycles, which implies that J contains at most $n_i - 1$ edges with both endpoints in the component C_i, for every $i = 1, \ldots, p$. Because $|J| = |I| + 1$, it follows that there is at least one edge $e \in J \setminus I$ whose endpoints lie in different components C_i and $C_j (i \neq j)$. It follows that $I \cup \{e\}$ is an independent set, since $(X, I \cup \{e\})$ also contains no cycles.

From Problem 6.20 one can deduce that the bases of the matroid $M(G)$ coincide with the edge sets of spanning trees of G. If (X, S) has p components containing respectively $m_1, \ldots, m_p$ vertices, then the fact that independent sets of edges contain no cycles implies that

$$\rho(S) = \sum_{i=1}^{p} (m_i - 1) = n - p$$

CHAPTER 7

7.1 Suppose that the sign of a negative edge is changed so that it becomes positive. It follows that the signs of the $n - 2$ triangles which contain this edge are also changed.

Suppose that r positive triangles become negative and s negative triangles become positive, so that $r + s = n - 2$. In the graph thereby obtained the number of negative triangles is equal to

$$n(f) + r - s \equiv n(f) + r + s = n(f) + n - 2,$$

where the congruence is taken modulo 2.

It follows that by changing, in turn, the sign of all the negative edges, one obtains zero negative triangles, and hence

$$n(f) + p(n - 2) \equiv 0,$$

that is,

$$n(f) \equiv np.$$

7.2 Suppose that the three colors are a, b, c. Each triangular face for which the vertex set is colored with all three colors has an edge whose endpoints are colored a and b respectively. All the other faces contain 0 or 2 edges with this property. Separate all the faces of the planar graph, and count the number of edges with endpoints of colors a and b. It follows that

$$\underbrace{1 + 1 + \cdots + 1}_{f_3} + 0 + 0 + \cdots + 0 + 2 + 2 + \cdots + 2 \equiv 0 \pmod{2}.$$

Thus f_3 represents the number of faces whose vertices are colored with all the colors a, b, and c, since each edge with colors a, b is counted twice (in both faces adjacent with it). Thus $f_3 \equiv 0 \pmod{2}$.

This property is a special case of Sperner's lemma of algebraic topology, which is equivalent to an analogous property for n-dimensional triangulations. The result remains valid for an arbitrary number of colors.

7.3 The necessity of the condition is immediate, since if a graph G without isolated vertices has an Eulerian circuit, then it is connected. In fact each two vertices, x and y, being incident to an arc which belongs to an Eulerian circuit, are joined by a walk.

On the other hand, an Eulerian circuit uses all the arcs which originate at and terminate in every vertex x. The Eulerian circuit uses a unique arc which ends at x and a unique arc which starts at x each time the vertex x is traversed, and hence $d^-(x) = d^+(x)$. In order to show the sufficiency, let C be a circuit of the graph G which contains a maximal number of arcs. The graph G is connected and $d^-(x) = d^+(x)$ for every vertex x. If C does not contain all the arcs of G, then it follows from the fact that G is connected that there is an arc (x, y) which has a vertex in common with the circuit C. Suppose, for example, that $x \in C$. One can make this assumption because the circuit C uses the same number of entry and exit arcs in each vertex, but the indegree of the vertex y is equal to the outdegree of the vertex y. Therefore if $y \in C$, there will exist an arc of the form (y, t).

Let G_C denote the spanning subgraph of G induced by the arcs which do not belong to the circuit C. Since $d^-(x) = d^+(x)$ for every vertex x, and since the circuit C uses the same number of entry and exit arcs at every vertex, it follows that all the vertices of G_C have equal indegrees and outdegrees.

Leave by the arc (x, y) of G_C, and move along the arcs of G_C using each arc exactly once. Continue this process as long as possible. One cannot end in a vertex $z \neq x$, since the vertices of G_C have equal indegrees and outdegrees and each traversal of a vertex $z \neq x$ uses an entry arc and an exit arc. Thus if one has arrived at a vertex $z \neq x$, one can also leave this vertex on an exit arc. Since the number of vertices of G_C is finite, one must terminate at the vertex x, and this produces a circuit C_1 in G_C.

The union of the arcs of the circuits C and C_1 is a circuit which is longer than C, which contradicts the assumption made. Thus C is an Eulerian circuit.

7.4 One can suppose that the graph G has at least one Eulerian circuit C, since otherwise the property is evident. The Eulerian circuit passes through every arc of G and hence passes through the vertex x with $d^+(x) \geq 3$.

Move along the circuit C by leaving from x and returning to x. At each traversal of x one obtains a circuit. Let these circuits be $C_1, C_2, \ldots, C_m$, where $m = d^+(x)$. Every permutation of the circuits $C_1, \ldots, C_m$ determines an order of passing through the arcs of the graph G once, and hence an Eulerian circuit. Two Eulerian circuits obtained in this way are identical if and only if the permutations of the cycles $C_1, \ldots, C_m$ are identical as cyclic permutations. The number of cyclic permutations of m objects is equal to $m!/m = (m-1)!$, since each cycle with m elements can be expressed in m different ways, by taking as first element each of the m elements. Thus one can obtain exactly $(m-1)!$ Eulerian circuits starting with the circuits $C_1, \ldots, C_m$ into which the circuit C is decomposed.

If two families of circuits $\{C_1, \ldots, C_m\}$ and $\{C'_1, \ldots, C'_m\}$ are different as sets, then the $(m-1)!$ Eulerian circuits obtained in this way from the first family will be different from all the $(m-1)!$ Eulerian circuits obtained from the second family. It follows that the total number of Eulerian circuits of the graph G is divisible by $(m-1)!$, which is an even number, since $m \geqslant 3$.

If G is a connected digraph with vertex set $X = \{x_1, \ldots, x_n\}$ such that $d^-(x_i) = d^+(x_i) = r_i$ for every $i = 1, \ldots, n$, then the theorem of van Aardenne-Ehrenfest and deBruijn states that the number of Eulerian circuits of G is equal to $\Delta_1 \prod_{k=1}^{n} (r_k - 1)!$, where Δ_1 is the number of spanning arborescences of G with the root x_1. [T. van Aardenne-Ehrenfest, N. G. deBruijn, *Simon Stevin*, **28** (1951), 203–217.]

7.5 First we show that every graph without isolated vertices and whose vertices have even degree can be expressed as the union of cycles without common edges. The union of two graphs

$$G_1 = (X_1, U_1) \quad \text{and} \quad G_2 = (X_2, U_2)$$

is defined to be

$$G_1 \cup G_2 = (X_1 \cup X_2, U_1 \cup U_2).$$

In the solution of Problem 6.3, it was shown that $d(x) \geqslant 2$ for every vertex x implies the existence of an elementary cycle C_1 in G. By suppressing the edges of the cycle C_1 one obtains a spanning graph of G which may contain some isolated vertices. After suppressing the isolated vertices one obtains a spanning subgraph G_1 of G whose vertices have even degrees and which does not contain isolated vertices. In fact, by suppressing the edges of the cycle C_1 some vertices of G have their degree reduced by 2. One can write

$$G = C_1 \cup G_1,$$

where G_1 has at least three edges less than G. G has all vertex degrees even and lacks isolated vertices.

By continuing this procedure the edges of G_1 are eventually exhausted and one finally obtains a single cycle C_k, that is, $G = C_1 \cup \cdots \cup C_k$ such that $C_1, \ldots, C_k$ do not have an edge pairwise in common.

Now let G be a graph all of whose vertices have even degree. Excluding isolated vertices, it has been shown that G can be expressed as the union of elementary cycles $C_1, \ldots, C_k$ which do not contain an edge pairwise in common. By selecting a sense for traversing each cycle C_p, one directs the edges of the cycle in the sense of their traversal. Thus the indegree and the outdegree of every vertex on the cycle C_p increase by one. Finally, after all the edges of the

Solutions

cycles C_p have been directed for $p=1,\ldots,k$, one obtains a directed graph which satisfies the given conditions.

7.6 If the graph G has an Eulerian cycle, then it follows that G is connected and has vertices of even degrees, since each traversal of a vertex uses two edges. Conversely, if G is connected and has even degrees, then by the preceding problem its edges can be directed so as to obtain a directed graph which satisfies $d^+(x)=d^-(x)$ for every vertex x. It follows from Problem 7.3 that the digraph so obtained has an Eulerian circuit, which corresponds to an Eulerian cycle in the graph G.

Suppose that G is connected and has $2k$ vertices of odd degree ($k\geqslant 1$). Let G_1 be the graph obtained from G by adding a new vertex which is joined by edges to all the $2k$ vertices of odd degree in G. It follows that G_1 is connected and has even degrees, and thus has an Eulerian cycle. After suppressing the additional vertex and the $2k$ edges incident with it, the Eulerian cycle decomposes into k walks which are disjoint with respect to their edges, do not use the same edge twice, and cover all the edges of G.

It can be seen that the sum of the degrees of the vertices of the graph is even, and hence the number of vertices of odd degree is also even.

7.7 Let K_n^* denote the complete directed graph with vertex set X of cardinality n and arc set of the form $\{(x,y)\mid x,y\in X \text{ and } x\neq y\}$. It follows that the graph K_n^* has $n(n-1)$ arcs and the set of arcs of the graph $\bar{G}$ is the complement of the set of arcs of the graph G with respect to the set of arcs of K_n^*. The number of Hamiltonian paths of K_n^* is equal to $n!$, since there is a bijection from the set of Hamiltonian paths of K_n^* onto the set of permutations of X. Let $u_1,\ldots,u_m$ denote the arcs of the graph $\bar{G}$, and let A_i denote the set of Hamiltonian paths of K_n^* which contain the arc u_i.

One can use the Principle of Inclusion and Exclusion (Problem 2.2) to obtain

$$h(G) = n! - |A_1 \cup \cdots \cup A_m|$$
$$= n! + \sum_{p=1}^{m} (-1)^p \sum_{1\leqslant i_1 < \cdots < i_p \leqslant m} |A_{i_1} \cap \cdots \cap A_{i_p}|. \qquad (1)$$

The term $|A_{i_1} \cap \cdots \cap A_{i_p}|$ represents the number of Hamiltonian paths of K_n^* which contain the arcs $u_{i_1},\ldots,u_{i_p}$. This term is nonzero only in the case of arcs $u_{i_1},\ldots,u_{i_p}$ which form elementary paths which are pairwise disjoint with respect to vertices. In this last case suppose that there exist r elementary paths which contain $p_1, p_2, \ldots, p_r$ vertices respectively. The number of connected components of the spanning subgraph of K_n^* generated by the arcs $u_{i_1},\ldots,u_{i_p}$ will be equal to

$$r + n - (p_1 + \cdots + p_r) = n - p,$$

since the number of arcs satisfies $p = (p_1 - 1) + \cdots + (p_r - 1) = p_1 + \cdots + p_r - r$.

Every Hamiltonian path which passes through the arcs $u_{i_1},\ldots,u_{i_p}$ defines a permutation of these $n-p$ components and conversely, and this correspondence is bijective. Thus if the arcs $u_{i_1}, \ldots, u_{i_p}$ form elementary paths without common vertices, then

$$|A_{i_1} \cap \cdots \cap A_{i_p}| = (n-p)!.$$

Finally, the numbers $|A_{i_1} \cap \cdots \cap A_{i_p}|$ are equal to 1 if and only if $p = n-1$ and the arcs $u_{i_1}, \ldots, u_{i_p}$ form a Hamiltonian path in the graph $\bar{G}$. Otherwise these numbers are even.

By using (1) one can conclude that $h(G) \equiv h(\bar{G}) \pmod 2$.

Let G be a graph, and consider the complete graph K_n which has $\binom{n}{2}$ edges and $n!/2$ Hamiltonian walks. Denote by $u_1, \ldots, u_m$ the edges of the graph $\bar{G}$, and by A_i the set of Hamiltonian walks of K_n which contain the edge u_i.

A formula analogous to (1) for $h(G)$ can be obtained immediately by replacing $n!$ with $n!/2$, which is even for every $n \geq 4$.

One can also show that

$$|A_{i_1} \cap \cdots \cap A_{i_p}| = \frac{(n-p)!}{2} 2^q,$$

if the edges $u_{i_1}, \ldots, u_{i_p}$ form q elementary walks without common vertices. The right-hand side is equal to zero otherwise. In fact, if one considers the $n-p$ connected components reduced to a single point, then one can form

$$\frac{(n-p)!}{2}$$

Hamiltonian walks, but each walk L from among the q elementary walks of the form

$$L = [x_1, \ldots, x_p]$$

can be inserted in each of the $(n-p)!/2$ Hamiltonian walks in two ways. One can choose the form $[x_1, \ldots, x_p]$ or $[x_p, \ldots, x_1]$. This results in distinct Hamiltonian walks which belong to the set $A_{i_1} \cap \cdots \cap A_{i_p}$.

Thus $|A_{i_1} \cap \cdots \cap A_{i_p}|$ is equal to 1 if $p = n-1$ and $q = 1$ (in other words $u_{u_1}, \ldots, u_{i_p}$ form a Hamiltonian walk in the graph G) and is even otherwise. In view of the fact that $n \geq 4$, one can conclude that $h(\bar{G}) \equiv h(G) \pmod 2$.

7.8 It can be shown that if the orientation of a unique arc $u = (a, b)$ of the tournament G is inverted, then a tournament G_1 is obtained such that $h(G_1) \equiv h(G)$. Let G_2 be the graph obtained from G by suppressing the arc u, and let G_3 be the graph obtained from G by adding the arc (b, a) with the same endpoints, but the opposite orientation to u.

Since $\bar{G}_2$ is obtained from G_3 by changing the direction of all its arcs, it follows that

$$h(\bar{G}_2) = h(G_3)$$

The results of the preceding problem then imply that

$$h(\bar{G}_2) \equiv h(G_2) \pmod 2.$$

Let $h_3(a, b)$, $h_3(b, a)$, and $h_3(0)$ denote respectively the number of Hamiltonian paths of G_3 which use the arc (a, b), the arc (b, a), or neither of these two arcs. Then

$$h_3(0) = h(G_2) \equiv h(G_3) = h_3(0) + h_3(a, b) + h_3(b, a),$$

from which one can conclude that $h_3(a, b) \equiv h_3(b, a)$ modulo 2. Hence

$$h(G) = h_3(0) + h_3(a, b) \equiv h_3(0) + h_3(b, a) = h(G_1).$$

Now let G be an arbitrary tournament. According to Problem 9.5, G contains a Hamiltonian path; say $(x_1, x_2, \ldots, x_n)$. By changing the direction of some arcs, it is possible to insure that the resulting tournament contains only arcs of the form (x_i, x_j) with $i < j$, i.e., it becomes a transitive tournament with a unique Hamiltonian path.

It has been shown that if the direction of an arc is changed, the parity of the number of Hamiltonian paths remains constant. This observation implies that $h(G) \equiv 1 \pmod{2}$. [L. Rédei, *Acta Litt. Szeged*, **7** (1934), 39–43.]

7.9 Let L be a longest elementary walk which begins at a vertex x_0 of the graph G:

$$L = [x_0, \ldots, x_k].$$

It follows that the vertex x_k is adjacent to x_{k-1} and to the other $d(x_k) - 1$ vertices which belong to the walk L, since otherwise L could be extended to a longer walk.

Suppose that x_k is adjacent to a vertex x_j such that $1 \leq j \leq k-2$. The walk

$$L_1 = [x_0, x_1, \ldots, x_j, x_k, x_{k-1}, \ldots, x_{j+1}]$$

is likewise a longest elementary walk which begins at x_0. The walk L_1 will be called a transformation of L. If L_1 is a transformation of L, then L is also a transformation of L_1. There are exactly $d(x_k) - 1$ transformations if x_k is a terminal vertex of L.

Let Y be the set of vertices of even degree in G, and let x_0 be an arbitrary vertex in G. It will be shown that there exists an even number of longest elementary walks L which begin at x_0 and have their last vertex in Y.

In order to prove this, let H be the graph defined as follows: The vertex set of H is the set of longest elementary walks which originate at x_0 in the graph G. Two vertices of H which correspond to two walks L_1 and L_2 are adjacent if and only if L_2 is a transformation of L_1.

The degree of a vertex of H which corresponds to the walk $L = [x_0, \ldots, x_k]$ is $d(x_k) - 1$. It follows that the set of longest elementary walks which originate at x_0 and terminate at a vertex of Y corresponds to the set of vertices of odd degree of the graph H. The number of vertices of odd degree is even for all graphs. This observation completes the proof of the property previously stated.

One can now prove that there exist an even number of Hamiltonian cycles which use a given edge $[x_0, y]$ in a graph G with all vertices of odd degree. To this end, consider the graph G_1 obtained from G by suppressing the edge $[x_0, y]$. In G_1 only the vertices x_0 and y have even degrees. Thus, by applying the previous property, it is found that there exist an even number of longest elementary walks which have endpoints x_0 and y. It follows that G has no Hamiltonian cycles which contain the edge $[x_0, y]$ (and thus an even number of such cycles) or else contains a positive even number of them. In fact, suppose that G contains a Hamiltonian cycle which passes through the edge $[x_0, y]$. One can then conclude that all maximal elementary walks with endpoints x_0 and y are Hamiltonian walks.

By the property just demonstrated, the number of these walks is even. But each of them generates, when taken together with the edge $[x_0, y]$, a Hamiltonian cycle which satisfies the given condition. [A. G. Thomason, *Annals of Discrete Math.*, 3 (1978), 259–268.]

7.10 First let G_1 be a spanning graph of G of the form $G_1 = (X, V)$ which has all its vertices of even degree, and let C be an elementary cycle of G with edge set W. It follows that the spanning graph $G_2 = (X, V \triangle W)$ also has all of its vertices of even degree. ($A \triangle B$ is the symmetric difference of A and B.) In fact, if the cycle C does not pass through the vertex x, then the degree of x in G_2 is even.

Suppose that $(u, x) \in W$, $(x, v) \in W$. If:

(1) $(u, x) \notin V$, $(x, v) \notin V$, then $d_{G_2}(x) = d_{G_1}(x) + 2$;
(2) $(u, x) \in V$, $(x, v) \notin V$ or $(u, x) \notin V$, $(x, v) \in V$, then $d_{G_2}(x) = d_{G_1}(x)$;
(3) $(u, x) \in V$, $(x, v) \in V$, then $d_{G_2}(x) = d_{G_1}(x) - 2$ for every $x \in V$.

The property will be established by induction on the number m. Since G is connected, it follows that $m \geqslant n - 1$. If $m = n - 1$, then G is a tree. The number of spanning trees of G of even degree is in this case equal to $2^{n-1-n+1} = 1$, since the unique spanning graph with this property is $(X, \varnothing)$.

In fact, suppose that there exists a spanning graph G_1 of G of the form (X, V) with $V \neq \varnothing$ and all vertices of even degree. It has been seen in the solution to Problem 7.5 that by suppressing the isolated vertices of G_1 the resulting graph can be expressed as a union of cycles without common vertices. This implies that G contains cycles, which contradicts the fact that G is a tree.

Suppose now that the property is true for all connected graphs with n vertices and at most p edges ($p \geqslant n - 1$). Let G be a connected graph with n vertices and $p+1$ edges. If by suppressing an arbitrary edge of the graph G, it becomes disconnected, then G is a tree and $p+1 = n-1$, which contradicts the inequality $p \geqslant n-1$. It follows that G contains an edge u whose elimination produces a connected graph G_u with p edges. The spanning graphs of even degree of G which do not contain the edge u coincide with the spanning graphs of even degree of G_u, and by the induction hypothesis their number is equal to 2^{p-n+1}.

It will now be shown that G contains the same number of spanning graphs of even degree which contain the edge u. Since G_u is connected, it follows that there exists an elementary walk which joins the endpoints of the edge u. This walk, together with the edge u, forms an elementary cycle C in G which contains the edge u. Let $G_1 = (X, U_1)$ be a spanning graph of G of even degree. It has been

Solutions

seen that $G_2 = (X, U_1 \triangle W)$ has even degree, where W is the set of edges of the cycle C. Also, if G_1 contains the edge u, then since $u \in W$, it follows that G_2 does not contain the edge u and hence is a spanning tree of even degree of G_u. The correspondence f which associates G_2 to G_1 is injective. Since

$$(A \triangle B) \triangle B = A$$

for every two sets A and B, it follows that f is also an involution, that is, $f(f(G_1)) = G_1$.

It follows that for every spanning graph G_2 of G_u of even degree there exists a spanning graph $G_1 = f(G_2)$ of even degree of G which contains u and is such that $G_2 = f(G_1)$. The mapping f is therefore surjective and hence bijective. Thus the number of spanning graphs of even degree of G is equal to $2^{p-n+1} + 2^{p-n+1} = 2^{(p+1)-n+1}$. This completes the proof by induction of the property.

7.11 The property will be proved by induction on the number n of vertices of the graph G. For $n = 1$ let $X_1 = \{x_1\}$ and $X_2 = \varnothing$, and for $n = 2$ let $X_1 = \{x_1\}$ and $X_2 = \{x_2\}$. Suppose that the property is true for all graphs with at most n vertices, and let G be a graph with $n+1$ vertices.

If all the vertices of G have even degree, then let $X_1 = X$ and $X_2 = \varnothing$. Otherwise let a be a vertex of odd degree in G, and denote by A the set of vertices in G adjacent to a. Define the graph G_1 as having the vertex set $Y = X \setminus \{a\}$. The pair $[x, y]$ is an edge in G_1 if and only if $x, y \in A$ and $[x, y]$ is not an edge of the graph G, or at least one of the vertices x and y does not belong to the set A and $[x, y]$ is an edge of the graph G. By the induction hypothesis there is a partition $Y = Y_1 \cup Y_2$ such that Y_1 and Y_2 both induce subgraphs of G_1 of even degree. But

$$|A \cap Y_1| + |A \cap Y_2| = |A| \equiv 1 \pmod{2},$$

so that one can suppose, for example, that $|A \cap Y_1|$ is even and $|A \cap Y_2|$ is odd.

Let

$$Z_1 = Y_1 \cup \{a\}, \quad Z_2 = Y_2.$$

We show that Z_1 and Z_2 induce subgraphs of even degree of G.

Let x be a vertex in Z_1. If $x \notin A$ and $x \neq a$, then its degree $d_{Z_1}(x)$ in the subgraph of G induced by Z_1 is even, by the definition of an edge in the graph G_1. If $x = a$, then its degree $d_{Z_1}(a) = |A \cap Y_1|$, which has been assumed to be an even number. Let $x \in A$; denote by $d_1(x)$ the degree of x in the subgraph of G_1 induced by Y_1, and by $d_2(x)$ the degree of x in the subgraph of G_1 induced by $Y_1 \cap A$. Further suppose that $d_3(x)$ is the degree of x in the subgraph of G induced by $Y_1 \cap A$. It follows from the definition of the graph G_1 that $x \in A$ is adjacent to $y \in Y_1 \cap A$ in the graph G if and only if x is not adjacent to y in the graph G_1, and hence

$$d_3(x) = |Y_1 \cap A| - 1 - d_2(x). \tag{1}$$

On the other hand, it also follows from the definition of G_1 that

$$d_{Z_1}(x) = d_1(x) - d_2(x) + d_3(x) + 1. \qquad (2)$$

The last term is 1 because $x \in A$ is adjacent to $a \in Z_1$ in the graph G.

From (1) and (2) it follows that

$$\begin{aligned}d_{Z_1}(x) &= d_1(x) - d_2(x) + \{|Y_1 \cap A| - 1 - d_2(x)\} + 1 \\ &= d_1(x) - 2d_2(x) + |Y_1 \cap A|,\end{aligned}$$

and in this sum each term is even.

If $x \in Z_2$, it can be shown analogously that if $x \notin A$ then $d_{Z_2}(x)$ is even. If $x \in A$, then

$$d_{Z_2}(x) = d_4(x) - d_5(x) + d_6(x)$$

where $d_4(x)$ is the degree of x in the subgraph of G_1 induced by $Y_2 = Z_2$. The number $d_5(x)$ is the degree of x in the subgraph of G_1 induced by $Y_2 \cap A$, and $d_6(x)$ is the degree of x in the subgraph of G induced by $Y_2 \cap A$. A similar argument shows that

$$d_6(x) = |Y_2 \cap A| - 1 - d_5(x),$$

which implies the equation

$$d_{Z_2}(x) = d_4(x) - 2d_5(x) + (|Y_2 \cap A| - 1).$$

This number is always even, since $d_4(x)$ is even by the induction hypothesis and $|Y_2 \cap A|$ is odd. The property is thus found to be true for every n.

It will now be shown that there exists a partition

$$X = X_1 \cup X_2$$

such that the degrees of the vertices of the subgraph generated by X_1 are even and the degrees of the vertices of the subgraph induced by X_2 are odd. To this end add to the graph G a new vertex y, which is adjacent to all the vertices of X. Let G_1 be the graph thus obtained. By the previous result, there exists a partition $Y_1 \cup Y_2$ of the set of vertices $X \cup \{y\}$ where Y_1 and Y_2 induce subgraphs of even degree in G_1. If, for example, $y \in Y_2$, then by denoting $X_1 = Y_1$ and $X_2 = Y_2 \setminus \{y\}$ one obtains the desired partition of the vertex set of G. [W. K. Chen, *SIAM J. Appl. Math.*, **20** (1971), 526–529.]

7.12 It is clear that if C contains only nonempty subsets, then $C_1 = C \cup \{\varnothing\}$ satisfies the given condition, and hence one can assume that $\varnothing \in C$.

It follows that $C \neq \{X\}$, since otherwise there exists a proper subset $Y \subset X$ which has elements in common with X, and hence with an odd number of subsets from C. Thus there exists a subset A in C such that $|A| = a$ is minimum and $a \geq 1$. Because $a \leq n-1$, it follows that $X \setminus A$ is a proper subset of X, and $X \setminus A$ intersects all sets of C but does not intersect A. By the hypothesis $|C| - 1$ is an even number, and hence C contains an odd number of subsets of X.

If $x \in X$ and $\{x\} \notin C$, it follows that $X \setminus \{x\}$ intersects all sets in C, and hence an odd number of sets, which contradicts the hypothesis. It follows that C contains all one-element subsets of X. Suppose that every subset $B \subset X$ such that $1 \leq |B| < k < n$ has the property $B \in C$. Let $A \subset X$ be such that $|A| = k$. If $A \notin C$, one can conclude that $X \setminus A$ intersects all but $2^{|A|} - 2 = |P(A) \setminus \{A, \varnothing\}|$ subsets from C.

Since $k \geq 1$, it follows that $2^{|A|} - 2 = 2^k - 2 \equiv 0 \pmod{2}$, and hence $X \setminus A$ intersects an odd number of subsets from C, which contradicts the hypothesis. This implies that $A \in C$. It has thus been proved by induction that every subset $A \subset X$ satisfying $1 \leq |A| \leq n-1$ belongs to C.

Suppose that $X \notin C$. Then $|C| = 2^n - 2 \equiv 0 \pmod{2}$, which is a contradiction. It follows that $C = P(X) \setminus \{\varnothing\}$ and $C_1 = P(X)$ are the solutions of the problem, since both satisfy the condition in the statement of the problem.

In fact, if $Y \subset X$ and $|Y| = k$, $1 \leq k \leq n-1$, then Y intersects $2^n - 2^{n-k}$ subsets in both collections C and C_1, and this number is even. [A. Adelberg, Problem E 2887, *American Mathematical Monthly*, **88**(5) (1981), 349.]

CHAPTER 8

8.1 Suppose that G is not connected, and let C_1 be a component which does not contain the vertex x_n. Let $|C_1| = k$, and let $x_{i_1}, \ldots, x_{i_k}$ be vertices which it contains, where

$$1 \leq i_1 < \cdots < i_k < n.$$

The component which contains x_n also contains all vertices adjacent to x_n, and therefore it contains at least $d_n + 1$ vertices. It follows that

$$k = |C_1| \leq n - (d_n + 1).$$

Since $k \leq i_k$, one can show that $d_k \leq d_{i_k} \leq k-1$, so that the vertices adjacent to x_{i_k} are all found in the component C_1. This is a contradiction, since by hypothesis $k \leq n - d_n - 1$ implies that $d_k \geq k$, and the property is therefore established.

8.2 It will first be shown that G contains a spanning graph A which is a tree. If G does not contain a cycle, then G itself is a tree and one can take $A = G$. Otherwise G contains at least one cycle C_1.

Now suppress an arbitrary edge u_1 of the cycle C_1. The result is a spanning graph G_1 of G. If G_1 does not contain a cycle, one can take $A = G_1$, since in this case G_1 is connected and does not contain a cycle and is hence a tree. Otherwise suppress an edge u_2 of a cycle C_2 of G_1, and so on. This process cannot continue indefinitely, since G contains at most $\binom{n}{2}$ edges. Finally one obtains a connected graph G_r without cycles, and A is defined to be equal to G_r.

It has been seen that the tree A has at least one vertex x_1 of degree 1 (see the solution of Problem 6.3). If $k = n$ then choose $H = G$. Otherwise suppress the vertex x_1 and the edge incident to x_1 in the tree A. This produces a new tree A_1, since it follows immediately that A_1 is connected and contains no cycles. Repeat this process so that the vertices $x_1, x_2, \ldots, x_{n-k}$ are eliminated. The result is a tree A_{n-k} with k vertices. The subgraph H is defined to be the subgraph induced by the k vertices of the tree A_{n-k}. The graph H is connected, since it contains the connected spanning tree A_{n-k} with the same vertex set.

8.3 Denote the number of edges and vertices of the p components by $m_1, \ldots, m_p$ and $n_1, \ldots, n_p$ respectively. Since each component is in fact a connected graph, it follows that

$$m_i \geq n_i - 1 \qquad (1)$$

for $i = 1, \ldots, p$. Equality holds only for components which are trees.

By adding inequalities (1) for $i = 1, \ldots, p$, one finds that $m \geq n - p$, since no two connected components have a common vertex and hence have no common edge.

8.4 Suppose that there exist two elementary walks of maximal length L_1 and L_2 which have no common vertex. Since G is connected, there are two vertices $x_1 \in L_1$ and $x_2 \in L_2$ which are joined by a walk Q which has in common with L_1 and L_2 only the endpoints x_1 and x_2. The vertex x_i divides the walk L_i into two subwalks L_{i1} and L_{i2} (one of which may possibly be empty) for $i = 1, 2$.

Let $l(L)$ be the length of the walk L, that is, the number of its edges. One can assume that $l(L_{i1}) \geq l(L_{i2})$ and $l(L_{11}) \geq l(L_{21})$. It follows that

$$l(L_{11}, Q, L_{21}) > l(L_{11}) + l(L_{21}) \geq 2l(L_{21}) \geq l(L_{21}) + l(L_{22}) = l(L_2),$$

and hence the walk (L_{11}, Q, L_{21}) is longer than L_2, which contradicts the hypothesis.

Suppose that G is a tree, and let L_1, L_2 be two walks of maximal length in G. These walks have at least one vertex in common. In fact the common vertices of L_1 and L_2 form a walk Q (which may reduce to a single vertex). Thus vertices of $L_1 \cup L_2$ generate a subgraph of the following form: There exist two walks L'_1, L'_2 which originate at one terminal vertex of Q and two walks L''_1 and L''_2 which originate at the other terminal vertex of Q, such that

$$L_1 = (L'_1, Q, L''_1) \quad \text{and} \quad L_2 = (L'_2, Q, L''_2).$$

It follows that $l(L'_1) = l(L'_2)$, since if, for example, $l(L'_1) > l(L'_2)$, then the walk (L'_1, Q, L''_2) would be longer than L_2, which would contradict the hypothesis.

In the same way it is seen that $l(L''_1) = l(L''_2)$.

Since (L'_1, L'_2) is a walk, its length is at most equal to the length of a maximal walk, and hence $2l(L'_1) \leq l(L_1)$ or $l(L'_1) \leq \frac{1}{2}l(L_1)$; analogously, $l(L''_1) \leq \frac{1}{2}l(L_1)$. Thus the median point or the two adjacent median points of the walk L_1 belong to the walk Q and hence to L_2. Since L_2 can be chosen arbitrarily, it follows that the median point(s) of L_1 [depending on whether $l(L_1)$ is even or odd] belongs to the intersection of all the maximal walks of the tree G.

If the connected graph G contains cycles, this property does not hold.

Solutions

8.5 In order to prove the necessity, let G be a bipartite graph which has a bipartition $X = A \cup B$. Every elementary cycle of G has the form

$$[a_1, b_1, a_2, b_2, \ldots, a_k, b_k, a_1]$$

where $a_i \in A$ and $b_i \in B$ for $1 \leq i \leq k$. The length of this cycle is $2k$, and thus an even number.

Now suppose that every elementary cycle of G has an even number of vertices. Color the vertices of G with two colors so that each two vertices joined by an edge have different colors. The coloring is performed as follows: Start with a vertex x_1 which is colored a. The vertices adjacent to x_1 are colored b. Then the vertices adjacent to the vertices which are colored b will be colored a, and so on. In this way no vertex in the connected component which contains x_1 will be colored once with a and once with b, which would be a contradiction.

In fact, suppose that the vertex z is colored a in a walk $L_1 = [x_1, y_1, \ldots, y_k, z]$ of even length and is colored b in a walk $L_2 = [x_1, t_1, \ldots, t_s, z]$ of odd length. If L_1 and L_2 have only endpoints in common, then the union of the edges of L_1 and L_2 forms an elementary cycle with an odd number of edges, which contradicts the hypothesis. Otherwise L_1 and L_2 have a given number of vertices in common, which implies that the union of the edges of L_1 and L_2 generates a spanning subgraph of G consisting of walks and cycles. It will be shown by induction on the number of edges of L_1 that the fact that G does not contain odd elementary cycles implies that the lengths of L_1 and L_2 must have the same parity.

If $l(L_1) = 1$ and $l(L_2)$ is even, then the elementary cycle consisting of the edge of L_1 and the edges of L_2 is odd, which contradicts the hypothesis. It follows that $l(L_2)$ is odd. Suppose that the property is true for every two walks L_1 and L_2 of the indicated form such that $l(L_1) \leq t$, and let L_1 and L_2 be two walks with the same endpoints such that $l(L_1) = t + 1$.

It has been assumed that L_1 and L_2 also have a vertex y in common other than x_1 and z. Let L_1' and L_2' be subwalks of L_1 and L_2 respectively contained between x_1 and y. Similarly let L_1'', L_2'' be subwalks of L_1 and L_2 which are located between y and z. Since $l(L_1') \leq t$, it follows from the induction hypothesis that $l(L_1')$ and $l(L_2')$ have the same parity. If x_1 is replaced by y, then since $l(L_1'') \leq t$, one can conclude similarly that $l(L_1'')$ and $l(L_2'')$ have the same parity. Since $l(L_1) = l(L_1') + l(L_1'')$ and $l(L_2) = l(L_2') + l(L_2'')$, it follows that $l(L_1)$ and $l(L_2)$ have the same parity. Thus the vertices of the connected component which contains x_1 can be colored in this way by two colors. By continuing this process for all connected components of G one obtains a coloring by two colors of the vertices of G such that each two adjacent vertices have different colors. It follows that G is bipartite by taking the set A to be the set of vertices colored a and the set B to be the set of those vertices colored by b.

It also follows that every graph G without odd cycles, which has at least one edge, is bichromatic, that is, $\chi(G) = 2$.

8.6 The vertex of degree 9 must be adjacent to all the other vertices of the graph and hence adjacent to the vertices of degree 1. It follows that the vertex of degree 7 can be adjacent to only $9-3=6$ vertices, which establishes a contradiction. Thus there does not exist a graph with the desired property.

8.7 The necessity of the condition is immediate, since $d_1 + \cdots + d_n$ is twice the number of edges and hence an even number. Also $d_n \leqslant d_1 + \cdots + d_{n-1}$, since every edge which is incident to x_n is also incident to one of the remaining vertices $[d_n = d(x_n)]$.

The sufficiency will be proven by induction on $d_1 + \cdots + d_n$. If $d_1 + \cdots + d_n = 2$, it follows that $d_1 = \cdots = d_{n-2} = 0$ and $d_{n-1} = d_n = 1$, and thus the desired graph contains one edge and $n-2$ isolated vertices.

Suppose the property is true for all sequences of numbers $d_1, \ldots, d_n$ which satisfy (1) and (2) and also satisfy $d_1 + \cdots + d_n \leqslant 2d$ ($d \geqslant 2$). Now suppose that $d_1 + \cdots + d_n = 2d + 2$ and conditions (1) and (2) are satisfied. Two cases will be considered.

(a) Suppose that $d_{n-2} < d_n$. It then follows that $d_n - 1$ is the largest number in the sequence $d_1, d_2, \ldots, d_{n-2}, d_{n-1} - 1, d_n - 1$ and it remains to show that:

(3) $d_1 + \cdots + d_{n-2} + (d_{n-1} - 1) + (d_n - 1) \equiv 0 \pmod{2}$;

(4) $d_1 + \cdots + d_{n-2} + d_{n-1} - 1 \geqslant d_n - 1$.

These two properties follow immediately from (1) and (2).

(b) Now let $d_{n-2} = d_n$. It follows that $d_{n-1} = d_n$. Condition (3) is again satisfied, and condition (2) becomes:

(5) $d_1 + \cdots + d_{n-3} + (d_{n-1} - 1) + (d_n - 1) \geqslant d_{n-2}$,

since $d_{n-2} = \max(d_1, \ldots, d_{n-2}, d_{n-1} - 1, d_n - 1)$. However, condition (5) also holds. In fact if $d_{n-2} = 1$, condition (1) implies that the left-hand side of inequality (5) is odd and thus is at least equal to 1, since in this case $d_{n-1} = d_n = 1$. If $d_{n-2} \geqslant 2$, then $d_{n-1} - 1 + d_n - 1 \geqslant d_{n-2}$ because $d_{n-2} = d_{n-1} = d_n$.

It follows that the numbers $d_1, \ldots, d_{n-2}, d_{n-1} - 1, d_n - 1$ whose sum is $2d$ satisfy conditions (1) and (2), and by the induction hypothesis there is a multigraph with n vertices whose degrees are equal to the given sequence. By joining the vertices of degrees $d_{n-1} - 1$ and $d_n - 1$ by a new edge one obtains a graph whose degrees are exactly $d_1, \ldots, d_n$.

8.8 Suppose that there exists a regular graph G of degree k with n vertices. Then $kn = 2m$, where m is the number of edges of G and the degree k cannot be greater than $n-1$. Thus two necessary conditions have been found which the number n must satisfy:

(a) $nk \equiv 0 \pmod{2}$;

(b) $n \geqslant k+1$.

It will be shown that if (a) and (b) are satisfied, then there exists a regular graph of degree k with n vertices.

Two cases will be considered:

Solutions

(1) k is even. Consider a regular polygon with n vertices. Join by an edge neighboring vertices, vertices which are two apart, ..., vertices which are $k/2$ apart. Since $k < n$, it follows that $k/2 < n/2$, and thus by construction each vertex has degree k and there are no multiple edges (see Figure 8.2).

(2) k is odd. It follows from condition (a) that n is even. The regular graph of degree $k-1$ with n vertices can be drawn as in case (1). Condition (b) assures that

$$\frac{k-1}{2} \leqslant \frac{n-2}{2} = \frac{n}{2} - 1,$$

and thus the endpoints of the longest diagonals of the polygon are not joined by an edge. Thus by joining diametrical vertices of the polygon by an edge, each vertex will have degree k, and hence a regular graph of degree k with n vertices is obtained.

8.9 Let x be a vertex of the graph G which is adjacent to the vertices $x_1, \ldots, x_m$ in G. Since G does not contain a complete subgraph with three vertices, it follows that the vertices $x_1, \ldots, x_m$ are not pairwise adjacent. It is also the case that every two vertices x_i, x_j with $1 \leqslant i, j \leqslant m$ and $i \neq j$ are adjacent to a vertex $z \neq x$. Hence z and x are not adjacent, since G does not contain a triangle.

The mapping which associates with every pair $\{x_i, x_j\}$ the vertex z which is not adjacent to x is injective. In fact, if such a z corresponds to two different pairs, then there are at least three vertices adjacent to both z and x, which contradicts the hypothesis. Since the number of pairs $\{x_i, x_j\}$ is equal to $\binom{m}{2}$, it follows that the number of vertices which are not adjacent to x is greater than or equal to $\binom{m}{2}$.

But the hypothesis implies that for each vertex z which is not adjacent to x there are two vertices from $x_1, \ldots, x_m$ which are adjacent to z and to x. Call

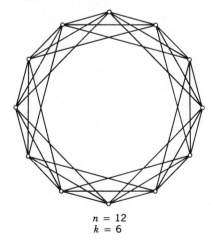
$n = 12$
$k = 6$

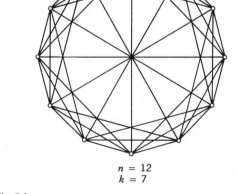
$n = 12$
$k = 7$

Fig. 8.2

these vertices x_r and x_s. The mapping which associates to each vertex z which is nonadjacent to x the pair of vertices $\{x_r, x_s\}$ is also injective.

Suppose that the same pair $\{x_r, x_s\}$ corresponds to two vertices $z_1 \neq z_2$. It follows that there exist at least three vertices x, z_1, z_2 adjacent to x_r and x_s, which again contradicts the hypothesis.

Thus the number of vertices z which are not adjacent to x is at most equal to $\binom{m}{2}$. From the two opposite inequalities it follows that the number of vertices which are not adjacent to x is equal to $\binom{m}{2}$, and hence

$$n = 1 + m + \binom{m}{2}.$$

This implies that

$$m = \frac{-1 + \sqrt{8n-7}}{2}.$$

It follows that $8n - 7 = k^2$ and thus $n = (k^2 + 7)/8$, which implies that $k = 2p + 1$, $n = \binom{p+1}{2} + 1$, and $m = p$. Thus G is a regular graph of degree p. For $p = 2$ one has $n = 4$ and $G = K_{2,2}$.

A necessary condition for the existence of this graph is that np is an even number, since it represents twice the number of edges in the graph.

8.10 Carry out the following construction: Suppose that a regular graph $G(r, g)$ of degree r and girth g has been constructed. Consider also a graph $G(r', g-1)$, where r' is equal to the number of vertices of the graph $G(r, g)$. Replace each vertex of the graph $G(r', g-1)$ by r' vertices of degree 1 (see Figure 8.2).

Now identify these r' vertices with the vertices of a copy of the graph $G(r, g)$. In Figure 8.3 the graph $G(r, g)$ is C_5, which is obtained for $r = 2$ and $g = 5$. Denote by G_1 the graph obtained in this way. The graph G_1 is regular of degree $r + 1$ by construction. It will now be shown that $g(G_1) = g$. Consider an elementary cycle of minimal length in a copy of the graph $G(r, g)$. Such a cycle has

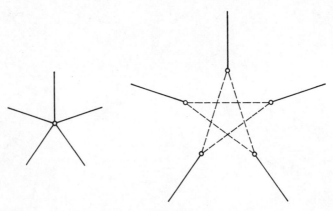

Fig. 8.3

length g, and hence $g(G_1) \leqslant g$. Since every other elementary cycle in a copy of $G(r, g)$ has length greater than or equal to g, one can let C be an elementary cycle of minimal length ($=s$), which is not contained in any copy of the graph $G(r, g)$. Replace each copy of $G(r, g)$ by a single vertex while preserving the edges incident to the vertices in $G(r, g)$. In this way the graph G_1 is transformed into the graph $G(r', g-1)$. The cycle C is transformed into a nonempty spanning subgraph of $G(r', g-1)$ with all its vertices of even degree. This is because the cycle C has a number of entries equal to the number of exits of each copy of $G(r, g)$ if one fixes a sense of traversal for the cycle C.

Every nonempty spanning subgraph all of whose vertices have even degrees contains an elementary cycle. In fact, if this were not the case, then a connected component of this subgraph would have the degrees of its vertices even and would not contain a cycle. It would thus be a tree. But this contradicts the fact that every tree has at least two vertices of degree one, that is, two vertices of odd degree.

Let C_1 be the elementary cycle contained in the image of the cycle C obtained by contracting each copy of $G(r, g)$ to a single vertex. It can also be seen that C contains at least one edge from a copy of $G(r, g)$ by the construction of the graph G_1. Thus C_1 has at most $s-1$ edges. One can deduce from this that

$$s-1 \geqslant l(C_1) \geqslant g-1,$$

since the graph $G(r', g-1)$, which contains the cycle C_1, has girth equal to $g-1$.

The fact that every cycle of G_1 has length at least equal to g implies that $s \geqslant g$ or $G_1 = G(r+1, g)$.

The existence of the graph $G(r, 3)$ for every $r \geqslant 2$ is demonstrated by considering the complete graph K_{r+1}. Examples of graphs of the form $G(2, g)$ where $g \geqslant 2$ are given by elementary cycles with g vertices.

By using the construction just described one can prove the existence of the graphs $G(3, 4), G(4, 4), G(5, 4), \ldots, G(r, 4)$ for every $r \geqslant 2$; $G(3, 5), G(4, 5), G(5, 5), \ldots$, $G(r, 5)$ for every $r \geqslant 2$; and so on. Thus one can show by induction that there exists a graph $G(r, g)$ for every $r \geqslant 2$ and $g \geqslant 3$.

8.11 Suppose that g is odd. Let x be an arbitrary vertex of the graph G, and denote by S_i the set of vertices which are found at a distance i from x, for

$$i = 0, 1, \ldots, \frac{g-1}{2}.$$

For each vertex z in the set S_i there is exactly one edge which joins it to a vertex in S_{i-1}. In fact, there exists at least one edge by the definition of the sets S_i. The existence of two such edges would lead to the existence of two walks from z to x, each of length i. But these form a cycle, and hence there is an elementary cycle of length less than g, and this contradicts the hypothesis. Since the degree of each vertex is equal to r, it can be seen that

$$|S_{i+1}| = (r-1)|S_i| \qquad \text{for} \quad i = 1, \ldots, \frac{g-3}{2}.$$

This then leads to the fact that

$$n \geq |S_0| + |S_1| + \cdots + |S_{(g-1)/2}|$$
$$= 1 + r + r(r-1) + \cdots + r(r-1)^{(g-3)/2}.$$

Suppose now that g is even, and consider two adjacent vertices x and y. Let S_i be the set of vertices at a distance i from the set $\{x, y\}$ for $i = 1, \ldots, g/2 - 1$. It follows that $|S_1| = 2(r-1)$ and $|S_{i+1}| = (r-1)|S_i|$ for $i = 1, \ldots, g/2 - 2$. From this one can conclude that

$$n \geq |S_0| + |S_1| + \cdots + |S_{g/2-1}| = 2 + 2(r-1) + \cdots + 2(r-1)^{g/2-1}.$$

8.12 (a) Let x_1 be a vertex of a regular graph G of degree 3 and girth $g(G) = 4$, and suppose that x_2, x_3, x_4 are adjacent to x_1. The vertices x_2, x_3, x_4 are not pairwise adjacent, since in that case one would have $g(G) = 3$. Denote by x_5, x_6 the vertices which, together with x_1, are adjacent to x_2. It follows that the number of vertices in G is at least equal to 6. Since G must be regular of degree 3, the vertices x_3 and x_4 are adjacent to x_5 and x_6, that is, $G = K_{3,3}$.

(b) Let $[x_1, x_2, \ldots, x_5, x_1]$ be a shortest elementary cycle of the graph G. Denote by y_i the third vertex adjacent to x_i for $i = 1, \ldots, 5$.

The vertices y_i are pairwise distinct, and are all different from the vertices x_j, since otherwise one would have $g(G) \leq 4$. It follows that the graph G has at least 10 vertices. If G has exactly 10 vertices then the only vertices at distance 4 from y_1 are y_3 and y_4 and hence y_1 must be adjacent to y_3 and y_4, in addition to x_1, because otherwise $g(G) \leq 4$. Similarly, it is the case that y_2 is adjacent to y_4 and y_5, y_3 is adjacent to y_1 and y_5, y_4 is adjacent to y_1 and y_2, and y_5 is adjacent to y_2 and y_3. In this way one obtains the Petersen graph (Figure 8.4).

8.13 Suppose that for every vertex x the subgraph G_x is not connected. Let $L = [x_1, \ldots, x_m]$ be an elementary walk of maximal length in the graph G. By hypothesis G_{x_1} is not connected. Denote by C_1 a component of this subgraph which does not contain the walk $[x_2, \ldots, x_m]$.

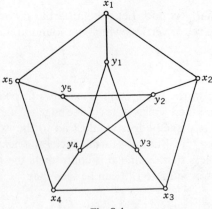

Fig. 8.4

Solutions

Since G is connected, there is an edge which joins a vertex $y \in C_1$ with the vertex x_1. This can be seen because the vertices of C_1 cannot be joined to the vertices of the other connected components of the subgraph G_{x_1}. It follows that $y \notin \{x_2, \ldots, x_m\}$ and thus the walk

$$[y, x_1, \ldots, x_m]$$

is an elementary walk which is longer than L, which contradicts the hypothesis that L is an elementary walk of maximal length.

Thus there exists a vertex x for which the subgraph G_x is connected. One can also see that x may be any terminal vertex of a spanning tree of G.

If G is an elementary circuit, then by suppressing an arbitrary vertex one obtains a subgraph which no longer is strongly connected, although G has this property.

8.14 Suppose that G is strongly connected and has vertex set X. Let $A \neq \varnothing$, $a \in A$, and $b \in X \setminus A$. It follows that there is a path in G of the form $(a, \ldots, b)$. Since $a \in A$ and $b \notin A$, there will exist at least one arc of this path of the form (x, y) where $x \in A$ and $y \notin A$.

It will now be shown that if for every nonempty set A there is at least one arc of the form (x, y) with $x \in A$ and $y \notin A$, then G is strongly connected.

Let a, b be two distinct vertices of G, and suppose that there is no path of the form $(a, \ldots, b)$. Denote by A the set of vertices of G which are endpoints of paths which originate at a. It follows that $b \notin A$ and there does not exist an arc of the form (x, y) with $x \in A$ and $y \notin A$. For otherwise every path of the form $(a, \ldots, x)$, extended by the arc (x, y) would produce a path from a to y. But in this case $y \in A$, which contradicts the hypothesis. Thus there does not exist any arc of the form (x, y), and this again contradicts the hypothesis. It follows that G is strongly connected.

Denote by G_1 the graph obtained from G by changing the direction of all the arcs of G. It follows from the definition of strong connectedness that G is strongly connected if and only if G_1 is strongly connected. It follows that G is strongly connected if and only if for each nonempty subset of vertices A there exists at least one arc of the form (y, x) where $y \notin A$ and $x \in A$.

8.15 Suppose that G contains a circuit, and hence an elementary circuit. Further assume that the following elementary circuit of G has a minimal number of arcs:

$$C = (x_1, x_2, \ldots, x_k, x_1),$$

where the vertices $x_1, \ldots, x_k$ are pairwise distinct. The property is clear for $k = 3$.

Suppose now that $k \geq 4$. Since G is complete and antisymmetric, there exists an arc (x_1, x_{k-1}), and hence a circuit with three vertices (x_1, x_{k-1}, x_k, x_1), which contradicts the hypothesis $k \geq 4$; or else there exists an arc (x_{k-1}, x_1), in which case one obtains a circuit which is shorter than C:

$$C_1 = (x_1, x_2, \ldots, x_{k-1}, x_1),$$

which contradicts the hypothesis that C contains a minimal number of arcs. The property is thus established by contradiction.

8.16 There is a unique arc between each two vertices of a tournament T with n vertices. It follows from this that $r_i + s_i = n - 1$ for $i = 1, \ldots, n$. Similarly $\sum_{i=1}^{n} r_i = \sum_{i=1}^{n} s_i = \binom{n}{2}$, since each sum represents the number of arcs of T and each arc (x, y) contributes exactly 1 to each sum. Thus

$$\sum_{i=1}^{n} r_i^2 = \sum_{i=1}^{n} (n - 1 - s_i)^2$$

$$= n(n-1)^2 - 2(n-1) \sum_{i=1}^{n} s_i + \sum_{i=1}^{n} s_i^2 = \sum_{i=1}^{n} s_i^2.$$

8.17 Let x be a vertex of G with maximal outdegree, that is,

$$d^+(x) = \max_{t \in X} d^+(t), \tag{1}$$

where X is the vertex set of the graph G. It will be shown that this vertex satisfies the conditions of the statement of the problem.

Let $y \in X$ and $y \neq x$. If G does not contain the arc (x, y), then it must contain (y, x). This follows from the fact that a tournament is complete. Suppose now that the vertex y cannot be reached by a path of length 2 which originates at x. Thus for every arc (x, z) there is an arc (y, z), since otherwise there would exist an arc (z, y) and hence a path of length 2 from x to z, namely (x, y, z), which is contrary to the hypothesis.

Finally, for every arc which originates at x [of the form (x, z)], there is an arc (y, z) which originates at y and is also an arc (y, x) which originates at y and terminates at x.

It follows that $d^+(y) > d^+(x)$, and this contradicts (1). The vertex x which was defined in (1) therefore satisfies the given condition.

8.18 The property will be established by induction on the number of vertices of the graph G. If G consists of two isolated vertices, let $S = \{x, y\}$. Otherwise there is an arc (y, x) and $S = \{y\}$.

Suppose that the property is true for all graphs with at most n vertices, and take G to be a graph with $n + 1$ vertices. Let x be a vertex of G, and denote by A the set $\{z | (x, z)$ is an arc of the graph $G\}$. Denote by G_1 the subgraph obtained from G by suppressing the vertices of the set $A \cup \{x\}$. It follows from the induction hypothesis that G_1 contains a set of pairwise nonadjacent vertices S_1 with the following property: Every vertex $z \notin S_1$ can be reached by starting from a vertex $y \in S_1$ and traversing a path of G_1 of length at most 2. Two cases will be studied.

(a) The vertices of $S_1 \cup \{x\}$ are not pairwise adjacent. Let $S = S_1 \cup \{x\}$. Every vertex z in G_1 such that $z \notin S_1$ can be reached by starting at a vertex $y \in S_1$ and traversing a path of length at most 2. If $z \in A$, then there exists a path (x, z) of length equal to 1.

(b) There is a vertex $z \in S_1$ which is adjacent to x. Since $z \notin A$, it follows that (z, x) is an arc in the graph G. Now let $S = S_1$. If y is a vertex of G_1 and $y \notin S_1$, then y can be reached by traversing a path of length at most 2 which originates at a vertex in S_1. If $y \in A$, there exists a path (z, x, y) of length two. If $y = x$, it has been shown that there is an arc (z, x) where $z \in S_1 = S$.

This completes the inductive proof of the property. [V. Chvátal, L. Lovász, *Hypergraph Seminar*, Lecture Notes in Math., 411, Springer-Verlag, 1974, p. 175.]

Since every tournament is complete, it follows that $|S| = 1$. This provides another proof of the previous problem.

8.19 The sequence of outdegrees of T written in increasing order will be called the score sequence of T. It will first be shown that $S: 0, 1, \ldots, n-1$ is the score sequence of a transitive tournament. Let T be a tournament with vertex set $V(T) = \{v_1, \ldots, v_n\}$ and arc set $E(T) = \{(v_i, v_j) | 1 \leq j < i \leq n\}$. It follows that $d^+(v_i) = i - 1$ for $i = 1, \ldots, n$; hence S is the score sequence of the transitive tournament T. Conversely, assume that T is a transitive tournament. It follows that $S: 0, 1, \ldots, n-1$ is the score sequence of T. To show this, it suffices to prove that no two vertices of T have the same outdegree. Suppose that u and v are distinct vertices of T such that $d^+(u) = d^+(v)$. Since T is a tournament, either (u, v) or (v, u) is an arc of T, say the former. Let W be the set of vertices of T adjacent from v; in this case $d^+(v) = |W|$. Since $(v, w) \in E(T)$ for each $w \in W$ and $(u, v) \in E(T)$, it follows that $(u, w) \in E(T)$ for each $w \in W$, since T is transitive. However, one then has $d^+(u) \geq 1 + |W| = 1 + d^+(v)$, which is a contradiction.

8.20 It is clear that $C(2) = 1$, since the vertices x_1 and x_2 are joined by an edge if the graph is connected. For $n \geq 3$ it must be shown that

$$n \, 2^{\binom{n}{2}} = \sum_{k=1}^{n} k \binom{n}{k} 2^{\binom{n-k}{2}} C(k).$$

Observe that $2^{\binom{n}{2}}$ represents the number of graphs with vertex set $X = \{x_1, \ldots, x_n\}$. In any such graph it is possible to label an arbitrary vertex in n ways, and hence $n \, 2^{\binom{n}{2}}$ is equal to the number of graphs with vertex set X in which one vertex is labeled.

It will now be shown that the right-hand side of the formula represents the same quantity. In order to do this consider a graph G with vertex set X which contains a labeled vertex x_i, where $1 \leq i \leq n$. The vertex x_i belongs to a connected component C of G. Let k be the number of vertices in C. It follows that $1 \leq k \leq n$ and the vertices of C are not joined by any edge to vertices in $X \setminus C$. The component C can be chosen in $\binom{n}{k}$ ways, the labeled vertex in C can be chosen in k ways, and the number of connected graphs with vertex set C is equal to $C(k)$. At the same time the number of graphs with vertex set $X \setminus C$ is equal to $2^{\binom{n-k}{2}}$.

Consider in turn $k = 1, \ldots, n$. One can see that in this way one generates all the $n \, 2^{\binom{n}{2}}$ graphs with vertex set X which have a marked vertex, and this observation completes the proof.

This recurrence relation yields the following values for $C(n)$:

$n =$	1	2	3	4	5	6	7
$C(n) =$	1	1	4	38	728	26,704	1,866,256

8.21 Consider the $2^{\binom{n}{2}}$ graphs with vertex set $X = \{x_1, \ldots, x_n\}$. Let G_{ij} be the set of graphs among them which have the following properties:

(1) x_i and x_j are not adjacent.
(2) Let x_k be another vertex such that x_i and x_k are adjacent. Then x_k and x_j are not adjacent.

Thus for $k \neq i, j$ there are three possible ways of joining the vertices x_i, x_j, and x_k. It follows that

$$|G_{ij}| = 3^{n-2} \, 2^{\binom{n-2}{2}} \quad \text{for every} \quad i \neq j.$$

The set of graphs with vertex set X and diameter at least 3 is $\bigcup_{i<j} G_{ij}$, and hence

$$\left| \bigcup_{i<j} G_{ij} \right| \leq \sum_{i<j} |G_{ij}| = \binom{n}{2} 3^{n-2} \, 2^{\binom{n-2}{2}},$$

from which it follows that

$$\lim_{n \to \infty} \binom{n}{2} \frac{3^{n-2} \, 2^{\binom{n-2}{2}}}{2^{\binom{n}{2}}} = \lim_{n \to \infty} \frac{8}{9} \binom{n}{2} \frac{3^n}{4^n} = 0.$$

Thus almost all graphs with n vertices have diameter equal to 1 or 2 as $n \to \infty$. But there exists a unique graph with diameter equal to 1, namely K_n. Thus almost all graphs with n vertices have diameter equal to 2 as $n \to \infty$. From this it follows that almost all graphs with n vertices are connected as $n \to \infty$.

It can be shown similarly that almost all directed graphs with n vertices have the property that for each two vertices x and y there is a path of length 1 or 2 from x to y as $n \to \infty$.

8.22 It follows immediately from the definition that this binary relation is reflexive and symmetric on the set U. In order to prove the transitivity it will be shown that if the edges u_1 and u_2 are found in the same elementary cycle C_1 and if u_2 and u_3 are found on the same elementary cycle C_2, then there is an elementary cycle C_3 which contains u_1 and u_3.

Traverse C_2 in both directions by starting at the endpoints of the edge u_3; terminate at the first vertex which is found on the cycle C_1. Let x and y be these vertices on the cycle C_1 (see Figure 8.5). It can happen that x or y are endpoints of u_3. However $x \neq y$, since the edge u_2 is found on the part of the cycle C_1 delimited by x and y and which does not contain u_1. The cycle C_3 (which is indicated by the heavy line in Figure 8.4) is obtained by taking the union of (1) the elementary walk on the cycle C_1 which joins x and y and which contains the

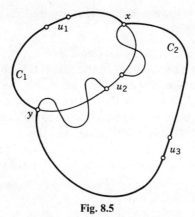

Fig. 8.5

edge u_1, and (2) the elementary walk in C_2 which connects x and y and contains the edge u_3.

8.23 It will be shown that $(1)\Rightarrow(3)\Rightarrow(2)\Rightarrow(1)$.

In order to show that $(1)\Rightarrow(3)$, let G be a 2-connected graph, and u and v two edges which have the endpoint x in common. Let y and z be their endpoints, which are different from x. Since G_x is connected, it follows that there exists an elementary walk which joins y and z in G_x. This elementary walk, together with the edges u and v, forms an elementary cycle in the graph G. The previous problem now implies that every two edges with a common endpoint are equivalent. Since G is connected, it follows that this equivalence relation has a single equivalence class which consists of the edge set of the graph G. Thus every two edges of the graph G lie on an elementary cycle. Also, the fact that G is connected implies that G has no isolated vertices.

$(3)\Rightarrow(2)$: Let x, y be two distinct vertices of the graph G. By hypothesis G has no isolated vertices, and thus there exist two distinct edges u and v which are incident with x and y respectively. For otherwise there exists an edge $[x, y]$, and the vertices x and y are no longer adjacent with other vertices of G. If $n=3$, this fact would imply that G has an isolated vertex, which contradicts the hypothesis. If $n \geqslant 4$, then since G has no isolated vertices, the set of remaining vertices contains at least one edge u. The edges $[x, y]$ and u do not belong to an elementary cycle, since $d(x)=d(y)=1$, and this contradicts the hypothesis. Thus there exist two distinct edges u and v which are incident with x and y respectively. In view of (3), there is an elementary cycle C which contains u and v and hence x and y.

$(2)\Rightarrow(1)$: Since every two vertices of G belong to an elementary cycle, it follows that there is an elementary walk which connects them, and thus G is connected. Suppose now that G is not 2-connected and hence that there is a vertex x such that G_x is not connected. Let a, b be two vertices which lie in different components of the subgraph G_x. Since each elementary walk between a and b in the graph G passes through x, it follows that there does not exist an elementary cycle in

G which contains the vertices a and b. This contradicts (2) and completes the proof that G is 2-connected.

8.24 Let $A=[a_0, a_1,\ldots, a_{r-1}, a_0]$ and $B=[b_0, b_1,\ldots, b_{r-1}, b_0]$ be two elementary cycles of maximal length in the graph G. Suppose that these cycles have at most one vertex in common, and let $a_0 = b_0$ be that common vertex. Since G is 2-connected, it follows that there also exists an elementary walk which does not pass through a_0 and which has one endpoint in the set of vertices $\{a_1,\ldots, a_{r-1}\}$ and the other endpoint in the set $\{b_1,\ldots, b_{r-1}\}$. Let $[a_p, x_1,\ldots, x_k, b_q]$ be such a walk where $k \geq 0$, $p, q \neq 0$, and x_i does not belong to the cycles A or B for $1 \leq i \leq k$. Suppose that $p \geq q$. One has therefore obtained an elementary cycle $[a_0,\ldots, a_p, x_1,\ldots, x_k, b_q,\ldots, b_{r-1}, a_0]$ which is longer than A, which contradicts the hypothesis.

Now let A and B be cycles which have no vertex in common. Since G is 2-connected, there exist two elementary walks without common vertices which join a vertex of A and a vertex of B. There will be an elementary walk of length greater than or equal to $r/2$ which is part of the cycle A and has as its endpoints the endpoints of the two walks. An analogous result can be obtained for the cycle B. The two parts of cycles A and B, together with the two elementary walks which join a vertex in A and a vertex in B, form an elementary cycle which is longer than A, and this again contradicts the hypothesis.

8.25 Let $m(G)$ denote the number of edges in the graph G. It follows that

$$\sum_{x \neq y} m(G_i - x - y) = \binom{n-2}{2} m(G_i), \tag{1}$$

since on the left-hand side each edge u contributes 1 for each pair $\{x, y\}$ of vertices which are different from the endpoints of u. The fact that there exist $\binom{n-2}{2}$ such pairs of vertices implies (1) for $i = 1, 2$.

By hypothesis, for $x \neq y$, the graph $G_1 - x - y$ is isomorphic to $G_2 - x - y$. Thus they have the same number of edges, and hence (1) implies that $m(G_1) = m(G_2)$. Now consider the sum

$$\sum_{y \neq x_0} m(G_i - x_0 - y), \tag{2}$$

where x_0 is a fixed vertex. If u is an edge incident with x_0, then its contribution to the sum (2) is zero. Otherwise its contribution to this sum is equal to $n-3$. This corresponds to the case in which y is different from x_0 and from the endpoints of the edge u. Thus it follows that

$$(n-3)m(G_i) - \sum_{y \neq x_0} m(G_i - x_0 - y) = (n-3)d_{G_i}(x_0).$$

Since the left-hand side of this identity is independent of i, one can see that $d_{G_1}(x_0) = d_{G_2}(x_0)$ for every vertex x_0.

In the same way one can show that

$$m(G_i) - m(G_i - x - y) = d_{G_i}(x) + d_{G_i}(y) - \alpha_i(x, y). \tag{3}$$

Here $\alpha_i(x, y) = 0$ if the vertices x and y are not adjacent in the graph G_i, and $\alpha_i(x, y) = 1$ if x and y are adjacent in G_i. It follows from (3) that for every vertex $x \neq y$ one has $\alpha_1(x, y) = \alpha_2(x, y)$, that is, the graphs G_1 and G_2 coincide.

8.26 Let G be the graph with n^2 vertices which correspond to the squares of a chessboard. Two vertices are considered to be adjacent if one can be reached from the other by moving a knight. Since a knight always moves from a black square to a white square or vice versa, it follows that this graph is bipartite. One can conclude from Problem 8.5 that this graph does not contain an elementary cycle with an odd number of vertices and hence there does not exist an elementary cycle with n^2 vertices. It is therefore impossible for a knight to visit all the squares of the chessboard in the manner described.

8.27 The set M of perfect matchings of G_n may be written as the union of two pairwise disjoint sets:

$$M = M_{yu} \cup M_{us},$$

where M_{yu} is the set of the perfect matchings of G_n that contain $[y, u]$ and M_{us} is the set of perfect matchings containing $[u, s]$ (see Figure 8.6). If a perfect matching of G_n contains $[y, u]$, then it follows that it also contains $[r, s]$, $[x, t]$, and $[v, z]$, and hence $|M_{yu}| = K(n-2)$. Thus this is equal to the number of perfect matchings of the graph obtained from G_n by deleting both hexagons α and β. If a matching belongs to M_{us}, then one can show that it contains the edge $[t, r]$, and hence $|M_{us}| = K(n-1)$. This is the number of perfect matchings of the graph obtained from G_n by deleting the hexagon β. Since $K(1) = 2$, $K(2) = 3$, and $K(n) = K(n-1) + K(n-2)$ for $n \geqslant 3$, it follows that $K(n) = F_{n+1}$ for any $n \geqslant 1$. [M. Gordon, W. H. T. Davison, *J. Chem. Phys.*, **20** (1952), 428–435.]

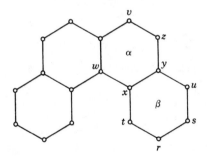

Fig. 8.6

8.28 As in Problem 6.9, one can show that for any three vertices x, y, z of a tree $d(x, y) + d(y, z) + d(x, z) \equiv 0 \pmod 2$ holds, and for any four vertices x, y, z, t of a tree the numbers $d(x, y) + d(z, t)$, $d(x, z) + d(y, t)$, $d(x, t) + d(y, z)$ are not all distinct.

(a) Suppose that G satisfies the hypothesis and contains an odd cycle with

$2k+1$ vertices. Let x, y, z be three distinct vertices on this cycle such that $d(x, y) = d(x, z)$ and $d(y, z) = 1$. In this case

$$d(x, y) + d(y, z) + d(x, z) = 2k + 1 \equiv 1 \pmod{2},$$

which is a contradiction.

On the other hand, if G contains an even cycle with $2k$ vertices ($k > 2$), let x, y, z, t be four vertices on the cycle such that

$$d(x, y) = d(z, t) = 1 \quad \text{and} \quad d(x, z) = d(y, t) = k - 1.$$

Then

$$d(x, y) + d(z, t) = 2, \quad d(x, z) + d(y, t) = 2k - 2,$$
$$d(x, t) + d(y, z) = 2k,$$

which is a contradiction. It is also clear that a cycle with four vertices is not itself embeddable in a tree.

(b) Let $K_{1,3}$ be a star with vertex set $\{a, b, c, d\}$ and edge set $\{[a, d], [b, d], [c, d]\}$. Define the graph $S(p)$ to be the graph obtained from $K_{1,3}$ by inserting new vertices on the edges of $K_{1,3}$ such that $d_{S(p)}(a, d) = d_{S(p)}(b, d) = d_{S(p)}(c, d) = p$. Suppose G is bipartite. Let $S = \{x, y, z\}$ be any set of three vertices of G. It will be shown that S is isometrically embeddable in the tree $S(d(G) - 1)$, where $d(G)$ is the diameter of G. Consider the identity

$$d(x, y) + d(y, z) + d(x, z) \equiv 0 \pmod{2}.$$

First note that if there exists a subgraph spanned by shortest paths between the vertices x, y, z which is a tree, then the equation follows from a previous result. Otherwise, one can assume that the vertices are distributed as in Figure 8.7 and

$$d(x, y) = d(x, u) + d(u, v) + d(v, y),$$
$$d(x, z) = d(x, u) + d(u, w) + d(w, z),$$
$$d(y, z) = d(y, v) + d(v, w) + d(w, z).$$

It follows that

$$d(x, y) + d(y, z) + d(x, z)$$
$$= 2d(x, u) + 2d(y, v) + 2d(z, w) + d(u, w) + d(w, v) + d(v, u)$$
$$\equiv d(u, w) + d(w, v) + d(v, u) \pmod{2}.$$

But this last sum is even, since it is the length of a cycle in a bipartite graph. To complete the proof one must find vertices x', y', z' in $S(d(G) - 1)$ such that if $d(x', d) = a$, $d(y', d) = b$, $d(z', d) = c$, then $a + b = d_G(x, y)$, $a + c = d_G(x, z)$, $b + c = d_G(y, z)$. Let $a = \sigma(x, y, z) - d_G(y, z)$, $b = \sigma(x, y, z) - d_G(x, z)$, $c = \sigma(x, y, z) - d_G(x, y)$, where $\sigma(x, y, z) = \{d_G(x, y) + d_G(x, z) + d_G(y, z)\}/2$. This is the solution of the

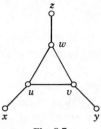

Fig. 8.7

system, and, for example, $a = \{d_G(x, y) + d_G(x, z) - d_G(y, z)\}/2 \leqslant \{2d(G) - 1\}/2$, which implies that $a \leqslant d(G) - 1$.

(c) Suppose that G is not bipartite. Then G contains an odd cycle with $2k + 1$ vertices and, as in the proof of (a), one can choose three vertices x, y, z such that

$$d(x, y) + d(y, z) + d(x, z) \equiv 1 \pmod{2},$$

which contradicts (b).

One can also show that three vertices x, y, z of a graph G are isometrically embeddable in a tree if and only if $d(x, y) + d(y, z) + d(x, z) \equiv 0 \pmod{2}$. [R. A. Melter and I. Tomescu, *Ars Combinatoria*, **12** (1981), 111–115.]

8.29 There is a round trip by at least one A_i which contains an odd number of stops. For $n = 1$ the statement is obvious, since one airline serves at least three cities C_1, C_2, C_3 and hence $[C_1, C_2, C_3, C_1]$ is a cycle with three landings. Use induction on n, and suppose that $n \geqslant 2$. One may assume that all round trips by A_n consist of an even number of stops, since otherwise there is nothing more to prove. Because the graph of service by A_n has no odd cycles, it follows from Problem 8.5 that this graph is bipartite. Then one can find a partition of the cities into two nonempty classes $\{Q_1, \ldots, Q_r\}$ and $\{R_1, \ldots, R_s\}$ where $r + s = N$, such that each flight by A_n runs between a Q-city and an R-city. Since $r + s = N \geqslant 2^n + 1$, at least one of r, s is greater than 2^{n-1}, say $r \geqslant 2^{n-1} + 1$. But $\{Q_1, \ldots, Q_r\}$ are only served by $A_1, \ldots, A_{n-1}$, and hence by the induction hypothesis at least one of these airlines flies a round trip with an odd number of landings. If there are $N = 2^n$ cities, there is a schedule with n airlines which contain no odd round trip on any of them. Let the cities be C_k, $k = 0, 1, \ldots, N - 1 = 2^n - 1$. Write k as an n-digit number in the binary system (possibly starting with one or more zeros). Link C_i and C_j by A_1 if the first digits of i and j are distinct, by A_2 if the first digits are the same but the second digits are different, ..., under A_n if the first $n - 1$ digits are the same but the nth digits are different.

All round trips under A_i are even, since the ith digit alternates for the vertices of such a cycle. Equivalently, the graph of service by A_i is bipartite for every $1 \leqslant i \leqslant n$. [Problem proposed to the jury of the 24th International Mathematical Olympiad, Paris, 1983.]

CHAPTER 9

9.1 Let V be a set consisting of $\rho(G)$ edges which cover all the vertices of the graph G. Since V is a minimal system of edges with this property, it follows that after suppressing an arbitrary edge u of V one vertex x will remain uncovered, and thus x is an endpoint only for u. Hence V induces a spanning graph of G whose components are stars of the form $K_{1,p}$ with $1 \leqslant p \leqslant n-1$. The number of these components is equal to $n - |V| = n - \rho(G)$, since each component $K_{1,p}$ contains p edges from V and $p+1$ vertices.

Select an edge from each component of the spanning graph induced by V. One obtains a matching with $n - \rho(G)$ edges, and it follows that

$$v(G) \geqslant n - \rho(G), \quad \text{or} \quad v(G) + \rho(G) \geqslant n.$$

In order to prove the opposite inequality let W be a matching consisting of $v(G)$ edges which do not have an endpoint pairwise in common. These $v(G)$ edges thus cover $2v(G)$ vertices.

Select an edge which is incident with each of the $n - 2v(G)$ vertices which are not covered by W. This procedure is possible because G does not contain isolated vertices. It is also the case that the vertices uncovered by W are pairwise non-adjacent, because this would contradict the maximality of W. The $n - 2v(G)$ edges are thus pairwise distinct. These edges, together with the set of edges W, cover all the vertices of G, and their number is equal to

$$v(G) + n - 2v(G) = n - v(G).$$

It thus follows that

$$\rho(G) \leqslant n - v(G), \quad \text{or} \quad v(G) + \rho(G) \leqslant n.$$

[T. Gallai, *Ann. Univ. Sci. R. Eötvös, Sectio Math.*, **2** (1959), 133–138.]

9.2 Let X be the vertex set of the graph G. There are $d(x)\{n-1-d(x)\}$ triples $\{x, y, z\}$ which are not triangles in G or $\bar{G}$ and which have a unique edge in G with $x \in X$ as an endpoint.

Each triple $\{x, y, z\}$ which is not a triangle in G or $\bar{G}$ contains one or two edges of G. Suppose that $[x, y]$ is an edge of G and that $[x, z]$ and $[y, z]$ are edges of $\bar{G}$. In the sum

$$\sum_{x \in X} d(x)\{n-1-d(x)\}$$

the triple $\{x, y, z\}$ is counted twice: once with respect to x, and once with respect to y. If $[x, y]$ and $[y, z]$ are edges of G and $[x, z]$ is an edge of $\bar{G}$, then in the summation the triple $\{x, y, z\}$ is also counted twice: once with respect to x and once with respect to z. It follows that the number of triangles in G and in $\bar{G}$ is equal to

$$\binom{n}{3} - \frac{1}{2} \sum_{x \in X} d(x)\{n-1-d(x)\}.$$

Solutions

(a) If G is regular of degree k, this formula becomes

$$\binom{n}{3} - \frac{nk}{2}(n-k-1).$$

(b) Thus $d(x)\{n-1-d(x)\} \leq \{(n-1)/2\}^2$ and hence the number of triangles in G and $\bar{G}$ is bounded below by

$$\binom{n}{3} - \frac{n}{2}\left(\frac{n-1}{2}\right)^2 = \frac{n(n-1)(n-5)}{24}.$$

This bound is positive for $n \geq 6$ and vanishes for $n=5$. The cycle C_5 with five vertices is such that neither C_5 nor $\bar{C}_5$ contains a triangle. It can be shown that the lower bound is attained if $n=4p+1$ where p is a natural number. [A. W. Goodman, Amer. Math. Monthly, **66** (1959), 778–783.]

9.3 Let $G=(X, U)$ where $|X|=n$ and $|U|=m$. Let $x, y \in X$, and denote by $A(x)$ the set of vertices which are adjacent to x. In similar fashion $A(y)$ will denote the set of vertices which are adjacent to y. It follows that

$$|A(x) \cap A(y)| = |A(x)| + |A(y)| - |A(x) \cup A(y)| \geq d(x) + d(y) - n.$$

If $[x, y] \in U$, then there are at least $d(x)+d(y)-n$ vertices which are adjacent to x and to y. Hence at least $d(x)+d(y)-n$ triangles in the graph G contain the edge $[x, y]$. From this one can conclude that G contains at least

$$\frac{1}{3} \sum_{[x,y] \in U} \{d(x) + d(y) - n\}$$

triangles, since each triangle is counted relative to each of its sides. In this sum $d(x)$ occurs exactly $d(x)$ times for every $x \in X$. It follows that the sum is equal to

$$\frac{1}{3}\left(\sum_{x \in X} d(x)^2 - mn\right).$$

By applying the Cauchy–Schwartz inequality and noting that $\sum_{x \in X} d(x) = 2m$, one can see that the last sum is bounded below by

$$\frac{1}{3}\left\{\frac{1}{n}\left(\sum_{x \in X} d(x)\right)^2 - mn\right\} = \frac{4m}{3n}\left(m - \frac{n^2}{4}\right).$$

9.4 Consider a set of three vertices $\{x, y, z\}$ in the graph. Either the triple is a circuit or there are two arcs of the form (x, y) and (x, z). In the latter case each would be the direction of the arc between y and z, the three vertices induce a transitive subgraph of the tournament. Let s_i denote the number of arcs which originate at the vertex x_i of the graph for $i=1, \ldots, n$. It follows that the number of transitive triangles is equal to

$$\sum_{i=1}^{n} \binom{s_i}{2} \geq n\binom{(n-1)/2}{2} = \frac{n(n-1)(n-3)}{8}.$$

To see this one uses Jensen's inequality and the fact that $\sum_{i=1}^{n} s_i = \binom{n}{2}$. Thus the number of circuits with three vertices is bounded above by

$$\binom{n}{3} - \frac{n(n-1)(n-3)}{8} = \frac{(n+1)n(n-1)}{24} = \frac{1}{4}\binom{n+1}{3}.$$

The upper bound is attained if n is odd, since there exists a tournament with n vertices such that $s_i = (n-1)/2$ for $i = 1, \ldots, n$. This will be established by induction on n.

For $n = 3$ one has only to consider a circuit with three vertices. Suppose that the property holds for a tournament with n vertices (n odd) and vertex set X. Orient the arcs so that $(n-1)/2$ arcs originate at each vertex of X.

Let y and z be two new vertices, and construct the arc (y, z). Let $X = Y \cup Z$ be a partition of the set X such that $|Y| = (n-1)/2$, $|Z| = (n+1)/2$. Now construct all the arcs of the form (y, x) and (x, z) for $x \in Y$ and all the arcs of the form (z, x) and (x, y) for $x \in Z$. One has thus obtained a tournament with $n+2$ vertices for which $s_i = (n+1)/2$ for $i = 1, \ldots, n+2$.

9.5 A tournament is a complete and antisymmetric graph. One can show that in general every complete directed graph has a Hamiltonian path.

The proof of this property uses induction on the number n of vertices in the graph. For $n = 2$ the graph has two vertices x and y. Since it is complete, there is an arc between these two vertices, for example (x, y). This arc is the desired Hamiltonian path.

Suppose that the property is true for all complete graphs with $n-1$ vertices. It will be shown that it is also true for complete graphs with n vertices.

If G has vertices $x_1, x_2, \ldots, x_n$, then the subgraph with vertices $x_1, \ldots, x_{n-1}$ is complete, and hence by the induction hypothesis it contains a Hamiltonian path. Denote this path by $(x_1, \ldots, x_{n-1})$. Since the graph is complete, there exists at least one arc between x_1 and x_n. If this arc is (x_n, x_1), then $(x_n, x_1, \ldots, x_{n-1})$ is a Hamiltonian path. In the opposite case only arc (x_1, x_n) exists.

By repeating this argument for a pair of vertices x_n and x_{n-1}, it is seen that if there exists an arc (x_{n-1}, x_n), then one can form a Hamiltonian path $(x_1, \ldots, x_{n-1}, x_n)$, and otherwise one has only the arc (x_n, x_{n-1}). Suppose only the arcs (x_1, x_n) and (x_n, x_{n-1}) exist. In view of the fact that x_n is connected by arcs to all vertices $x_2, \ldots, x_{n-2}$, there will exist among them two adjacent vertices x_k and x_{k+1} for which there exist arcs (x_k, x_n) and (x_n, x_{k+1}). In fact there is an arc from x_1 to x_n, and there is an arc from x_n to x_{n-1}, the last vertex on the path. Thus at a given moment, by leaving from x_1 towards x_{n-1}, the direction of the arc towards x_n must be changed.

In this way one constructs a Hamiltonian path for G, namely $(x_1, \ldots, x_k, x_n, x_{k+1}, \ldots, x_{n-1})$.

In order to obtain the upper bound, consider $m = [n/2]$ arcs $a_1, a_2, \ldots, a_m$ which have no vertices pairwise in common and are selected from a tournament with n vertices. Suppose that there exists a Hamiltonian path whose $(2i-1)$st arc is a_i for $i = 1, \ldots, m$. In this case the Hamiltonian path is uniquely determined.

Solutions

Thus the number of Hamiltonian paths is at most equal to the number of ways in which one can choose m arcs without common vertices. The arc a_1 can be chosen in $\binom{n}{2}$ ways, a_2 in $\binom{n-2}{2}$ ways, and so on. It follows that the number of ways of choosing the sequence of arcs $a_1, \ldots, a_m$ is equal to

$$\binom{n}{2}\binom{n-2}{2}\cdots\binom{n-2m+2}{2} = \frac{n!}{2^m} \leq \frac{n!}{2^{n/2}}.$$

Let $t(n)$ denote the maximum number of Hamiltonian paths in the class of tournaments with n vertices. The following values are known: $t(3)=3$, $t(4)=5$, $t(5)=15$, $t(6)=45$, and $t(7)=189$. If

$$\alpha = \lim_{n \to \infty} \left(\frac{t(n)}{n!}\right)^{1/n},$$

then α exists and satisfies the inequalities $0.5 = 2^{-1} \leq \alpha \leq 2^{-3/4} < 0.6$. Szele's conjecture states that $\alpha = 0.5$, but a proof of this has not yet been given. [T. Szele, *Mat. Fiz. Lapok*, **50** (1943), 223–256.]

9.6 Suppose that the graph G has n vertices and m edges, and does not contain an elementary cycle with four vertices. Count the number of pairs of vertices $\{x, y\}$ in which both members are adjacent to a third vertex z. If the vertex z is fixed, then there are $\binom{d(z)}{2}$ such pairs. Each pair $\{x, y\}$ is counted at most once, since if it was counted with respect to z_1 and z_2 with $z_1 \neq z_2$, then $[x, z_1, y, z_2, x]$ would form a cycle with four vertices, contrary to hypothesis.

Hence it follows that

$$\sum_{z \in X} \binom{d(z)}{2} \leq \binom{n}{2},$$

where X is the vertex set of the graph G. Since the function $x(x-1)/2$ is convex, it follows from Jensen's inequality that

$$\binom{n}{2} \geq \sum_{z \in X} \binom{d(z)}{2} \geq n \binom{2m/n}{2} = \frac{m(2m-n)}{n},$$

and thus

$$m^2 - \frac{nm}{2} - \frac{n^3 - n^2}{4} \leq 0.$$

From this one can see that

$$m \leq \frac{n}{4} - \sqrt{\frac{n^3}{4} - \frac{3n^2}{16}} = \frac{n}{4}(1 + \sqrt{4n-3}),$$

which contradicts the hypothesis. Thus G must contain an elementary cycle with four vertices.

If $n = q^2 + q + 1$ and q is a power of 2, then the maximum number of edges of a graph G with n vertices which contains no C_4 is $(q/2)(q+1)^2$ [Z. Füredi, *J. Combinatorial Theory*, B, **34**(2) (1983), 187–190], and this result also holds if q is a prime power.

9.7 Let $n=(k-1)q+r$ with $0\leq r\leq k-2$ and $(k-2)n=(k-1)p+s$, where $0\leq s\leq k-2$ and r, s are integers. It follows that if $r=0$ then $s=0$ and if $r\geq 1$ then $s=k-r-1$.

Suppose that the number of vertices x of degree $d(x)\leq p$ is smaller than m, that is, the other vertices y have degree $d(y)\geq p+1$. Thus one can assume that there exists a partition $X\cup Y$ of the set of vertices such that $d(x)\leq p$ for every vertex $x\in X$, $d(y)\geq p+1$ for every vertex $y\in Y$, and also $|X|<m$, $|Y|>n-m$.

Let y_1 be a vertex in Y, and denote by $A(y_1)$ the set of vertices which are adjacent to y_1. In the set $A(y_1)\cap Y$ choose another vertex y_2. It follows that $A(y_1)\cap A(y_2)$ will contain at least $|A(y_1)\cap A(y_2)|=|A(y_1)|+|A(y_2)|-|A(y_1)\cup A(y_2)|\geq 2(p+1)-n$ vertices.

Suppose that the vertex sets $A(y_1),\ldots,A(y_{k-2})$ with $y_j\in \bigcap_{i<j}A(y_i)\cap Y$ have been constructed for $j=1,\ldots,k-2$. Let $B=\bigcap_{i=1}^{k-2}A(y_i)$. It will be shown that $B\cap Y$ is nonempty. This property will demonstrate the possibility of choosing the vertices $y_2,\ldots,y_{k-1}$. In fact it can easily be shown by induction that $|B|\geq (k-2)(p+1)-(k-3)n$.

Consider two cases:

(1) $r=s=0$. In this case $(k-2)(p+1)-(k-3)n=(k-2)\{(k-2)q+1\}-(k-3)(k-1)q=q+k-2\geq q$.

(2) $r\geq 1$. It follows that

$$(k-2)(p+1)-(k-3)n=(k-1)p-p+k-2$$
$$-(k-1)p-s+n$$
$$=n-p+k-s-2$$
$$=\frac{n}{k-1}+\frac{(k-2)n}{k-1}-p+k-s-2$$
$$=q+\frac{r}{k-1}+p$$
$$+\frac{s}{k-1}-p+k-s-2$$
$$=q+\frac{r+s}{k-1}+k-s-2$$
$$\geq q+1,$$

since $r+s=k-1$ and $k-s-2\geq 0$.

Thus in both cases B contains at least $m=\{n/(k-1)\}$ vertices.

If $B\cap Y=\emptyset$, then $B\subset X$, which is impossible, since $|X|<m$ and $|B|\geq m$. Thus there exists a vertex $y_{k-1}\in B\cap Y$ which has degree $d(y_{k-1})\geq p+1$. In this way one can see that

$$\left|\bigcap_{i=1}^{k-1} A(y_i)\right| \geq (k-1)(p+1) - (k-2)n = k-1-s > 0,$$

which implies the existence of a vertex $y_k \in \bigcap_{i=1}^{k-1} A(y_i)$.

It follows from the method of construction that the vertices $y_1, y_2, \ldots, y_k$ form a complete subgraph with k vertices and this contradicts the hypothesis. Thus there exist at least m vertices of degree less than or equal to p. [I. Tomescu, *Studii si Cercet. Mat.*, **31**(3) (1979), 353–358, which extends K. Zarankiewicz's lemma, *Colloquium Math.*, **1** (1947), 10–15.]

9.8 Define a graph with 1001 vertices $x_1, \ldots, x_{1001}$ as follows: The vertices x_i and x_j are joined by an edge if person i and person j of the set M do not know each other. Since each subset of 11 persons contains at least two persons who know each other, it follows that G does not contain a complete subgraph with 11 vertices.

Now apply the result of the previous problem with $n=1001$ and $k=11$. It follows that G contains at least $m=101$ vertices with degree less than or equal to $p=900$. Thus in the complementary graph $\bar{G}$ there are at least 101 vertices with degree greater than or equal to $1000-900=100$. This establishes the desired result, since in $\bar{G}$ the vertices x_i and x_j are adjacent if and only if persons i and j know each other.

9.9 The theorem will be proven by induction on n. For $n=1, \ldots, k-1$ the graph with n vertices which has a maximal number of edges and does not contain a complete subgraph with k vertices is the complete graph K_n, and it has the indicated form, that is, each class contains a single vertex. Suppose that the theorem is true for $n' \leq n-1$. If the graph G has n vertices and does not contain a complete subgraph with k vertices, then Problem 9.7 implies the existence of a vertex x with degree $d(x) \leq p = [(k-2)n/(k-1)]$.

Consider the subgraph G_x obtained from G by suppressing the vertex x and all the edges which have x as an endpoint. The subgraph G_x has $n-1$ vertices, does not contain a complete subgraph with k vertices, and may or may not contain a maximal number of edges with respect to this property.

If G_x does not contain a maximal number of edges, then replace it by a graph with $n-1$ vertices which does not contain a complete subgraph with k vertices and has a maximal number of edges. By the induction hypothesis, this graph contains $k-1$ classes of vertices. There are r' classes which each contain $t'+1$ vertices. The remaining classes each contain t' vertices, where $n-1 = (k-1)t' + r'$ and $0 \leq r' \leq k-2$. Each vertex y is joined by an edge to all the vertices which do not belong to the same class as y.

Add the vertex x to a class which contains t' vertices, and join it to all the vertices which do not belong to the same class as x. One thus obtains a graph with n vertices, which does not contain a complete subgraph with k vertices, and which is unique up to isomorphism. The degree of the vertex x in the graph thus obtained is equal to $n-1-t' = n-1-[(n-1)/(k-1)]$. But a simple calculation shows that $n-1-[(n-1)/(k-1)]=[(k-2)n/(k-1)]$. It follows that a graph with n vertices and without complete subgraphs with k vertices and which contains a maximal number of edges must have the desired structure. Thus

$$M(n,k) = \binom{n}{2} - r\binom{t+1}{2} - (k-1-r)\binom{t}{2}$$
$$= \tfrac{1}{2}\{n^2 - n - rt(t+1) - (k-1-r)t(t-1)\},$$

where $t=(n-r)/(k-1)$. One can further show that

$$M(n,k) = \tfrac{1}{2}\{n^2 - n - t(2r + t(k-1) - k + 1)\}$$
$$= \frac{1}{2}\left(n^2 - n - \frac{n-r}{k-1}(2r+n-r-k+1)\right)$$
$$= \frac{1}{2}\left(n^2 - n - \frac{(n-r)(n+r-(k-1))}{k-1}\right)$$
$$= \frac{k-2}{k-1} \cdot \frac{n^2 - r^2}{2} + \binom{r}{2}.$$

9.10 First we show that for every choice of four points A, B, C, D from M, there exist at least two points which are at a distance less than or equal to $1/\sqrt{2}$. If three of the points are collinear, then one of the distances is less than or equal to $\tfrac{1}{2} < 1/\sqrt{2}$ and the property is seen to hold. Otherwise it will be shown that the configuration formed from the four points contains a triangle with an angle of at least $90°$. There are two possible cases:

(1) Three points, say A, B, C, form a triangle with the point D in its interior. The sum of the angles from D is equal to $360°$, and thus at least one angle is greater than or equal to $120°$.

(2) The four points form a convex quadrilateral. The sum of the angles of the quadrilateral is $360°$, and hence at least one angle is greater than or equal to $90°$.

Let ABC be the triangle with $A \geqslant 90°$. It follows that

$$a^2 \geqslant b^2 + c^2 \geqslant 2\min(b^2, c^2).$$

If $b \geqslant c$, it can be seen that $a^2 \geqslant 2c^2$ or $c^2 \leqslant a^2/2 \leqslant \tfrac{1}{2}$ and hence $c \leqslant 1/\sqrt{2}$.

Now define a graph which has as its vertices the $3n$ points; two vertices are joined by an edge if the distance between them is greater than $1/\sqrt{2}$. From the property just established, one can conclude that this graph does not contain a complete subgraph with four vertices. Thus it follows from the preceding problem that the number of edges is at most equal to $M(3n, 4) = 3n^2$. The $3n^2$ distances which are larger than $1/\sqrt{2}$ can be chosen in the interval $(1-\varepsilon, 1)$ for every $\varepsilon > 0$, by grouping each n points sufficiently close to the vertices of an equilateral triangle of side 1.

Solutions

9.11 The proof follows directly from Turán's theorem, since $M(2n, 3) = n^2$. The graph which realizes this maximum number of edges and does not contain a triangle is the complete bipartite graph $K_{n,n}$.

9.12 It will be shown that for every $n \geq 2$ the maximum number $f(n)$ of maximal complete subgraphs (cliques) in the class of graphs with n vertices takes on the following values: $f(n) = 3^{n/3}$ for $n \equiv 0 \pmod{3}$; $f(n) = 4 \times 3^{(n-1)/3 - 1}$ for $n \equiv 1 \pmod{3}$, and $f(n) = 2 \times 3^{(n-2)/3}$ for $n \equiv 2 \pmod{3}$.

For $2 \leq n \leq 4$ the expression for $f(n)$ can be obtained by a simple counting argument. In fact, if $n = 2$ or $n = 3$, the maximal number of cliques is obtained for graphs consisting of isolated vertices. In the case of $n = 4$ one must consider both a graph which only has isolated vertices and the complete bipartite graph $K_{2,2}$.

Let G be a graph with $n \geq 5$ vertices which contains $f(n)$ cliques. G contains at least two nonadjacent vertices x, y, since otherwise G would be the complete graph on n vertices K_n, which contains a unique clique; this would contradict the maximality of G.

Let $V(x)$ be the set of vertices adjacent to x in G. Denote by $G(x, y)$ the graph obtained from G by suppressing all the edges incident with x and replacing them with edges from the vertex x to each vertex in the set $V(y)$.

The symbol G_x will designate the subgraph obtained from G by suppressing the vertex x, and $a(x)$ will denote the number of complete subgraphs contained in $V(x)$ and which are maximal with respect to the subgraph G_x. Finally, $c(x)$ is the number of cliques of G which contain x.

By suppressing the edges incident with x, one causes $c(x) - a(x)$ cliques to disappear. But joining x by an edge to all the vertices of $V(y)$ creates $c(y)$ cliques. Thus if $c(G)$ denotes the number of cliques in G, one has

$$c(G(x, y)) = c(G) + c(y) - c(x) + a(x).$$

It can be assumed that $c(y) \geq c(x)$, because otherwise one only has to consider the graph $G(y, x)$.

Since $c(G)$ is maximal, it follows that $c(G(x, y)) \leq c(G)$ or $c(y) = c(x)$ and $a(x) = 0$. This implies that $c(G(x, y)) = c(G)$, and the graph $G(x, y)$ contains the same maximal number of cliques $f(n)$ as the graph G. It can be proved similarly that

$$c(G(y, x)) = c(G) + c(x) - c(y) + a(y).$$

Thus, in view of the fact that $c(x) = c(y)$, it follows that $a(y) = 0$, and hence $c(G(x, y)) = c(G(y, x)) = c(G)$.

Let x be an arbitrary vertex of the graph G, and let $y_1, \ldots, y_p$ be vertices which are not adjacent to x. Transform the graph G into the graph $G_1 = G(y_1, x)$, followed by the transformation of G_1 into $G_2 = G_1(y_2, x)$, and so on, until one finally transforms G_{p-1} into $G_p = G_{p-1}(y_p, x)$, in every case preserving the number $f(n)$ of cliques. The graph G_p has the property that the vertices $x, y_1, \ldots, y_p$, are not joined to each other by an edge, and $V(x) = V(y_1) = \cdots =$

$V(y_p)$. If $V(x)=\emptyset$, the process terminates. Otherwise consider a vertex in $V(x)$ and repeat the construction for the subgraph obtained from G by suppressing the vertices $x, y_1, \ldots, y_p$.

Finally a multipartite complete graph G^* is obtained with the property that its vertices can be partitioned into k classes which contain $n_1, \ldots, n_k$ vertices respectively. Two vertices are adjacent if and only if they do not belong to the same class. It is also the case that $c(G^*)=f(n)$. But $c(G^*)=n_1 \cdots n_k$, from which it follows that

$$f(n) = \max_k \max_{n_1+\cdots+n_k=n} n_1 n_2 \cdots n_k.$$

Suppose that $f(n)=m_1 m_2 \cdots m_p$, where $m_1+\cdots+m_p=n$. It is clear that $\max(m_1,\ldots,m_p) \leq 4$, since if, for example, $m_1 \geq 5$, then $3(m_1-3) > m_1$ or $2m_1 > 9$, and thus the product $m_1 \cdots m_p$ would not be a maximum. There cannot exist two factors equal to 4, since $4 \times 4 < 3 \times 3 \times 2$; also one cannot have three factors equal to 2, since $2 \times 2 \times 2 < 3 \times 3$.

Similarly, it follows that $-1 \leq m_i - m_j \leq 1$ for every $i,j=1,\ldots,p$. If, for example, $m_i \geq m_j + 2$, then $m_i m_j < (m_i-1)(m_j+1) = m_i m_j + m_i - m_j - 1$. Thus all the factors $m_1, \ldots, m_p$ are equal to 3 if $n \equiv 0 \pmod 3$. If $n \equiv 1 \pmod 3$, then a single factor is equal to 4 (or two factors are equal to 2), and the rest are equal to 3. When $n \equiv 2 \pmod 3$, one factor is equal to 2 and the rest are equal to 3. [J. W. Moon, L. Moser, *Israel J. Mathematics*, **3**(1) (1965), 23–28.]

9.13 First we show that after the final application of the algorithm (when one obtains a set of $n-1$ edges which does not contain a cycle) there is in fact a tree with n vertices. For suppose that the graph obtained is not connected and has $p \geq 2$ connected components which contain $n_1, \ldots, n_p$ vertices, respectively. Since these connected components do not contain cycles, each is a tree. It follows that the number of edges in each component is equal to $n_1-1, \ldots, n_p-1$, respectively, and hence

$$n_1 - 1 + \cdots + n_p - 1 = n - p = n - 1,$$

from which it follows that $p=1$. Thus the graph is connected and without cycles, that is, a tree.

Suppose now that the spanning tree A obtained in the final application of the algorithm is not minimal, and thus there is another tree A_1 such that $c(A_1) < c(A)$, where $c(A)$ denotes the sum of the costs of the edges of A.

Let $u_1, u_2, \ldots, u_{n-1}$ denote the edges of A. The indices are to correspond to the order in which the edges are obtained in the algorithm. Suppose that the first edge of the tree A (in this sequence) which is not an edge of A_1 is the edge u_k. Add the edge u_k to the tree A_1.

Let G_1 be the graph which is thus formed. It will have a unique cycle, consisting of u_k and the unique walk in the tree A_1, which joins the endpoints of u_k. This cycle contains at least one edge v_i which does not belong to the tree A, since otherwise A would contain a cycle. By suppressing the edge v_i, a tree A_2 results, since the graph obtained from G_1 by suppressing the edge v_i does not

Solutions

contain cycles and has $n-1$ edges. By the preceding argument it is therefore a tree. The cost of this tree is equal to

$$c(A_2) = c(A_1) + c(u_k) - c(v_i).$$

It follows from the definition of the algorithm that u_k is an edge of minimal cost which does not form a cycle with the edges $u_1, u_2, \ldots, u_{k-1}$. But no edge v_i forms a cycle with these edges, since A_1 is a tree. Thus $c(u_k) \leq c(v_i)$ and hence $c(A_2) \leq c(A_1)$. By repeating this procedure one replaces all the edges of the tree A_1 by edges of A and obtains a sequence of trees $A_2, A_3, \ldots, A_r = A$, which by construction satisfy

$$c(A_1) \geq c(A_2) \geq \cdots \geq c(A_r) = c(A),$$

and thus $c(A_1) \geq c(A)$. This is a contradiction, because it has been assumed that $c(A_1) < c(A)$ and thus A is a minimal tree in G. This observation justifies the algorithm.

9.14 Suppose that there are two trees of minimal cost, A and A_1. It follows that one can find an edge u in A which is not an edge in A_1. Consider the walk in A_1 which joins the endpoints of the edge u. There is an edge v of this walk which joins two vertices located in the two components of the graph obtained from A by suppressing the edge u. It follows that the graph A_2 obtained from A by suppressing the edge u and inserting the edge v is a tree. This is also true for the graph A_3 obtained from A_1 by suppressing the edge v and inserting the edge u. Since $c(u) \neq c(v)$, one can conclude that

$$\min(c(A_2), c(A_3)) < c(A),$$

since $c(A_2) = c(A) - c(u) + c(v)$ and $c(A_3) = c(A) + c(u) - c(v)$. These observations contradict the hypothesis that $c(A)$ is a minimum.

Hence the minimal spanning tree is unique.

9.15 Let C be a complete subgraph of G of maximal cardinality, and choose $x_i = 1/k$ for $i \in C$ and $x_i = 0$ for $i \notin C$, where $X = \{1, \ldots, n\}$. It follows that

$$\sum_{[i,j] \in U} x_i x_j = \frac{1}{k^2} \binom{k}{2} = \frac{1}{2}\left(1 - \frac{1}{k}\right),$$

and thus $f(G) = \max \sum_{[i,j] \in U} x_i x_j \geq \frac{1}{2}(1 - 1/k)$.

In order to prove the opposite inequality we use induction on n. If $n = 1$, then $k = 1$ and $f(G) = 0$, which implies that equality holds. Suppose that $f(G) \leq \frac{1}{2}(1 - 1/k)$ for all graphs with at most $n-1$ vertices, and let G be a graph with n vertices. If the maximum, $f(G)$, is attained for $x_i = 0$ where $1 \leq i \leq n$, then $f(G) = f(G')$. Here G' is the graph obtained from G by suppressing the vertex i and the edges incident with it. By using the induction hypothesis with respect to G' one can conclude that

$$f(G) = f(G') = \frac{1}{2}\left(1 - \frac{1}{k'}\right) \leq \frac{1}{2}\left(1 - \frac{1}{k}\right),$$

since $k' \leq k$.

Suppose now that $f(G)$ is attained for all variables $x_i > 0$. If G is not complete, then there exist two nonadjacent vertices, say 1 and 2. Let $F(G) = F(x_1, \ldots, x_n) = \sum_{[i,j] \in U} x_i x_j$, and $0 < c \leqslant x_1$. Then

$$F(x_1 - c, x_2 + c, x_3, \ldots, x_n) = F(x_1, \ldots, x_n) + c\left(\sum_{i \in A} x_i - \sum_{i \in B} x_i\right),$$

where A represents the set of vertices in G which are adjacent to vertex 2, but not adjacent to vertex 1. Similarly, B denotes the set of vertices in G which are adjacent to vertex 1, but not adjacent to vertex 2. It remains to show that $\sum_{i \in A} x_i = \sum_{i \in B} x_i$. For if $\sum_{i \in A} x_i > \sum_{i \in B} x_i$ then $F(x_1 - c, x_2 + c, x_3, \ldots, x_n) > F(x_1, \ldots, x_n)$, which is a contradiction. But if the opposite inequality held, then $F(x_1 + c, x_2 - c, x_3, \ldots, x_n) > F(x_1, \ldots, x_n)$ for every $0 < c \leqslant x_2$, which contradicts the maximality of $F(x_1, \ldots, x_n)$.

If $c = x_1$, then

$$F(0, x_1 + x_2, x_3, \ldots, x_n) = F(x),$$

and hence the maximum is attained for the subgraph G' obtained from G by suppressing the vertex 1. This reduces to the previous case if one of the variables takes the value zero. It thus follows from the induction hypothesis applied to G' that

$$f(G) \leqslant \frac{1}{2}\left(1 - \frac{1}{k}\right).$$

If $G = K_n$, it can be seen that

$$F(x_1, \ldots, x_n) = \sum_{1 \leqslant i < j \leqslant n} x_i x_j = \frac{1}{2}\left((x_1 + \cdots + x_n)^2 - \sum_{i=1}^{n} x_i^2\right)$$

$$= \frac{1}{2}\left(1 - \sum_{i=1}^{n} x_i^2\right) \leqslant \frac{1}{2}\left(1 - \frac{1}{n}\right),$$

and hence the inequality is also established in this case. [T. S. Motzkin, E. G. Straus, *Canadian J. Mathematics*, **17**(4) (1965), 533–540.]

9.16 Suppose that the arcs $u_1, \ldots, u_m$ of G are numbered so that $c(u_1) \geqslant c(u_2) \geqslant \cdots \geqslant c(u_m)$. Let k be the smallest index with the property that the set of arcs $\{u_1, \ldots, u_k\}$ contains the arcs of a path $D_0 = (a, \ldots, b)$. Since $\{u_1, \ldots, u_{k-1}\}$ does not contain all the arcs of any path from a to b, it follows that $C_0 = \{u_k, u_{k+1}, \ldots, u_m\}$ is an (a, b)-cut with $C_0 \cap A(D_0) = \{u_k\}$. [Here $A(D_0)$ denotes the set of arcs of the path D_0.] It follows that

$$\min_{u \in D_0} c(u) = c(u_k) \quad \text{and} \quad \max_{u \in C_0} c(u) = c(u_k). \tag{1}$$

Let C be an arbitrary (a, b)-cut. Then $C \cap A(D_0) \neq \emptyset$. Now let $u_i \in A(D_0) \cap C$. Since $A(D_0) \subset \{u_1, \ldots, u_k\}$, one can conclude that $i \leqslant k$ and hence

$$\max_{u \in C} c(u) \geqslant c(u_i) \geqslant c(u_k). \tag{2}$$

If $D=(a,\ldots,b)$ is any path from a to b, then $A(D) \cap C_0 \neq \emptyset$. Now let $u_i \in A(D) \cap C_0$. In view of the definition of C_0 it follows that $i \geq k$ and hence

$$\min_{u \in D} c(u) \leq c(u_k). \tag{3}$$

It follows from (1) and (3) that $\max_D \min_{u \in D} c(u) = c(u_k)$. From (1) and (2) one can conclude that $\min_C \max_{u \in C} c(u) = c(u_k)$, which establishes the property in question. [J. Edmonds, D. R. Fulkerson, *J. Combinatorial Theory*, **8** (1970), 299.]

9.17 Let g be a function which satisfies the two conditions in the statement of the problem, and let $D = (a, x_1, \ldots, x_k, b)$ be a path in the graph G. It follows that

$$c(D) = c(a, x_1) + c(x_1, x_2) + \cdots + c(x_k, b)$$
$$\geq g(x_1) + \{g(x_2) - g(x_1)\} + \cdots + \{g(b) - g(x_k)\} = g(b),$$

which implies that $\min_D c(D) \geq \max_g g(b)$. If there is a path $D = (a, \ldots, b)$ and a function g which satisfies the two conditions such that $c(D) = g(b)$, then the opposite inequality also holds and hence the given equality is shown to be valid.

An inductive procedure will now be given for constructing a function g on the vertices of the graph G. Let $g(a) = 0$. Suppose that g has been defined on a set of vertices $S \subset X$ with $a \in S$, and for every vertex $x \in S$ there is a path $D_x = (a, \ldots, x)$ in the subgraph generated by S such that $c(D_x) = g(x)$.

Consider all the arcs of the form (x, y) with $x \in S$ and $y \notin S$. Choose one for which the sum $g(x) + c(x, y)$ is a minimum. Let $g(y) = g(x) + c(x, y)$; the function g is now defined on the set of vertices $S \cup \{y\}$. It will be shown that every arc (z, u) in the subgraph generated by $S \cup \{y\}$ satisfies the inequality

$$g(u) - g(z) \leq c(z, u). \tag{1}$$

If $u, z \in S$, the validity of the inequality follows from the induction hypothesis. If $z \in S$ and $u = y$, then the method of choosing the arc (x, y) implies that

$$g(y) = g(x) + c(x, y) \leq g(z) + c(z, y),$$

and (1) is satisfied.

Let $u \in S$ and $z = y$. It will be shown that in this case $g(u) \leq g(y)$ and hence $g(u) - g(y) \leq 0 \leq c(y, u)$, and thus (1) is satisfied.

In order to prove this it will be shown by induction that the function g increases with an increase in the number of vertices chosen in G. This property holds for the first vertex c selected after a, since $g(c) = g(a) + c(a, c) = c(a, c) \geq 0$, and hence $g(c) \geq g(a)$.

Suppose that this property holds for every set of vertices of cardinality less than or equal to $|S|$ constructed by the indicated procedure. Suppose further that there exists a vertex $u \in S$ such that $g(y) < g(u)$.

Since $g(y) = g(x) + c(x, y) \geq g(x)$, it follows that $g(x) \leq g(y) < g(u)$. The induction hypothesis now implies that the vertex x was selected in the set S before the vertex u. Let $M \subset S$ be the largest set (with respect to inclusion) which contains x and not u constructed by the indicated procedure. At the next step one adds the vertex u to M. Thus there exists $v \in M$ such that

$$g(u) = g(v) + c(v, u).$$

Since $x, v \in M$, the inequality

$$g(y) = g(x) + c(x, y) < g(v) + c(v, u) = g(u)$$

contradicts the choice of u as the next vertex for which the function g is defined by this procedure. Thus it has been shown that the function g takes increasing values, and therefore (1) is valid and the function f satisfies both of the given conditions. By the induction hypothesis there exists a path $D_x = (a, \ldots, x)$ with $c(D_x) = g(x)$ in the subgraph generated by S. It follows that by adding the arc (x, y) to the path D_x one obtains a path $D_y = (a, \ldots, x, y)$, whose value is equal to

$$c(D_y) = c(D_x) + c(x, y) = g(x) + c(x, y) = g(y).$$

Thus g has been defined on $S \cup \{y\}$, and satisfies the two conditions on this set. Also, for every $x \in S \cup \{y\}$ there exists a path $D_x = (a, \ldots, x)$ which satisfies $c(D_x) = g(x)$ in the subgraph generated by $S \cup \{y\}$. This completes the inductive argument. It follows that g can be defined on the set of vertices X in only one way. Thus, in particular, g is defined at b, and a path $D = (a, \ldots, b)$ and a function g which satisfies $c(D) = g(b)$ have been constructed.

Let $c(u)$ be the length of the arc $u \in U$. Then $\min_D c(D)$ is the minimum distance between vertices a and b in the graph G, where $D = (a, \ldots, b)$ runs over the set of paths from a to b. The described procedure provides a construction for a unique function $g: X \to \mathscr{R}$ with the property that $g(x) \geq 0$. This is because $g(a) = 0$ and because it has been seen that g takes on only increasing values. Also, for every vertex $b \neq a$ one has

$$\min_D c(D) = g(b).$$

Thus the indicated procedure is also an algorithm for finding the minimal distances from a fixed vertex a to all vertices $b \neq a$ in the graph G.

9.18 The following sum will be calculated in two different ways:

$$\sum_{x \in X} \left(\sum_{u \in \omega^-(x)} f(u) - \sum_{u \in \omega^+(x)} f(u) \right). \qquad (*)$$

In view of the condition (C) for the conservation of the flow for every vertex $x \neq a, b$, it follows that the term corresponding to $x \neq a, b$ in the sum $(*)$ is zero. Thus $(*)$ reduces to $\sum_{u \in \omega^-(b)} f(u) - \sum_{u \in \omega^+(a)} f(u)$, since $\omega^-(a) = \omega^+(b) = \emptyset$.

But since every arc $(x, y) \in U$ belongs to both of the sets $\omega^+(x)$ and $\omega^-(y)$, it follows that by regrouping terms one can express $(*)$ in the form

$$\sum_{u \in U} \{f(u) - f(u)\} = 0.$$

From this (a) follows easily.

Solutions

In order to prove (b) consider the summation:

$$\sum_{x \in A} \left(\sum_{u \in \omega^-(x)} f(u) - \sum_{u \in \omega^+(x)} f(u) \right). \quad (**)$$

It can be seen that for every vertex $x \neq b$, the corresponding term in (**) is zero and hence summation (**) reduces to $\sum_{u \in \omega^-(b)} f(u) = f_b$, since $\omega^+(b) = \varnothing$ and $a \notin A$. By regrouping terms one can write (**) in the form

$$\sum_{u \in U_A} \{f(u) - f(u)\} + \sum_{u \in \omega^-(A)} f(u) - \sum_{u \in \omega^+(A)} f(u) = \sum_{u \in \omega^-(A)} f(u) - \sum_{u \in \omega^+(A)} f(u),$$

where U_A is the set of arcs u which have both endpoints in the set A. On the other hand, if the arc $u \in \omega^-(A)$, then the flow $f(u)$ appears in (**) with a plus sign, and if $u \in \omega^+(A)$ then the flow $f(u)$ appears in (**) with a minus sign, since there will exist vertices $x_1, x_2 \in A$ such that $u \in \omega^-(x_1)$ and $u \in \omega^+(x_2)$.

Comparing the two expressions for the sum (**), one sees that

$$f_b = \sum_{u \in \omega^-(A)} f(u) - \sum_{u \in \omega^+(A)} f(u) \leq \sum_{u \in \omega^-(A)} f(u) \leq \sum_{u \in \omega^-(A)} c(u) = c(\omega^-(A))$$

This is because condition (B) bounds the flow on each arc of the network and the flow takes on non-negative values.

9.19 Let v be a walk which joins the source a and the sink b. Denote by v^+ the set of arcs of v directed in the same sense as the direction determined in traversing the walk v from a to b. Similarly v^- is the set of arcs oriented in the opposite direction.

Let

$$\varepsilon = \min \left(\min_{u \in v^+} \{c(u) - f(u)\}, \min_{u \in v^-} f(u) \right).$$

For $\varepsilon > 0$ one can increase the flow f_b as follows: Increase the flow on each arc $u \in v^+$ by ε, and diminish the flow by ε on each arc $u \in v^-$. It follows from the method of defining ε that one obtains a new flow f', which satisfies

$$0 \leq f'(u) \leq c(u)$$

for each arc $u \in U$. The conservation condition is also satisfied at each vertex $x \neq a, b$. The flow at the sink increases, since $f'_b = f_b + \varepsilon > f_b$, and the last arc of the walk v which terminates in b belongs to the set v^+.

A walk for which $\varepsilon = 0$ is said to be saturated. Thus if a flow f realizes max f_b, there will not exist a nonsaturated walk from a to b. Otherwise the flow could be increased at b.

Let f be a maximal flow in the network G, and consider all the elementary walks from a to b. Suppress all the arcs u for which there exists an elementary walk v such that $u \in v^+$ and $f(u) = c(u)$ or $u \in v^-$ and $f(u) = 0$. Here u is the first arc encountered with this property in traversing the walk v from a to b.

The spanning graph thus obtained has at least two connected components, since otherwise there would exist a nonsaturated walk from a to b, which contradicts the maximality of the flow f. One of these components consists of a subset of vertices A which contains the sink b and does not contain the source a.

Thus A defines a cut $\omega^-(A)$. All the arcs $u \in \omega^-(A)$ have the same orientation as the walks from a to b, so that $f(u)=c(u)$. It is also the case that all the arcs $u \in \omega^+(A)$ are directed opposite to the direction of the respective walks from a to b which use these arcs. Thus $f(u)=0$ for every $u \in \omega^+(A)$.

The arcs from $\omega^-(A)$ and $\omega^+(A)$ respectively must have been suppressed from the network, since otherwise the connected component A would not be maximal with respect to inclusion, which contradicts the definition of a connected component.

It follows from the preceding problem that

$$f_b = \sum_{u \in \omega^-(A)} f(u) - \sum_{u \in \omega^+(A)} f(u) = \sum_{u \in \omega^-(A)} c(u) = c(\omega^-(A)).$$

But it has been seen that for every flow f and every cut induced by the set $A \subset X$ with $a \notin A$ and $b \in A$ one has the inequality

$$f_b \leq c(\omega^-(A)).$$

This also follows from the preceding problem.

Thus a maximal flow f and a cut $\omega^-(A)$ have been found for which the inequality becomes an equality. Hence the cut $\omega^-(A)$ has a minimal capacity which is equal to the maximal flow at the sink.

9.20 It is clear that the given algorithm has a finite number of steps, which is bounded above, for example, by $\sum_{u \in \omega^-(b)} c(u)$. In fact, it follows from Problem 9.18 that $\max_f f_b \leq c(\omega^-(A))$, where $\omega^-(A)$ is any cut of the network. If $A=\{b\}$, then $\omega^-(A)=\omega^-(b)$. From this one can deduce that $\max_f f_b \leq c(\omega^-(A)) = \sum_{u \in \omega^-(b)} c(u)$. Start with a zero flow on each arc, so that $f_b = 0$. At each step the flow f_b increases by ε. Recall that the capacity of an arc is a non-negative integer. Thus the values of the flow and of ε itself are integers, and $\varepsilon > 0$ implies that $\varepsilon \geq 1$. Therefore, at each step the flow f_b increases by at least 1, and hence the number of steps of the algorithm is bounded above by the capacity of a cut.

It will now be shown that when one can no longer mark the sink b, then the flow obtained has the maximal value at the sink. The set of arcs which join a marked vertex to an unmarked vertex constitutes a cut with minimal capacity.

To show this let A be the set of vertices which cannot be labeled by the given algorithm. It follows that $a \notin A$ and $b \in A$, since it has been assumed that the sink of the network cannot be labeled. Thus $\omega^-(A)$ is a cut of the network. One has $f(u)=c(u)$ for every arc $u \in \omega^-(A)$, since if $u=(x, y)$ then $x \notin A$ and $y \in A$. If $f(u)<c(u)$, the vertex y could have been labeled, which contradicts the fact that $y \in A$. Also one has $f(u)=0$ for every arc $u \in \omega^+(A)$, since if $u=(y, x)$ then $y \in A$ and $x \notin A$. If $f(u)>0$, the vertex y could have been labeled by starting from the labeled vertex x, which contradicts the fact that $y \in A$.

By applying (b) of Problem 9.18 one can write

$$f_b = \sum_{u \in \omega^-(A)} f(u) - \sum_{u \in \omega^+(A)} f(u) = \sum_{u \in \omega^-(A)} c(u) = c(\omega^-(A)).$$

Solutions

But it has been shown that for every flow f and every cut $\omega^-(A)$ one has $f_b \leq c(\omega^-(A))$. For the given flow and cut this inequality becomes an equality. Thus the flow f_b is maximal, and the cut $\omega^-(A)$ has minimal capacity.

This reasoning also provides a new proof of the Ford–Fulkerson theorem of the preceding problem in the case when the capacities of the arcs are integers.

Observe that this algorithm allows one to find the maximal flow after a finite number of steps in every network G whose capacities are rational numbers.

Let the arcs of the network be $u_1, \ldots, u_m$ with $c(u_i) = p_i/q_i$ where p_i, q_i are non-negative integers and $q_i \geq 1$ for every $1 \leq i \leq m$. Multiply the capacity of all the arcs by the least common multiple $[q_1, \ldots, q_m]$ of the denominators of these fractions. One thus obtains a transport network G_1 which has the same graph and whose capacities are integers. Let f be a flow in the network G_1, and define the flow g by

$$g(u) = \frac{f(u)}{[q_1, \ldots, q_m]} \tag{1}$$

for every $u \in U$. The conservation condition at every vertex $x \neq a, b$ and the boundedness on each arc will be satisfied for the network G.

In fact, from $\sum_{u \in \omega^-(x)} f(u) = \sum_{u \in \omega^+(x)} f(u)$ it follows that $\sum_{u \in \omega^-(x)} g(u) = \sum_{u \in \omega^+(x)} g(u)$. Also, $0 \leq f(u) \leq [q_1, \ldots, q_m] c(u)$ implies that $0 \leq g(u) \leq c(u)$ for every arc $u \in U$. If f_b is maximal, then $g_b = f_b / [q_1, \ldots, q_m]$ is maximal in the network G.

In the opposite case, there is a flow h in the network G such that $h_b > g_b$. Define the flow with components equal to $[q_1, \ldots, q_m] h(u)$ for each arc $u \in U$ of the network G_1. One obtains a flow with value at the sink equal to $[q_1, \ldots, q_m] h_b > [q_1, \ldots, q_m] g_b = f_b$, which contradicts the maximality of the flow f.

It follows that the flow g defined by (1) is maximal in the network G if and only if the flow f is maximal in the network G_1. This maximal flow can be determined by the given algorithm in a finite number of steps, since in the network G_1 all the capacities are integers.

The following problem shows that the Ford–Fulkerson algorithm may not have a finite number of steps if the capacities of the arcs are irrational.

9.21 Let $v = [a, \ldots, b]$ be a walk in the network from the source to the sink, and let the arc $u \in v^+$. The reserve of flow on this arc will be the quantity $c(u) - f(u)$ by which the flow can be increased without becoming larger than the capacity of the arc.

It will be shown by induction that there exists a procedure for determining a walk from a to b on which one can increase the flow according to the algorithm of the preceding problem so that at step n the flow f_b increases by a_n. Start at step 0 with a zero flow on each arc, and consider the walk $[a, x_1, y_1, b]$. One finds that $\varepsilon_0 = \min(c, a_0, c) = a_0$. Define a new flow equal to $a_0 < c$ on the arcs $(a, x_1), (x_1, y_1), (y_1, b)$. On the other arcs the flow will be equal to zero.

Let $n = 1$. Let A'_1, A'_2, A'_3, A'_4 be a permutation of the arcs $A_1, A_2, A_3, A_4,$

so that A'_1 has flow reserve 0, A'_2 has reserve a_n, A'_3 has reserve a_{n+1}, and A'_4 has reserve a_{n+1}. Notice that for $n=1$ one can take $A'_i = A_i$ for $1 \leq i \leq 4$.

Henceforth, at step n, choose a walk from a to b which includes among the arcs A'_i only the arcs A'_2 and A'_3; for example,

$$[a, x'_2, y'_2, x'_3, y'_3, b].$$

It follows that $\varepsilon_1 = \min(c - f(a, x'_2), a_n, c - f(y'_2, x'_3), a_{n+1}, c - f(y'_3, b))$. Thus by the induction hypothesis the amount of flow sent to b up to the present on the arcs of the network is equal to $\sum_{i=1}^{n-1} a_i$, and thus $f(a, x'_2) \leq \sum_{i=1}^{n-1} a_i$. This implies that $c - f(a, x'_2) \geq \sum_{i=n}^{\infty} a_i > a_n > a_{n+1}$. Similarly one can deduce that $c - f(y'_2, x'_3) > a_n$ and $c - f(y'_3, b) > a_n$ and hence $\varepsilon_1 = a_{n+1}$. Increase the flow on the arcs (a, x'_2), (x'_2, y'_2), (y'_2, x'_3), (x'_3, y'_3), and (y'_3, b) by a_{n+1}. The flow at b then increases by a_{n+1}. The arcs A'_1, A'_2, A'_3, A'_4 have reserves of flow equal to $0, a_n - a_{n+1} = a_{n+2}, 0$ and a_{n+1} respectively.

Now choose a walk v from a to b on which one can increase the flow at b so that $A'_2 \in v^+$ and $v^- = \{A'_1, A'_3\}$. For example,

$$v = [a, x'_2, y'_2, y'_1, x'_1, y'_3, x'_3, y'_4, b]$$

with $\varepsilon_2 = \min(c - f(a, x'_2), a_{n+2}, c - f(y'_2, y'_1), c(A'_1), c - f(x'_1, y'_3), c(A'_3), c - f(x'_3, y'_4), c - f(y'_4, b))$. The arcs A'_1 and A'_3 have their reserve of flow equal to 0, and thus on these arcs the flow is equal to the capacity of the arc.

One first sees that $c - f(a, x'_2) \geq \sum_{i=n+2}^{\infty} a_i + a_n > a_{n+2}$, since the amount of flow transmitted to b up to now is equal to $\sum_{i=1}^{n-1} a_i + a_{n+1}$. All the differences which appear in the expression for ε_2 are by the same reasoning greater than a_{n+2}. Similarly $\min(c(A'_1), c(A'_3)) \geq a_2 > a_{n+2}$ for $n \geq 1$. It follows that $\varepsilon_2 = a_{n+2}$, and thus one increases by a_{n+2} the flow on the arcs $(a, x'_2), (x'_2, y'_2), (y'_2, y'_1), (x'_1, y'_3), (x'_3, y'_4)$, and (y'_4, b). Decrease the flow by a_{n+2} on the arcs A'_1 and A'_3. The flow at b is still increased by a_{n+2}.

The reserves of flow on the arcs A'_1, A'_2, A'_3, A'_4 are now equal to $a_{n+2}, 0, a_{n+2}, a_{n+1}$, respectively, and the flow f_b has increased following step n by $a_{n+1} + a_{n+2} = a_n$ and thus has the value

$$\sum_{i=0}^{n} a_i.$$

Observe that now one has a situation analogous to that existing before applying step n. The reserve of flow of the arcs A'_i leads to the following permutation of the arcs A'_i:

$$\begin{pmatrix} A'_1 & A'_2 & A'_3 & A'_4 \\ A'_2 & A'_4 & A'_1 & A'_3 \end{pmatrix}.$$

Now redefine $A'_1 \leftarrow A'_2, A'_2 \leftarrow A'_4, A'_3 \leftarrow A'_1, A'_4 \leftarrow A'_3, n \leftarrow n+1$, and return to step n.

One can now see that the algorithm of the preceding problem will not terminate, but after n steps one has

$$f_b = \sum_{i=1}^{n} a_i < c.$$

Solutions

However, max $f = 4c$, and the maximal flow can therefore be obtained as follows:

$$f(a, x_1) = f(x_1, y_2) = f(y_2, b) = f(a, x_2) = f(x_2, y_1)$$
$$= f(y_1, b) = f(a, x_3) = f(x_3, y_4) = f(y_4, b)$$
$$= f(a, x_4) = f(x_4, y_3) = f(y_3, b) = c,$$

and on the other arcs the flow will be equal to zero. [L. R. Ford, D. R. Fulkerson, *Flows in Networks*, Princeton University Press, 1962.]

9.22 Let $G = (A, B, U)$ be a bipartite graph where U is the set of edges. Each edge has one of its endpoints in A and the other in B. Now construct a network in the following manner.

Consider two new vertices, a source a and a sink b. Associate with all the arcs (a, x) where $x \in A$ a capacity equal to 1. The arcs (y, b) with $y \in B$ also have a capacity equal to 1. For each edge $[x, y]$ with $x \in A$ and $y \in B$, one will consider the arc (x, y) as existing in the associated network.

All of the arcs of the form (x, y) with $x \in A$ and $y \in B$ have a capacity equal to $C = |A| + 1$.

Thus to the bipartite graph of Figure 9.2 corresponds the network of Figure 9.3 for which $C = 5$. It follows from the method of defining this network that

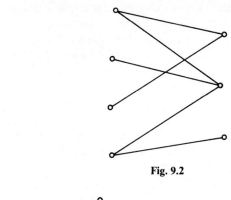

Fig. 9.2

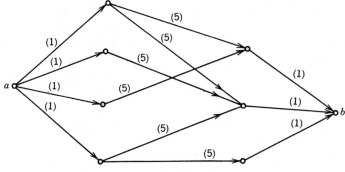

Fig. 9.3

no flow with integral components can have values other than 0 and 1 on the arcs of the network. The arcs of the form (x, y) with $x \in A$ and $y \in B$ which have flow equal to 1 determine a matching consisting of the corresponding edges $[x, y]$ of the graph G. In fact the arcs with flow equal to 1 cannot have a common endpoint, since the entry arcs and the exit arcs of the network both have capacity equal to 1. Thus they can have a maximal flow equal to 1. The number of arcs $u = (x, y)$ which have a flow $f(u) = 1$ is thus equal to the number of edges in the associated matching, which is, in turn, equal to the flow f_b at the sink.

The maximal flow f_b in the associated network corresponds to a maximal matching in the bipartite graph G obtained by considering the arcs between A and B which have a flow equal to 1. It follows that $v(G) = \max f_b$. The Ford–Fulkerson theorem implies that $\max f_b = \min_T c(\omega^-(T))$, where T is a set of vertices such that $a \notin T$ and $b \in T$. Note that for $T = A \cup B \cup \{b\}$ one has $c(\omega^-(T)) = |A|$ and $C = |A| + 1$. It follows that the minimal capacity of a cut in the network is attained for a cut which does not contain an arc of the form (x, y) with $x \in A$ and $y \in B$. Thus $\min_T c(\omega^-(T)) = c(\omega^-(T_0))$, where

$$\omega^-(T_0) = \{(a, x) | x \in A_0\} \cup \{(y, b) | y \in B_0\}, \qquad A_0 \subset A \text{ and } B_0 \subset B.$$

Now we show that $A_0 \cup B_0$ is a support for the bipartite graph G. To do this, consider an arbitrary edge $[x, y] \in U$ to which corresponds the path (a, x, y, b) in the network associated with G. Since every path from a to b contains at least one arc of a cut in the network, it follows that either $(a, x) \in \omega^-(T_0)$ and hence $x \in A_0$, or $(y, b) \in \omega^-(T_0)$ and hence $y \in B_0$. Thus one has shown that $A_0 \cup B_0$ is a support of G of cardinality $|A_0 \cup B_0| = |A_0| + |B_0| = |\omega^-(T_0)| = c(\omega^-(T_0)) = v(G)$, since each arc in $\omega^-(T_0)$ has a capacity equal to 1. It follows that $\tau(G) \leq |A_0 \cup B_0| = v(G)$.

On the other hand, if V is a set of edges which form a matching and S is a set of vertices which form a support of G, it follows that $|V| \leq |S|$. (This is because the edges in V do not have an endpoint pairwise in common, but each one has at least one vertex from S.) This implies that $v(G) \leq \tau(G)$, which, together with the previously established opposite inequality, establishes the equality of the two numbers for the bipartite graphs.

9.23 The matrix A can be considered to be the matrix of a bipartite graph $G = (X, Y, U)$, where $X = \{x_1, \ldots, x_n\}$, $Y = \{y_1, \ldots, y_m\}$, and $U = \{(x_i, y_j) | a_{ij} = 1\}$. The maximum number of elements equal to 1 which are found on different rows and columns corresponds to the edges of a maximal matching of G, which contains $v(G)$ edges. If the rows $i_1, \ldots, i_r$ and the columns $j_1, \ldots, j_s$ together contain all the elements equal to 1 in the matrix A, then $\{x_{i_1}, \ldots, x_{i_r}, y_{j_1}, \ldots, y_{j_s}\}$ is a support of the graph G.

The minimal number of rows and columns with this property is thus equal to $\tau(G)$. By the preceding problem, $v(G) = \tau(G)$, which completes the proof.

9.24 Let $E(n) = [\frac{1}{4}(n-1)^2]$. The proof is by induction on n. For $n \leq 3$ the result follows by enumerating all possible cases. Suppose that the property is valid for all graphs with at most n vertices, and let G be a graph with $n+1$

Solutions

vertices. Let G_x denote the subgraph obtained from G by deleting x and all edges incident to it. It follows from the induction hypothesis that $\delta_2(G_x) \leq E(n)$. Hence by using at most $E(n)$ operations (α) and (β), G can be transformed into $K_{n_1} \cup K_{n_2}$, where $n_1 \geq 0$, $n_2 \geq 0$, and $n_1 + n_2 = n$. If x is adjacent to p_1 vertices of K_{n_1} and to p_2 vertices of K_{n_2}, then G can be transformed into $K_{n_1} \cup K_{n_2+1}$ or $K_{n_1+1} \cup K_2$ either by p_1 operations (α) and $n_2 - p_2$ operations (β) or by p_2 operations (α) and $n_1 - p_1$ operations (β). Let $z_1 = p_1 + n_2 - p_2$ and $z_2 = p_2 + n_1 - p_1$. It follows that $z_1 + z_2 = n$, and hence $\min(z_1, z_2) \leq \frac{1}{2}n$ for even n and $\min(z_1, z_2) \leq \frac{1}{2}(n-1)$ for odd n. Finally one has $\delta_2(G) \leq \delta_2(G_x) + \min(z_1, z_2) \leq E(n+1)$, since $E(n) + \frac{1}{2}n = E(n+1)$ for even n and $E(n) + \frac{1}{2}(n-1) = E(n+1)$ for odd n.

In order to characterize the extremal graphs G with n vertices with this property, observe that if $\delta_2(G) = E(n)$, then for any vertex x of G one has $\delta_2(G_x) = E(n-1)$. The characterization of all graphs G such that $\delta_2(G) = E(n)$ can now be obtained by induction on n. For $n \leq 3$ it can be shown directly that all extremal graphs are complete bipartite graphs. Suppose that this property is true for all graphs with at most n vertices, and let G be a graph with $n+1$ vertices such that $\delta_2(G) = E(n+1)$. If the subgraph G_x with n vertices is composed only of isolated vertices, then x is also isolated or it is adjacent to all vertices of G_x, since otherwise there would exist a vertex $y \neq x$ such that G_y is not a complete bipartite graph, and hence $\delta_2(G_y) < E(n)$ by the induction hypothesis. But this would imply that $\delta_2(G) < E(n+1)$, which is a contradiction.

Thus G contains only isolated vertices (i.e., $G = K_{0,n+1}$) or $G = K_{1,n}$. A similar proof can be used when G_x is a graph $K_{p,q}$ where $p, q > 0$ and $p + q = n$. It remains to show that $\delta_2(K_{p,q}) = E(n)$ when $p, q \geq 0$ and $p + q = n$. For $K_{p,q}$ one may obtain a clique composed of x vertices from the set with p vertices and y vertices from the set with q vertices of $K_{p,q}$; the remaining vertices constitute the second clique. Hence the number of operations (α) and (β) is equal to $\binom{x}{2} + \binom{p-x}{2} + \binom{y}{2} + \binom{q-y}{2} + x(q-y) + y(p-x) = (x-y)^2 - (p-q)(x-y) - \frac{1}{2}n + \frac{1}{2}(p^2 + q^2)$. This expression has a minimum equal to $E(n)$ if $0 \leq x \leq p$, $0 \leq y \leq q$, $p + q = n$, and this minimum is reached only if $x - y = \frac{1}{2}(p-q)$ for even n and $x - y = \frac{1}{2}(p-q-1)$ [or $\frac{1}{2}(p-q+1)$] for odd n. For the graph $K_{0,n}$ with n isolated vertices one finds that $\binom{x}{2} + \binom{y}{2} \geq E(n)$ if $x + y = n$, and equality holds only for $x = y = \frac{1}{2}n$ for even n and $x = \frac{1}{2}(n-1)$, $y = \frac{1}{2}(n+1)$ for odd n. [M. Petersdorf, *Wiss. Z. Techn. Hochsch. Ilmenau* **12** (1966), 257–260, and I. Tomescu, *Math. et Sci. Humaines*, **42** (1973), 37–40.]

Let $\delta_k = \max_G \delta_k(G)$, where $\delta_k(G)$ represents the minimum number of operations (α) and/or (β) which transform G into the union of k disjoint cliques (some of them may be empty). It is known that $\delta_1 = \binom{n}{2}$ and $\delta_k = [\frac{1}{4}(n-1)^2]$ for any $2 \leq k \leq n$. [I. Tomescu, *Discrete Math.*, **10** (1974), 173–179.]

CHAPTER 10

10.1 The property will be established by induction on the number of vertices of the graph G.

If G has at most $k+1$ vertices, then it is evident that $\chi(G) \leqslant k+1$. Suppose that the property is true for all graphs with at most n vertices, let G be a graph with $n+1$ vertices such that the degrees of the vertices are bounded above by k, and let x be a vertex of G. Each vertex of the graph G_x obtained from G by suppressing the vertex x and the edges incident with x has degree at most equal to k.

It follows from the induction hypothesis that

$$\chi(G_x) \leqslant k+1.$$

Since x is adjacent to at most k vertices from G_x, one can color the vertex x with a color which does not appear among the colors of the vertices adjacent to x. Thus the total number of colors used to color G is not greater than $k+1$. It follows that $\chi(G) \leqslant k+1$.

10.2 It will first be shown by induction on n that $\chi(G) + \chi(\bar{G}) \leqslant n+1$. For $n=1$ and $n=2$ there is in fact an equality. Suppose now that the inequality holds for all graphs with at most $n-1$ vertices, and let G be a graph with n vertices, x a vertex of G, and G_x the graph obtained from G by suppressing the vertex x and all the edges incident with x. It is clear that:

$$\chi(G) \leqslant \chi(G_x) + 1, \qquad \chi(\bar{G}) \leqslant \chi(\bar{G}_x) + 1.$$

If at least one of these inequalities is strict, then the desired inequality follows, since by the induction hypothesis one has $\chi(G_x) + \chi(\bar{G}_x) \leqslant n$. However, if $\chi(G) = \chi(G_x) + 1$ and $\chi(\bar{G}) = \chi(\bar{G}_x) + 1$, then it follows that the vertex x is adjacent to at least one vertex colored with each of the $\chi(G_x)$ colors of G_x; thus $d_G(x) \geqslant \chi(G_x)$ and analogously $d_{\bar{G}}(x) \geqslant \chi(\bar{G}_x)$. It can thus be seen that $\chi(G_x) + \chi(\bar{G}_x) \leqslant d_G(x) + d_{\bar{G}}(x) = n-1$ and hence $\chi(G) + \chi(\bar{G}) \leqslant n+1$.

Let $\alpha(G)$ denote the maximal number of vertices of G which induce a subgraph consisting of isolated vertices. It follows that $\chi(G)\alpha(G) \geqslant n$, since each class of a coloring with $\chi(G)$ colors induces a subgraph which consists of pairwise nonadjacent vertices.

On the other hand, it is also the case that $\chi(\bar{G}) \geqslant \alpha(G)$ since each vertex in a set M of pairwise nonadjacent vertices in G must receive a color different from the other colors of M, for every coloring of the vertices of $\bar{G}$. Thus $\chi(G)\chi(\bar{G}) \geqslant n$. The other two inequalities now follow immediately from the inequality $(a+b)^2 \geqslant 4ab$ for $a, b \in \mathcal{R}$. [E. A. Nordhaus, J. W. Gaddum, *American Math. Monthly*, **63** (1956), 176–177.]

10.3 Consider a planar representation of the graph G. It will be shown that the faces in the interior of the Hamiltonian cycle C can be colored with two colors so that each two faces which have a common edge are colored differently.

Now construct the dual graph G_1^* of the faces which are found in the interior of the cycle C as follows: Each face is represented as a new vertex located in the interior of this face. Two vertices are adjacent if the corresponding faces have at least one edge in common. Thus it must be shown that the vertices of G_1^* can be colored with two colors so that each two adjacent vertices have different colors.

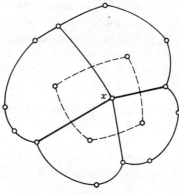

Fig. 10.1

The graph G_1^* does not contain cycles. For otherwise an elementary cycle of G_1^* would contain in its interior a vertex x of the graph G. But x is found on the Hamiltonian cycle C, whose edges incident with x are represented by heavy lines in Figure 10.1. This is a contradiction, since the vertices of G_1^* and hence the faces of G under consideration are not contained in the interior of the cycle C. Since G_1^* does not contain cycles, it follows that it does not contain elementary cycles. One can therefore conclude from Problem 8.5 that G_1^* is bipartite.

By coloring the vertices of each part A and B of G_1^* with the same color one obtains a coloring of the vertices with two colors. Since all the edges of G_1^* are of the form $[a, b]$ with $a \in A$ and $b \in B$, it follows that each two adjacent vertices have different colors.

In the same way, consider the dual graph G_2^* of the faces in the exterior of C. It can be seen that one can also color the faces outside of C with two other colors. One has thus found a coloring with four colors of the faces in a planar representation of the graph G, with the desired property.

10.4 Denote by $(x_1, y_1), \ldots, (x_n, y_n)$ the coordinates of the points of intersection with respect to a pair of perpendicular axes. One can assume that the directions of the axes are chosen so that the abscissas $x_1, \ldots, x_n$ are pairwise distinct, for example $x_1 < x_2 < \cdots < x_n$. Color the vertices of the graph in this order with three colors. If one has colored with three colors the vertices with abscissas $x_1, \ldots, x_{i-1}$, then the vertex (x_i, y_i) has at most two adjacent vertices which have already been colored, since there do not exist three concurrent lines. Thus there is a third color usable for the vertex (x_i, y_i) for $i = 2, \ldots, n$. The inequality under consideration is therefore satisfied.

10.5 The property will be established by induction on the number f of faces of the graph G. If $f = 1$, it can easily be seen that G is connected and does not contain cycles; it is thus a tree whose only face is the infinite face. In this case, the formula is satisfied, since $m = n - 1$. Suppose now that $f > 1$ and that the property is true for all planar connected graphs with at most $f - 1$ faces.

Let $[a, b]$ be an edge of a cycle of G. The edge $[a, b]$ is located on the boundary between two faces S and T. By suppressing the edge $[a, b]$ one obtains a new planar connected graph G_1 with n_1 vertices, m_1 edges, and f_1 faces, in which the faces S and T are joined to form a new face, while the other faces of G remain unchanged. Thus $n_1 = n$, $m_1 = m - 1$, and $f_1 = f - 1$. By the induction hypothesis one has $f_1 = m_1 - n_1 + 2$ and hence $f = f_1 + 1 = m_1 - n_1 + 3 = m - n + 2$.

10.6 Since G is a planar graph with m edges, there will exist $m_1 \leqslant m$ edges on the boundary between exactly two faces. If G has no vertices of degree 1, then $m_1 = m$. Cut out all the faces of the graph, and count in two different ways the $2m_1$ edges which are found on the boundary of all the faces. Since each face has at least three edges on its boundary, it follows from Euler's formula that

$$2m \geqslant 2m_1 \geqslant 3f = 3(m - n + 2),$$

or

$$m \leqslant 3n - 6.$$

If G does not contain triangles, then one can show similarly that

$$2m \geqslant 2m_1 \geqslant 4f = 4(m - n + 2),$$

or

$$m \leqslant 2n - 4.$$

10.7 Suppose that the complete graph with five vertices is planar. Then

$$f = m - n + 2 = 10 - 5 + 2 = 7.$$

It follows from the preceding problem that

$$20 = 2m \geqslant 3f = 21,$$

which is a contradiction. If the complete bipartite graph $K_{3,3}$ were planar, then one would have $f = m - n + 2 = 9 - 6 + 2 = 5$.

In the planar representation of the graph $K_{3,3}$, no face can be triangular, since the bipartite graphs do not contain odd cycles (Problem 8.5). Thus each face has a boundary consisting of at least four edges. It follows analogously that

$$18 = 2m \geqslant 4f = 20,$$

which is a contradiction.

Thus K_5 and $K_{3,3}$ are not planar.

10.8 Let G be a graph which is not a triangulation of the plane (each face a cycle with three vertices). Edges will be added so that each face of the resulting graph is triangular. If the graph obtained contains a vertex x with degree $d(x) \leqslant 5$, then it will follow that the graph G contains a vertex of degree at most 5, since by adding edges the degrees of the vertices increase.

Suppose therefore that G is a triangulation. In view of Euler's formula (Problem 10.5) one can write

$$v+f-m=2, \qquad (1)$$

where v is the number of vertices, f is the number of faces (including the infinite face), and m is the number of edges of the graph G. Since each face has three edges and each edge belongs to two faces, it follows that

$$3f = 2m. \qquad (2)$$

Suppose now that each vertex has degree $d(x) \geqslant 6$. In any graph one has the relation

$$\sum d(x) = 2m,$$

since every edge $[x, y]$ is counted twice, both in $d(x)$ and $d(y)$. It is also the case that

$$2m \geqslant 6v, \quad \text{or} \quad \frac{m}{3} \geqslant v.$$

Euler's formula and (2) lead to the conclusion that

$$2 = v + f - m = v + \frac{2m}{3} - m = v - \frac{m}{3} \leqslant 0,$$

which is a contradiction. It follows that every planar graph G contains at least one vertex x such that $d(x) \leqslant 5$.

This upper bound for the minimum degree of a planar graph cannot be improved, as can be seen from Figure 10.2, which represents a planar graph with 12 vertices which is regular of degree 5 (the graph of an icosahedron).

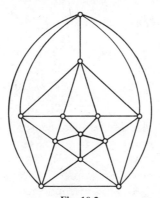

Fig. 10.2

10.9 It is sufficient to suppose that G is connected, since otherwise one could add edges between vertices located in different connected components in a planar representation of G. A new connected planar graph G_1 would be obtained such that $g(G_1) = g$.

It is also the case that the number of edges of G_1 is greater than the number of edges of G.

Thus let G be a planar connected graph with n vertices, m edges, and $g(G)=g$. The desired property will be established by induction on the number of vertices. If $n=g$, then G is a cycle with n vertices, and hence $m=n$ and the inequality becomes an equality. Suppose now that the inequality is true for all planar graphs with at most $n-1$ vertices, and let G be a planar graph with n vertices and $g(G)=g$.

Suppose that G contains an edge $[a, b]$ whose removal disconnects the graph G. Denote by H the spanning graph obtained from G by suppressing the edge $[a, b]$. H consists of two disjoint planar graphs G_1 and G_2 which have n_1 vertices and m_1 edges (n_2 vertices and m_2 edges) respectively. At least one of the two graphs G_1 and G_2 has girth equal to g; say $g(G_1)=g_1=g$. The other graph is a tree or has girth $g(G_2)=g_2 \geqslant g$. This property follows from the fact that since the edge $[a, b]$ does not belong to any cycle in the graph G, it can be concluded that $g(H)=g(G)$.

If the graph G_2 is a tree, then

$$m_2 = n_2 - 1 < \frac{g}{g-2}(n_2 - 2).$$

Otherwise, by using the induction hypothesis it follows that

$$m_2 \leqslant \frac{g_2}{g_2 - 2}(n_2 - 2) \leqslant \frac{g}{g-2}(n_2 - 2)$$

and similarly

$$m_1 \leqslant \frac{g}{g-2}(n_1 - 2).$$

Thus, since $n_1 + n_2 = n$, we can conclude that

$$m = m_1 + m_2 + 1 \leqslant \frac{g}{g-2}(n_1 + n_2 - 4) + 1$$

$$< \frac{g}{g-2}(n_1 + n_2 - 4) + \frac{2g}{g-2} = \frac{g}{g-2}(n-2).$$

Hence if G contains an edge whose removal causes G to become disconnected, then the inequality is proven and it has been seen that the inequality is in fact strict.

It remains to consider the case in which G does not contain an edge $[a, b]$ whose elimination transforms G into a disconnected graph. Let f_i be the number of faces which contain i edges for $i \geqslant g$, and let f be the total number of faces in a planar representation of G. It follows that

$$\sum_{i \geqslant g} f_i = f \quad \text{and} \quad \sum_{i \geqslant g} i f_i = 2m,$$

since in this case each edge $[a, b]$ lies on the boundary between two faces. Thus

$$2m = \sum_{i \geqslant g} i f_i \geqslant \sum_i g f_i = gf.$$

By applying Euler's formula (Problem 10.5) one can conclude that

$$m+2 = n+f \leqslant n+\frac{2}{g}m,$$

and hence

$$m \leqslant \frac{g}{g-2}(n-2).$$

Observe that equality holds only for $f_{g+1} = f_{g+2} = \cdots = 0$, and thus when all the faces of G are cycles with g vertices. This case can be realized for certain values of n, for example, for $n = g$.

10.10 The property will be proven by induction on the number of vertices of the graph G.

If G has at most five vertices, the property is immediate.

Suppose that every planar graph with at most n vertices has chromatic number less than or equal to 5, and let G be a planar graph with $n+1$ vertices. It follows from Problem 10.8 that there is a vertex x in G with degree $d(x) \leqslant 5$.

Denote by G_x the subgraph obtained from G by suppressing the vertex x and the edges incident with x. By the induction hypothesis the vertices of G_x can be colored by using at most five colors so that each two adjacent vertices have different colors. If $d(x) \leqslant 4$ then one can find, for the vertex x, an available color, different from the colors of the vertices adjacent to x. Thus the vertices of G can be colored with at most five colors, that is, $\chi(G) \leqslant 5$.

Suppose that $d(x) = 5$ and that the vertices y_1, y_2, y_3, y_4, y_5 adjacent to x are colored with at most four colors in the coloring with at most five colors of G_x. One can find an available color for x, and the property is established. The only remaining case is that in which $d(x) = 5$ and the vertices $y_1, \ldots, y_5$ are colored with exactly five colors, say A, B, C, D, E, in the coloring of G_x (Figure 10.3). Consider the connected component M which contains y_1 of the subgraph of G_x consisting of vertices colored with either color A or color C. If the vertex y_3 does not belong to this component, one can interchange the colors A and C in M to obtain a coloring with five colors of the graph G_x in which y_1 has the color C.

Thus the color A has become available, and the vertex x can be colored with A to obtain a coloring with five colors of the vertices of the graph G. Otherwise, if y_1 and y_3 belong to M, it follows that there exists a walk which joins y_1 and y_3 and which contains alternatingly the colors A and C.

Consider the connected component N which contains y_2 in the subgraph G_x consisting of vertices colored B or D. In this case $y_4 \notin N$. In fact, there would otherwise be a walk with endpoints y_2 and y_4 whose vertices are colored alternatingly B and D. Since the graph is planar, it would turn out that this walk must have a vertex in common with the walk which joins y_1 and y_3 (see Figure 10.3). But this is impossible, since these two walks consist of sets of vertices for which the sets of colors are disjoint. Thus $y_4 \notin N$. By interchanging the colors

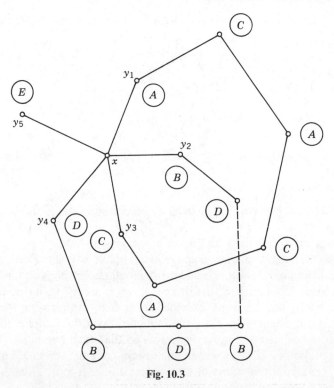

Fig. 10.3

B and D in the component N, one obtains a coloring of the subgraph G_x with five colors in which the vertex y_2 has color D. Thus the color B becomes available. One can color x with B, and this produces a coloring with five colors of the vertices of the graph G.

In 1976 K. Appel and W. Haken proved the Four-Color Theorem. It states that for every planar graph G one has the inequality $\chi(G) \leqslant 4$. They used an electronic computer to study more than 1900 configurations which occurred in the proof.

10.11 The graphs with the smallest number of vertices which do not contain triangles and which have chromatic numbers $\chi(G_1) = 3$ and $\chi(G_2) = 4$ respectively are illustrated in Figures 10.4 and 10.5.

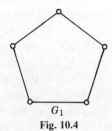

Fig. 10.4

Solutions

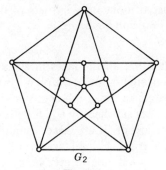

Fig. 10.5

In general, suppose the graph G has vertices $x_1, \ldots, x_n$, does not contain triangles, and has $\chi(G) = k$. Associate a new vertex y_i with each vertex x_i of G for $i = 1, \ldots, n$. Join the vertex y_i to all the vertices adjacent to x_i for $i = 1, \ldots, n$. Also add a new vertex z which is joined by an edge to all the vertices $y_1, \ldots, y_n$. It will be shown that the graph G_1 which is thereby obtained does not contain triangles and has $\chi(G_1) = k + 1$. Since G does not contain triangles and the vertices $y_1, \ldots, y_n$ are pairwise nonadjacent, it follows that every triangle in G_1 must have three vertices of the form x_i, x_j, y_k. This is a contradiction, since x_i and x_j are adjacent to x_k and thus G contains a triangle x_i, x_j, x_k, which contradicts the hypothesis.

It will also be shown that $\chi(G_1) = k + 1$. Suppose the vertices $x_1, \ldots, x_n$ have been colored with k colors so that each two adjacent vertices have different colors. The vertices $y_1, \ldots, y_n$ can be colored with k colors by coloring the vertex y_i with the same color as the vertex x_i for $i = 1, \ldots, n$. This is the minimum number of colors with which one can color the vertices $y_1, \ldots, y_n$, since if $p < k$ colors were sufficient to color the vertices $y_1, \ldots, y_n$, it would follow that $\chi(G) < k$ by coloring the vertex x_i with the color of the vertex y_i for $i = 1, \ldots, n$.

Since the vertex z is adjacent to all the vertices y_i, it follows that $\chi(G_1) = k + 1$.

By this construction, starting from the cycle with five vertices (C_5) of Figure 10.4, one obtains the graph represented in Figure 10.5. It has no triangles, and its chromatic number is equal to 4. By repeating the construction one finds that for every natural number $k \geq 3$ there exists a graph G with $\chi(G) = k$ which does not contain triangles.

10.12 Suppose that G contains three mutually adjacent vertices (a_1, b_1), (a_2, b_2), and (a_3, b_3) such that $a_2 = a_1 + b_1$, $a_3 = a_1 + b_1$, and $a_3 = a_2 + b_2$. This would imply that $a_3 = a_1 + b_1 + b_2 > a_1 + b_1$, which is a contradiction.

Let $x = c_1$ and $x = c_2$ be lines with $c_2 < c_1$. It follows that the point $(c_1, c_2 - c_1)$ on the line with equation $x = c_1$ is adjacent to all the points with positive integer coordinates on the line with equation $x = c_2$. Thus the set of colors of the points of G located on the line $x = c_1$ is different from the set of colors of the points of G located on the line $x = c_2$. Suppose that the chromatic number $\chi(G) = m < \infty$. It follows that the number of lines with equation $x = c$, $c > 0$, and $c \in \mathscr{Z}$ is at most equal to the number of nonempty subsets of the set of colors, that is, $2^m - 1$. This is a contradiction, and hence $\chi(G) = \infty$. [A. Gyárfás, *Discrete Math.*, **30**(2) (1980), 185.]

10.13 One can assume that all the faces in a planar representation of G are triangles, that is, G is a triangulation of the plane. In fact, if G is not a triangulation, one can add new edges to the graph G until one obtains a triangulation G_1. If the desired inequality is satisfied by G_1, it will be satisfied further for G, since the degrees of the vertices of G_1 are larger than the degrees of the vertices of G. Thus let G be a planar graph with all faces triangular for which the sum of the squares of the degrees is maximal. Let x be a vertex of minimal degree. It will be shown that $d(x) = 3$.

In fact, if $d(x) \leq 2$ it follows that $d(x) = 2$ and the graph G reduces to K_3, which contradicts the hypothesis that $n \geq 4$. Suppose that $d(x) \geq 4$, and let $x_1, x_2, \ldots, x_r$ be vertices adjacent to x such that $d(x_1) \leq d(x_i)$ for $i = 2, \ldots, r$, where $r = d(x) \geq 4$ (Figure 10.6).

Suppress the edge $[x, x_1]$ and add the edge $[x_2, x_r]$ to produce a new planar graph G_1 without multiple edges for $r \geq 4$.

Let S denote the sum of the squares of the degrees of G, and S_1 the sum of the squares of the degrees of G_1. It follows that

$$S_1 - S = \{d(x_r) + 1\}^2 + \{d(x_2) + 1\}^2 + \{d(x) - 1\}^2 + \{d(x_1) - 1\}^2$$
$$- \{d(x_r)\}^2 - \{d(x_2)\}^2 - \{d(x)\}^2 - \{d(x_1)\}^2$$
$$= 2d(x_r) + 2d(x_2) - 2d(x) - 2d(x_1) + 4 > 0,$$

since $d(x) \leq d(x_1) \leq d(x_i)$ for $i = 2, \ldots, r$. However, this inequality contradicts the maximality of the graph G, and thus every graph G for which S is maximal contains a vertex x of degree $d(x) = 3$.

The proof of the property will now be completed by induction on n. Let $n = 4$, and consider a planar representation of the graph K_4. In this case both sides of the inequality are equal to 36. Every other graph with four vertices has a sum of squares of its degrees less than 36, since the degrees of some vertices strictly decrease; the property is therefore established.

Suppose now that the property is true for all planar graphs with at most n vertices, and let G be a planar graph with $n + 1$ vertices for which the sum of the squares of the degrees is maximal and which has all of its faces triangular. By

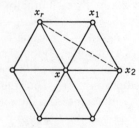

Fig. 10.6

the previous observation, there is a vertex x with degree $d(x) = 3$. Let a, b, c be vertices adjacent to x. Denote by G_1 the subgraph obtained from G by suppressing the vertex x and the three edges incident with x. Using the earlier notation, one sees that

$$S = S_1 + 9 + \{d(a)+1\}^2 + \{d(b)+1\}^2 + \{d(c)+1\}^2$$
$$\quad - \{d(a)\}^2 - \{d(b)\}^2 - \{d(c)\}^2$$
$$= S_1 + 2\{d(a)+d(b)+d(c)\} + 12.$$

Consider three vertices a, b, c which induce a complete subgraph with three vertices in a planar graph with n vertices. It will now be shown that

$$d(a) + d(b) + d(c) \leqslant 2n + 1. \tag{1}$$

In fact, if another vertex d is adjacent to all three vertices a, b, c, then an additional vertex e can be adjacent to at most two of the vertices $a, b,$ or c, say with a and with b. This follows from Problem 10.7 and the planarity of the graph G (Figure 10.7). Thus the number of edges which join vertices a, b, c to the other $n-3$ vertices of the graph G is bounded above by $2(n-3)+1 = 2n-5$. Recalling the contribution of the edges $[a, b], [b, c],$ and $[a, c]$ to the degrees $d(a), d(b), d(c)$, one obtains

$$d(a) + d(b) + d(c) \leqslant 2n - 5 + 6 = 2n + 1$$

and thus (1) is verified.

By the induction hypothesis $S_1 \leqslant 2(n+3)^2 - 62$ and thus $S \leqslant 2(n+3)^2 - 62 + 2(2n+1) + 12 = 2(n+4)^2 - 62$ which completes the inductive proof of the inequality. For $n = 4$ it has been seen that the inequality becomes an equality for the graph K_4 with degree sequence 3, 3, 3, 3. It will now be shown by induction on n that there exists a planar triangulated graph with n vertices and such that the degrees of the vertices which border the infinite face are equal to 3, $n-1$, and $n-1$ respectively. For this graph the inequality also become an equality.

In fact it is easily seen that this assertion is true for $n = 4$. Suppose that it is true for n, and let G be a planar graph with the desired property.

Suppose that the infinite face of G is bounded by the cycle $[a, b, c, a]$ with $d(a) = 3$, $d(b) = n-1$, and $d(c) = n-1$. Proceed to define a planar graph G_1 with the desired property as follows: Consider a new vertex x_{n+1} (on the infinite

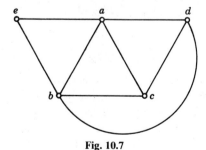

Fig. 10.7

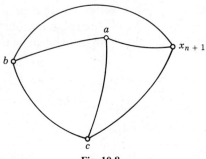

Fig. 10.8

face of the graph G) which is joined by edges to a, b, c as in Figure 10.8. The new infinite face of the graph G_1 is bounded by the cycle $[b, c, x_{n+1}, b]$. Also $d(x_{n+1}) = 3$, $d(a) = 4$, $d(b) = n$, and $d(c) = n$. G_1 is a triangulation, and $S_1 = S + 9 + 2n^2 - 2(n-1)^2 + 16 - 9 = S + 4n + 14$. However, by the induction hypothesis $S = 2(n+3)^2 - 62$, and thus $S_1 = 2(n+3)^2 - 62 + 4n + 14 = 2(n+4)^2 - 62$ and the property is established.

By using the fact that the function $x^{\alpha-1}: [0, \infty) \to \mathscr{R}$ is convex for $\alpha \geqslant 2$, it can be shown analogously that

$$\sum_{i=1}^{n} d_i^\alpha \leqslant 2(n-1)^\alpha + (n-4)4^\alpha + 2 \times 3^\alpha \tag{2}$$

for every planar graph G with $n \geqslant 4$ vertices and $\alpha \geqslant 2$. Equality holds for the planar graph constructed for $\alpha = 2$. It has degrees $d_1 = d_2 = n - 1$, $d_3 = \cdots = d_{n-2} = 4$, $d_{n-1} = d_n = 3$.

For $\alpha = 1$ inequality (2) becomes an equality for every triangulation of the plane and expresses the fact that the number of edges of a triangulation is equal to $3n - 6$ (by Problem 10.6).

10.14 Denote by $U = \{u_1, \ldots, u_m\}$ the set of edges of the graph G, and let $A_i = \{f : X \to \{1, \ldots, \lambda\} \mid f(x) = f(y) \text{ where } u_i = [x, y]\}$. The number of functions which satisfy the desired condition is equal to

$$P_G(\lambda) = \lambda^n - |A_1 \cup A_2 \cup \cdots \cup A_m|.$$

One can evaluate the cardinality of the set $A_1 \cup \cdots \cup A_m$ by using the Principle of Inclusion and Exclusion (Problem 2.2):

$$|A_1 \cup \cdots \cup A_m| = \sum_{\substack{V \subset U \\ V \neq \varnothing}} (-1)^{|V|-1} p(V, \lambda),$$

where $p(V, \lambda)$ represents the number of functions $f : X \to \{1, \ldots, \lambda\}$ which take on the same value for the endpoints of every edge in V. It follows that these functions have the same value from the set $\{1, \ldots, \lambda\}$ for each connected component of the spanning graph (X, V) of G. Thus $p(V, \lambda) = \lambda^{c(V)}$. This justifies the expression for the chromatic polynomial of the graph G in view of the fact that the term λ^n is obtained for $V = \emptyset$ when the spanning graph (X, V) consists only of isolated vertices and thus has n connected components.

Observe that a single term of degree n is obtained for $V = \emptyset$, namely λ^n, but terms of degree $n-1$ are obtained for $|V| = 1$. In this case there are $\binom{m}{1} = m$ terms equal to $-\lambda^{n-1}$, and hence

$$P_G(\lambda) = \lambda^n - m\lambda^{n-1} + \cdots,$$

where m is the number of edges of the graph G.

If $P_G(\lambda) = \lambda^n + a_1 \lambda^{n-1} + a_2 \lambda^{n-2} + \cdots + a_n$, then one can show similarly that $a_2 = \binom{m}{2} - c_3(G)$, where $c_3(G)$ represents the number of triangles of the graph G. Terms of degree $n-2$ are obtained for $c(V) = n-2$, when $|V| = 2$ and the spanning graph (X, V) of G contains only two edges, adjacent or not. This also occurs if $|V| = 3$, when the spanning graph contains three edges which form a triangle.

Since two edges can be chosen in $\binom{m}{2}$ ways, one has indeed obtained the expression for the coefficient a_2.

10.15 If a λ-coloring f of the graph $G - e$ has the property that $f(x) \neq f(y)$, where $e = [x, y]$, then f is also a λ-coloring for G and vice versa. It follows that $P_{G-e}(\lambda) - P_G(\lambda)$ is the number of those λ-colorings f of $G - e$ with the property that $f(x) = f(y)$.

A λ-coloring f of $G - e$ of this kind produces a λ-coloring g of $G|e$ by defining $g(z) = f(x) = f(y)$ and $g(t) = f(t)$ for every $t \neq z$. Conversely, every λ-coloring g of the graph $G|e$ induces a λ-coloring f of $G - e$ with the property $f(x) = f(y)$ by defining $f(x) = f(y) = g(z)$ and $f(t) = g(t)$ for $t \neq x, y$. Since both correspondences are injective, it follows that

$$P_{G-e}(\lambda) - P_G(\lambda) = P_{G|e}(\lambda)$$

for every natural number λ.

Since the equality holds for polynomials of a given degree, it follows that the chromatic polynomials are equal for every λ.

10.16 (a) Let K_n have vertex set $\{x_1, \ldots, x_n\}$. A λ-coloring f can be defined at x_1 in λ ways, at x_2 in $\lambda - 1$ ways, $\ldots$, at x_n in $\lambda - n + 1$ ways, so that there are

$$P_{K_n}(\lambda) = \lambda(\lambda - 1) \cdots (\lambda - n + 1)$$

possible ways of defining the function f so that it takes on pairwise different values for all the vertices $x_1, \ldots, x_n$.

(b) Let x be a vertex of degree 1 of T_n, and denote by $T_n - x$ the tree which is obtained from T_n by suppressing the vertex x and the edges incident with x. A λ-coloring of $T_n - x$ can be extended in $\lambda - 1$ ways to a λ-coloring of T_n, and thus

$$P_{T_n}(\lambda) = (\lambda - 1) P_{T_n - x}(\lambda).$$

By continuing in this manner one sees that
$$P_{T_n}(\lambda)=(\lambda-1)^{n-1}P_{T_1}(\lambda)=\lambda(\lambda-1)^{n-1},$$
since the chromatic polynomial of the tree with a single vertex is λ.

(c) Let e be an edge of the cycle C_n. Apply the property of Problem 10.15 to obtain
$$P_{C_n}(\lambda)=P_{C_n-e}(\lambda)-P_{C_n|e}(\lambda),$$
where C_n-e is a walk with n vertices. Thus from (b) it follows that $P_{C_n-e}(\lambda)=\lambda(\lambda-1)^{n-1}$ and $C_n|e$ is a cycle with $n-1$ vertices. Thus
$$P_{C_n}(\lambda)=\lambda(\lambda-1)^{n-1}-P_{C_{n-1}}(\lambda),$$
and a repetition of this argument yields
$$P_{C_n}(\lambda)=\lambda(\lambda-1)^{n-1}-\lambda(\lambda-1)^{n-2}+\cdots+(-1)^{n-2}\lambda(\lambda-1),$$
since $P_{C_3}(\lambda)=P_{K_3}(\lambda)=\lambda(\lambda-1)(\lambda-2)=\lambda(\lambda-1)^2-\lambda(\lambda-1)$. One can thus write $\lambda(\lambda-1)^{p-1}=(\lambda-1+1)(\lambda-1)^{p-1}=(\lambda-1)^p+(\lambda-1)^{p-1}$ for $2\leq p\leq n$, and hence
$$P_{C_n}(\lambda)=(\lambda-1)^n+(\lambda-1)^{n-1}-(\lambda-1)^{n-1}-(\lambda-1)^{n-2}+\cdots$$
$$+(-1)^{n-2}(\lambda-1)^2+(-1)^{n-2}(\lambda-1)$$
$$=(\lambda-1)^n+(-1)^n(\lambda-1).$$

10.17 Both properties will be proven by induction on the number of edges of the graph G. For the graph with n isolated vertices one has $P_G(x)=x^n$, which has the desired form.

Suppose now that the property of alternating signs of the coefficients of the chromatic polynomial is true for all graphs with n vertices and $m\leq p-1$ edges. It will be established for an arbitrary graph G with n vertices and $p\leq\binom{n}{2}$ edges.

Let e be an edge of G. It follows from the result of Problem 10.15 that
$$P_G(x)=P_{G-e}(x)-P_{G|e}(x). \tag{1}$$
Since the graphs $G-e$ and $G|e$ have at most $p-1$ edges, one can use the induction hypothesis to show that
$$P_{G-e}(x)=x^n-b_{n-1}x^{n-1}+b_{n-2}x^{n-2}-\cdots+(-1)^{n-1}b_1x,$$
$$P_{G|e}(x)=x^{n-1}-c_{n-2}x^{n-2}+c_{n-3}x^{n-3}-\cdots+(-1)^{n-2}c_1x,$$
where $b_i, c_i\geq 0$. Substituting in (1), one finds that
$$P_G(x)=x^n-(b_{n-1}+1)x^{n-1}+(b_{n-2}+c_{n-2})x^{n-2}-\cdots+(-1)^{n-1}(b_1+c_1)x,$$
which completes the proof of the first property.

The graph G, which has one more edge than $G-e$, has the coefficient $a_{n-1}=b_{n-1}+1$. But $a_{n-1}=0$ for the graph with n vertices and no edges. It follows that a_{n-1} is the number of edges in the graph G.

Now let G be a connected graph with n vertices. It will be shown that

Solutions

$a_i \geqslant \binom{n-1}{i-1}$ by induction on the number of edges of G. If G has the minimal number of edges, equal to $n-1$, then it is a tree, and by Problem 10.16 its chromatic polynomial is equal to

$$P_G(x) = x(x-1)^{n-1} = x^n - \binom{n-1}{1}x^{n-1} + \binom{n-1}{2}x^{n-2} - \cdots$$
$$+ (-1)^{n-1}\binom{n-1}{n-1}x.$$

This establishes the inequality $a_i \geqslant \binom{n-1}{i-1}$ for $1 \leqslant i \leqslant n-1$.

Suppose that the property is true for every connected graph with n vertices and m edges such that

$$n - 1 \leqslant m \leqslant p - 1.$$

The property will be proved for a connected graph G with n vertices and $p \leqslant \binom{n}{2}$ edges.

Suppose that G is not a tree. Then there exists an edge e such that $G - e$ is a connected graph. The graph $G|e$ obtained by identifying the endpoints of e is also connected, and, using the previous notation and the induction hypothesis, one can see that $a_i = b_i + c_i$ for $i = 1, \ldots, n-1$ and $c_{n-1} = 1$, and hence $a_i \geqslant \binom{n-1}{i-1}$.

Since the chromatic polynomial of a graph is equal to the product of the chromatic polynomials of its components, it follows that the smallest number s such that x^s has a nonzero coefficient in $P_G(x)$ is equal to the number of components in G.

10.18 If the graph G has components $G_1, \ldots, G_m$, then a λ-coloring can be defined on a component independent of its definition on the other components. It follows that

$$P_G(\lambda) = P_{G_1}(\lambda) \cdots P_{G_m}(\lambda).$$

Thus $P_G(\lambda)$ has no roots in the interval $(0, 1)$ if and only if each polynomial $P_{G_i}(\lambda)$ has this property for $i = 1, \ldots, m$.

Finally, one can suppose that G is connected and proceed to show that $(-1)^{n-1}P_G(\lambda) > 0$ if the graph G has n vertices and $0 < \lambda < 1$. This property will be established by induction on the number of edges of the graph G. If G has a minimal number $n - 1$ of edges, then it is a tree, $P_G(\lambda) = \lambda(\lambda - 1)^{n-1}$, and the property is satisfied.

Suppose now that the property is true for all connected graphs G with $m \geqslant n - 1$ edges. It will be shown that the property also holds for a graph G with n vertices and $m + 1$ edges.

Since G has a number of edges which is greater than $n - 1$, it follows that there is an edge e in G with the property that $G - e$ is connected. The graph $G|e$ obtained by identifying the endpoints of e is also connected; it has $n - 1$ vertices and at most m edges. Therefore, by the induction hypothesis

$$(-1)^{n-1}P_{G-e}(\lambda) > 0 \text{ and } (-1)^{n-2}P_{G|e}(\lambda) > 0 \quad \text{for } 0 < \lambda < 1.$$

Furthermore, it follows from Problem 10.15 that

$$(-1)^{n-1}P_G(\lambda) = (-1)^{n-1}\{P_{G-e}(\lambda) - P_{G|e}(\lambda)\}$$
$$= (-1)^{n-1}P_{G-e}(\lambda) + (-1)^{n-2}P_{G|e}(\lambda)$$
$$> 0 \quad \text{for every} \quad 0 < \lambda < 1.$$

Thus one can conclude that $P_G(\lambda)$ has no root in the interval $(0, 1)$.

The fact that $(3-\sqrt{5})/2 \in (0, 1)$ implies that $(3-\sqrt{5})/2$ is not a root of any chromatic polynomial $P_G(\lambda)$. Since the chromatic polynomials have rational coefficients, it follows that $(3+\sqrt{5})/2 = \tau + 1$ is not a root of any chromatic polynomial $P_G(\lambda)$ for any graph G.

10.19 If x is a vertex of maximum degree, then there are D edges incident with x, which thus have a common endpoint. It follows that

$$q(G) \geq D.$$

The proof of $q(G) \leq D+1$ will use induction on the number m of edges of the graph. If $m=1$ then $q(G)=1$ and $D=1$, so that the inequality is satisfied. Suppose that $q(G) \leq D+1$ for all graphs with at most $m-1$ edges, and let G be a graph with m edges and maximum degree D.

Now suppose that all the edges of G, with the exception of the edge $e_1 = [v, w_1]$, have been colored with $D+1$ colors so that every two edges with a common endpoint have different colors. It will be shown that under these circumstances there exists a coloring of the edges of G with $D+1$ colors, that is, $q(G) \leq D+1$.

In fact, by the induction hypothesis there is a coloring with $D'+1$ colors of the edges of the graph G_1 obtained from G by suppressing the edge $[v, w_1]$. But $D'=D$ or $D'=D-1$, and hence $D' \leq D$. There also exists a coloring of the edges of G_1 with at most $D+1$ colors which satisfies the given condition.

It will be shown that this coloring can be extended to the edge e_1. Since the maximum degree in G is D, it follows that for the edges incident with v and with w_1, at least one color from among the $D+1$ colors is missing.

If the same color is missing at v and w_1, then one can use it to color the edge e_1 and the proof is finished. Otherwise, let α be the color missing from among the edges incident with v, and let $\beta_1 \neq \alpha$ be the color missing from among the edges incident with w_1.

(a) Let $e_2 = [v, w_2]$ be the edge incident with v which is colored β_1. There exists such an edge, because otherwise this color would be missing from among the edges incident to v and to w_1, contrary to hypothesis. Delete the color of the edge e_2, and color e_1 with β_1. Suppose that the vertices v, w_1, w_2 belong to the same connected component of the spanning graph $H(\alpha, \beta_1)$ of G consisting of the vertices of G and the edges colored with α or β_1. If this were not possible, one could interchange the colors α and β_1 in the component which contains the vertex w_2, without changing the color β_1 of e_1. Since the edge e_2 is colored β_1, it follows that w_2 is not incident with another edge colored β_1. By inter-

Solutions

changing α and β_1 at w_2 it turns out that the color α becomes available and it is also missing from among the colors of the edges incident with v. It follows that the edge $e_2 = [v, w_2]$ can be colored with α and this yields a coloring of the edges of the graph G with at most $D+1$ colors.

It remains to consider the case in which the vertices v, w_1, w_2 belong to the same component of the graph $H(\alpha, \beta_1)$.

(b) Let $\beta_2 \neq \beta_1$ be a color which is missing from the edges incident with w_2. One can suppose that β_2 is the color of an edge incident with v, for otherwise the proof could be finished by coloring the edge e_2 with β_2. Let $e_3 = [v, w_3]$ be the edge colored β_2 which is incident with v. Delete the color β_2 from the edge e_3, and color e_2 with β_2, since β_2 does not occur at w_2. Following the same reasoning as in (a), one can consider only the case in which v, w_2, w_3 belong to the same component of the spanning graph $H(\alpha, \beta_2)$.

(c) The number of colors is finite, and the colors themselves $\beta_1, \beta_2, \beta_3, \ldots$ are pairwise distinct. It follows that either one can color all the edges incident with v (and thereby obtain a coloring of the edges of G with at most $D+1$ colors) or else one arrives at the following situation: One can no longer use operations of type (a) or (b) to recolor the last edge $[v, w_k]$ which has color β_{k-1}. This is because every color missing from w_k is a color β_i where $i < k-1$.

Similarly one can suppose that v, w_i, w_{i+1} belong to the same component C of the spanning graph $H(\alpha, \beta_i)$, and thus there is a walk with endpoints w_i and w_{i+1} which consists only of edges colored alternately α and β_i. The color α does not occur in the edges incident with v, and β_i does not occur in the edges incident with w_{i+1}. It follows that C is a walk from v to w_{i+1} which passes through w_i and which contains only edges colored α and β_i alternately (Figure 10.9).

This walk does not contain the vertex w_k, since β_i does not occur among the colors of the edges which have an endpoint at w_k. If C_1 is the component of $H(\alpha, \beta_i)$ which contains w_k, then C and C_1 are disjoint. This follows because

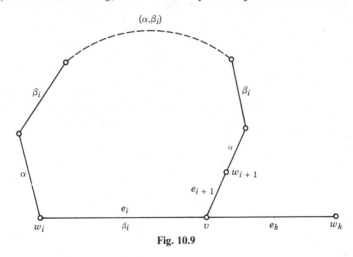

Fig. 10.9

otherwise they would form a single component, contrary to the definition of C and the fact that the color β_i is missing at w_k. Thus in the component C_1 one can interchange the colors α and β_i without causing the color α to occur at v for the edge $[v, w_i]$ of color β_i.

The color β_i is missing at w_k. It follows that after interchanging the colors in C_1, the color α will be missing at w_k and in v. Thus one can color the edge $[v, w_k]$ with color α. This yields a coloring of the edge set of G with $D+1$ colors.

Thus, for example, $q(G) = D+1$ for an odd cycle C_{2s+1}, for which $q(C_{2s+1}) = 3$ and $D = 2$.

10.20 The number of edges without a common endpoint which can be colored with the same color is less than or equal to $n/2$. Thus for n even there exist at most $n/2$ edges which can be colored with the same color, and for n odd there exist at most $(n-1)/2$ edges which can be colored with the same color. Since the complete graph K_n has $\binom{n}{2}$ edges, it follows that $q(K_n) \geq n-1$ for n even and $q(K_n) \geq n$ for n odd.

The opposite inequality will be established by means of a construction which provides a coloring of the edges of K_n with n colors for n odd and with $n-1$ colors for n even.

Represent the vertices of the graph K_5 by the vertices of a regular pentagon $ABCDE$. Color the sides of the pentagon with the colors $1, 2, \ldots, 5$, and all the diagonals parallel to a side with the same color as the respective side (Figure 10.10). One thus obtains a coloring of the edges of K_5 with five colors which satisfies the condition that two edges which have a common endpoint have different colors.

Observe that in this way one color becomes available at each vertex. For example, no edge incident with A has color 4. The set of colors which are available at all the vertices is the set of five colors used to color the edges of K_5.

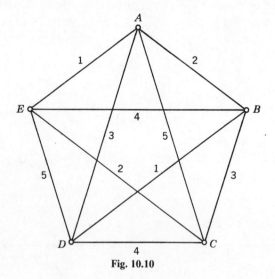

Fig. 10.10

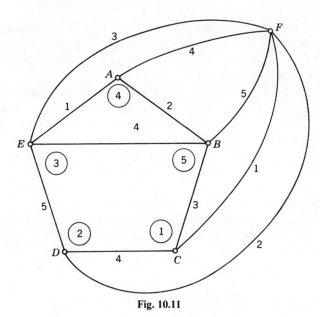

Fig. 10.11

It follows that by taking a new vertex F and joining it to all the vertices of the pentagon, the coloring of the edges of K_5 is uniquely extended to a coloring of the edges of K_6 with five colors. For example, the edge AF is given color 4, which is available at vertex A, and so on (Figure 10.11). This construction can be generalized to any two graphs K_n and K_{n+1} with n odd.

10.21 Represent the chess players by the vertices of a complete graph K_n. The matches played in one day can be represented by a set of edges which do not have an endpoint pairwise in common. It follows that the minimum number of days in which the tournament can end is equal to the minimum number of colors needed to color the edges of K_n so that each two edges with a common endpoint have different colors. In fact, one can use the same color to color edges which correspond to matches which are played in the same day. Thus the minimum number of days is $q(K_n)$, which (by the previous problem) is equal to $n-1$ for even n and to n for odd n.

10.22 The property will be proven by induction on the number n of vertices of the graph. For $n=1$ or 2 the result is immediate for $1 \leqslant k \leqslant n$. Suppose that the property is true for all graphs with at most $n-1$ vertices.

Let G be a graph with n vertices and chromatic number $\chi(G)=k$ ($1 \leqslant k \leqslant n$). Let x be a vertex of the graph, and let G_x be the subgraph obtained by suppressing the vertex x and the edges incident with x. The number of minimal colorings of the graph G with $\chi(G)$ colors will be denoted $C_m(G)$.

If $\chi(G_x)=k$, it follows that

$$C_m(G) \leqslant k C_m(G_x) \leqslant k \cdot k^{n-k-1} = k^{n-k}.$$

The inequality becomes an equality only when x is an isolated vertex and the subgraph G_x has a maximum number of minimal colorings. This follows from the fact that the vertex x can be added to a minimal partition of G_x with k classes in at most k different ways.

If $\chi(G_x)=k-1$, then a minimal coloring of G is given as follows: $\{x\}, C_1, \ldots, C_{k-1}$, where the sets $C_1, \ldots, C_{k-1}$ consist of pairwise nonadjacent vertices and there exist $k-1$ vertices $x_1 \in C_1, \ldots, x_{k-1} \in C_{k-1}$ which are adjacent to x. Otherwise one would have $\chi(G) \leqslant k-1$.

Let X be the vertex set of G. It follows that every k-coloring of G has a class which contains the vertex x and a subset of the set $X \setminus \{x_1, \ldots, x_{k-1}\}$. Since $X \setminus \{x, x_1, \ldots, x_{k-1}\}$ contains $n-k$ elements, it follows from the induction hypothesis that the number of k-colorings of X which contain in the same class the vertex x and r vertices of the set $X \setminus \{x, x_1, \ldots, x_{n-k}\}$ is bounded above by $\binom{n-k}{r}(k-1)^{n-k-r}$ for $0 \leqslant r \leqslant n-k$. In fact the r vertices can be chosen from the $n-k$ vertices in $\binom{n-k}{r}$ distinct ways. The maximal number of $(k-1)$-colorings of a graph G with $n-k-r$ vertices and $\chi(G)=k-1$ is equal to $(k-1)^{n-k-r}$. Thus

$$C_m(G) \leqslant \sum_{r=0}^{n-k} \binom{n-k}{r}(k-1)^{n-k-r} = k^{n-k}.$$

Equality holds only when $\{x, x_1, \ldots, x_{k-1}\}$ induces a complete subgraph and the remaining $n-k$ vertices are isolated. Thus it has been established by induction that $C_m(G) \leqslant k^{n-k}$ for every graph G with n vertices and chromatic number equal to k. The upper bound is attained only when G consists of a complete k-subgraph and $n-k$ isolated vertices. Observe that this graph is the unique graph G with $\chi(G)=k$ and the minimal number $\binom{k}{2}$ of edges. [I. Tomescu, C. R. Acad. Sci. Paris, **A272** (1971), 1301–1303.]

10.23 Let $C(n, k)$ be the desired number of colorings. It is clear that $C(n, n) = C(n, 2) = 1$ for every $n \geqslant 2$, and $C(n, k) = 0$ for $k > n$. It will be shown by induction on n that $C(n, k) = S(n-1, k-1)$ for every tree with n vertices. For $n=2$ the property is obviously true. Suppose that the property is true for every tree with at most $n-1$ vertices. It follows that $C(n, k) = S(n-2, k-2) + (k-1)S(n-2, k-1)$.

In fact every tree A with n vertices contains a vertex x of degree 1 such that the subgraph A_x obtained from A by suppressing the vertex x is a tree with $n-1$ vertices. The set of $C(n, k)$ colorings of A can be expressed as the union of the set of k-colorings for which the vertex x is alone in a class of the partition [there are $S(n-2, k-2)$ such colorings, i.e., the number of $(k-1)$-colorings of the tree A_x] and of the set of k-colorings in which the vertex x occurs in a class together with other vertices of the tree. Since x is adjacent to a unique vertex of the tree, it follows that x can be added to exactly $k-1$ classes of a k-coloring of a tree with $n-1$ vertices. Thus, by the induction hypothesis this set of colorings has cardinality $(k-1)S(n-2, k-1)$. It follows from the recurrence relation for Stirling numbers of the second kind (Problem 3.4) that $C(n, k) = S(n-1, k-1)$ for every tree with n vertices.

10.24 The property will be proven by induction on k. For $k=1$ the graph G

does not contain an edge, and it is sufficient to let $H=G$. In this case $d_H(x)=d_G(x)=0$ for every $x \in X$, and the graph H is monochromatic, since it consists of isolated vertices.

Suppose that the property is true for all graphs which do not contain a complete subgraph with $k+1$ vertices, and let G be a graph which does not contain a complete subgraph with $k+2$ vertices.

Let x be a vertex of maximum degree in G. Denote by X_1 the set of vertices adjacent to x, and let G_1 be the subgraph induced by the set of vertices X_1.

Since G_1 does not contain a complete subgraph with $k+1$ vertices, it follows from the induction hypothesis that there exists a graph H_1 with vertex set X_1, which is k-chromatic and satisfies

$$d_{H_1}(z) \geq d_{G_1}(z) \qquad \text{for every} \quad z \in X_1.$$

Let H be a graph with vertex set X. In H all the vertices of $X \setminus X_1$ are joined to all the vertices of X_1. Also adjacent are all pairs of vertices in X_1 which were adjacent in H_1. The graph H is $(k+1)$-chromatic by construction, since $\chi(H_1)=k$. If $y \notin X_1$ then

$$d_H(y) = |X_1| = d_G(x) \geq d_G(y),$$

and if $y \in X_1$, it follows that

$$d_H(y) = |X| - |X_1| + d_{H_1}(y)$$
$$\geq |X| - |X_1| + d_{G_1}(y) \geq d_G(y),$$

which establishes the property.

Now let G be a graph with n vertices and without a complete subgraph with $k+1$ vertices. Let $A_1, \ldots, A_k$ be monochromatic subsets of vertices of H.

Denote $|A_i|=a_i$, so that $a_1 + \cdots + a_k = n$, and let U, V be the edge sets of the graphs G and H respectively. It follows that

$$|U| = \frac{1}{2} \sum_{x \in X} d_G(x) \leq \frac{1}{2} \sum_{x \in X} d_H(x) = |V| \leq \sum_{1 \leq i < j \leq k} a_i a_j,$$

since there are edges in H only between vertices from distinct sets A_i and A_j.

The last sum is maximal if and only if $|a_i - a_j| \leq 1$. Thus $a_1 = \cdots = a_r = m+1$ and $a_{r+1} = \cdots = a_k = m$, where $m = [n/k]$ and $n \equiv r \pmod{k}$. Observe that this limit is attained only if the last inequality becomes an equality and hence in the class of complete multipartite graphs. (See Turán's theorem, Problem 9.9.) [P. Erdös, *Mat. Lapok*, **21** (1970), 249–251.]

10.25 Suppose that G possesses the maximum number of edges in the class of graphs with n vertices and chromatic number equal to k, and that the number of vertices with color i is equal to n_i for $1 \leq i \leq k$ ($n_1 + \cdots + n_k = n$). It is clear that any two vertices of different colors are adjacent, that is, G is a complete multipartite graph. If for example $n_1 \geq n_2 + 2$, then one can move one vertex from the class with n_1 vertices into the class with n_2 vertices, thus obtaining a new multipartite complete graph G_1 with n vertices and m_1 edges. It follows that

$$m_1 - m = n_1 - 1 - n_2 \geq 1,$$

which is a contradiction. Hence $|n_i - n_j| \leq 1$ for any $1 \leq i, j \leq k$, and G is isomorphic with the Turán graph defined in Problem 9.9.

10.26 Let G be a graph with n vertices, and let $P_G(\lambda) = \lambda^n - a_{n-1}\lambda^{n-1} + \cdots + (-1)^{n-1}a_1\lambda$. It follows that a_{n-1} is the number of edges of G, and $\chi(G) = t$ if and only if $P_G(1) = P_G(2) = \cdots = P_G(t-1) = 0$ and $P_G(t) > 0$. If $P_G(\lambda) = P_{T(n,k)}(\lambda)$, it follows that $\chi(G) = \chi(T(n,k)) = k-1$ and that G and $T(n,k)$ have the same number $M(n,k)$ of edges. From Problem 10.25 one can conclude that G is isomorphic to $T(n,k)$. [C.-Y. Chao, G. A. Novacky, *Discrete Math.*, 41 (1982), 139–143.]

10.27 One can show that $P_G(k) = \sum_{i=0}^{n} i!\binom{k}{i}C_i(G)$ holds, where $C_i(G)$ stands for the number of i-colorings of G, which are partitions of the vertex set of G. In fact, i colors can be chosen from the set of k colors in $\binom{k}{i}$ different ways, and a partition into i classes generates $i!$ colorings, taking into account the order of the classes. The formula for $C_k(G)$ results from the inverse binomial formulas (Problem 2.17) if $a_k = P_G(k)$ and

$$b_k = k!C_k(G) = \sum_{i=0}^{k}(-1)^{k-i}\binom{k}{i}P_G(i)$$
$$= \sum_{i=0}^{k}(-1)^i\binom{k}{i}P_G(k-i).$$

10.28 The set of points having coordinates $(0, 0)$, $(\frac{1}{2}, \frac{1}{2})$, $(\frac{1}{2}, -\frac{1}{2})$, and $(1, 0)$ is a 4-clique for G_4, and the points $(0, 0)$, $(0, 1)$, $(1, 0)$, and $(1, 1)$ induce a 4-clique for G_8, which implies that $\chi(G_4) \geq 4$ and $\chi(G_8) \geq 4$. It remains to define a 4-coloring for G_4 and G_8. In the case of G_4 consider all lines with slope 1 or -1 passing through the points of E^2 having integer coordinates (digital points). The intersection points of these lines are points $M(p, q)$ where $p, q \in \mathscr{Z}$ and points $N(r/2, s/2)$ where $r, s \in \mathscr{Z}$ and $r, s \equiv 1 \pmod{2}$. Let S denote the set of all such points. Color $M(p, q) \in S$ with color α if $p \equiv q \pmod 2$ and with color β if $p \equiv q+1 \pmod 2$. Color $N(r/2, s/2) \in S$ with the color γ if $r \equiv s \pmod 4$ and with the color δ if $r \equiv s+2 \pmod 4$. For $P(u, v) \in S$ consider the points $Q = (u - \frac{1}{2}, v + \frac{1}{2})$ and $R = (u - \frac{1}{2}, v - \frac{1}{2})$. If P is colored with the color $a \in \{\alpha, \beta, \gamma, \delta\}$, then all interior points of the segments PQ and PR will also be colored with the color a. In this way any square $ABCD$ having vertices in S and length of a side equal to $\sqrt{2}/2$ will have its four vertices colored with $\alpha, \beta, \gamma, \delta$, and the colors of the sides will be a, a, b, c, where $a, b, c \in \{\alpha, \beta, \gamma, \delta\}$. In this case color all interior points of $ABCD$ with the color a. Thus all points of E^2 will be colored with four colors. It is easy to see that if $d_4(E, F) = 1$ then E and F have different colors.

A 4-coloring of G_8 may be defined in a similar manner. Let S denote the set of digital points of E^2, and color the points of S in the following way: the point $M(p, q)$ with $p, q \in \mathscr{Z}$ will be colored with the color α if $p \equiv 1 \pmod 2$ and $q \equiv 1 \pmod 2$; β if $p \equiv 0 \pmod 2$ and $q \equiv 0 \pmod 2$; γ if $p \equiv 1 \pmod 2$ and $q \equiv 0 \pmod 2$; δ if $p \equiv 0 \pmod 2$ and $q \equiv 1 \pmod 2$. If $M(p, q) \in S$ is colored with the color a, then all interior points of the segments MQ and MR, where $Q = (p-1, q)$ and $R = (p, q-1)$ will also be colored with the color a. Any unit square $ABCD$ with its vertices in S has its four vertices colored with $\alpha, \beta, \gamma, \delta$, and the colors of the sides are $a, a, b, c \in \{\alpha, \beta, \gamma, \delta\}$. Color all interior points of $ABCD$ with the color a. Now if $d_8(E, F) = 1$, then points $E, F \in E^2$ will have different colors. Let G be the corresponding graph for Euclidean distance in E^2. The problem of determining $\chi(G)$ is an open problem in Euclidean Ramsey theory. It follows from Problem 14.1 that $\chi(G) \geq 4$. The existence of a 7-coloring of the plane covered by congruent regular hexagons of side $s \in (1/\sqrt{7}, \tfrac{1}{2})$ implies that $\chi(G) \leq 7$.

10.29 If G has no subgraph isomorphic to K_p, it is clear that G has at least $p+1$ vertices. Suppose that G has $p+1$ vertices. Because $\chi(G) = p$, it follows that there exists a partition of its vertex set of the form $\{x_1\}, \{x_2\}, \ldots, \{x_{p-1}\}, \{x_p, x_{p+1}\}$ where $\{x_1, \ldots, x_{p-1}\}$ spans a complete subgraph K_{p-1} and x_p and x_{p+1} are not adjacent. Since G has no p-clique, it follows that x_p is nonadjacent to at least one vertex x_i, and x_{p+1} is nonadjacent to at least one vertex x_j, where $1 \leq i, j \leq p-1$. But in this case $\{x_1\}, \ldots, \{x_i, x_p\}, \ldots, \{x_j, x_{p+1}\}, \ldots, \{x_{p-1}\}$ (for $i \neq j$) or $\{x_1\}, \ldots, \{x_i, x_p, x_{p+1}\}, \ldots, \{x_{p-1}\}$ (for $i=j$) is a $(p-1)$-coloring of G, which contradicts the hypothesis. Thus G contains at least $p+2$ vertices.

Now consider a $(p-1)$-clique C, two vertices $a, b \in C$, and three new vertices $x, y, z \notin C$. By definition x is adjacent to all vertices $c \in C$ such that $c \neq a$; y is adjacent to all vertices $c \in C$ such that $c \neq b$; z is adjacent to x, y and to all vertices $c \in C$ such that $c \neq a, b$. The graph G defined in this way satisfies all the conditions in the statement of the problem, and it reduces to the five-vertex cycle C_5 for $p = 3$.

CHAPTER 11

11.1 Denote the two vertex sets of $K_{n,n}$ by $X = \{x_1, \ldots, x_n\}$ and $Y = \{y_1, \ldots, y_n\}$; the edges of the graph are of the form $[x_i, y_j]$ with $1 \leq i, j \leq n$.

Since every Hamiltonian cycle passes through x_1, one can determine the number of ways of constructing a Hamiltonian cycle which originates and terminates at x_1. One can leave from x_1 towards one of the vertices in Y in n ways. From here one can return to a vertex of X, other than x_1, in $n-1$ ways. Now a continuation to a vertex in Y, other than the vertex which has already been traversed, in $n-1$ ways, and so on. When one arrives at the last nonvisited vertex in X, there will still exist a nonvisited vertex in Y which one can leave for and return to x_1. The number of cycles thus obtained is equal to

$$n(n-1)(n-1)(n-2)(n-2) \cdots (1)(1) = (n-1)! \, n!.$$

The family of cycles thus obtained contains every cycle exactly twice, corresponding to the two directions in which the cycle can be traversed. Thus $K_{n,n}$

contains exactly $\frac{1}{2}(n-1)!\,n!$ cycles which pass through each vertex exactly once.

11.2 For $h=0$ there are $(n-1)!/2$ Hamiltonian cycles in the complete graph K_n, and the formula is seen to be satisfied. In fact, every Hamiltonian cycle determines a cyclic permutation of the vertices of K_n; the number of cyclic permutations on n elements is equal to $(n-1)!$. It is possible to obtain two distinct circular permutations from the same Hamiltonian cycle by traversing the cycle in both directions. Thus the number of Hamiltonian cycles in K_n is equal to $\frac{1}{2}(n-1)!$. This formula can also be established by induction, since for $n=3$ the graph K_n is a Hamiltonian cycle. Suppose the formula holds for $n \leqslant m$, and let K_{m+1} be the complete graph with $m+1$ vertices: $x_1, \ldots, x_{m+1}$. Every Hamiltonian cycle in K_{m+1} can be obtained from a Hamiltonian cycle of K_m by inserting the vertex x_{m+1} between two adjacent vertices of the cycle. Each Hamiltonian cycle of K_m generates in this way m different cycles of K_{m+1}, and by the induction hypothesis K_m has $(m-1)!/2$ Hamiltonian cycles. It follows that the number of Hamiltonian cycles in K_{m+1} is equal to $\{(m-1)!/2\}m = m!/2$, and hence the property holds for every $m \geqslant 3$.

Let $h \geqslant 1$, and suppose that h edges are each replaced by a new vertex. One thereby obtains a complete graph with $n-h$ vertices which has $(n-h-1)!/2$ Hamiltonian cycles. Let z be a vertex which has replaced the edge $[x, y]$ in one of the $(n-h-1)!/2$ Hamiltonian cycles, and let u, v be adjacent to z in this cycle. The edge $[x, y]$ can replace the vertex z in this Hamiltonian cycle in exactly two distinct ways: The walk $[u, z, v]$ is replaced by $[u, x, y, v]$ or by $[u, y, x, v]$. After carrying out this operation for all the vertices which represent the h edges, one obtains $\{(n-h-1)!/2\}2^h = (n-h-1)!\,2^{h-1}$ pairwise distinct Hamiltonian cycles which pass through the h given edges.

Observe that every Hamiltonian cycle which passes through the h edges can be obtained without repetition in this way. The h edges have no vertices in common, and hence it follows that $2h \leqslant n$.

11.3 Let the vertices of K_n be numbered $0, 1, 2, \ldots, 2k$, and consider the following Hamiltonian cycle:

$$C_1 = [0, 1, 2, 2k, 3, 2k-1, 4, 2k-2, 5, \ldots, k+3, k, k+2, k+1, 0],$$

which is represented in Figure 11.1.

Now add 1 modulo $2k$ a total of $k-1$ times to all the nonzero numbers in this sequence, with the exception that $2k$ is not replaced by its residue. One thereby obtains a total of $k = (n-1)/2$ Hamiltonian cycles. In order to show that these Hamiltonian cycles are disjoint with respect to edges it is necessary to consider the sum modulo $2k$ of each two consecutive nonzero numbers in the sequence. For C_1 these sums form the set $\{2, 3\}$, and adding 2 modulo $2k$ yields

$$\{4, 5\}, \{6, 7\}, \ldots, \{2k-2, 2k-1\}, \{0, 1\}.$$

It can be seen that these k Hamiltonian cycles are disjoint with respect to edges. This implies that they cover all the edges of K_n, since K_n contains $n(n-1)/2$

Solutions

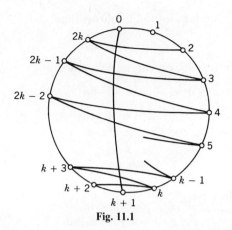

Fig. 11.1

edges, and the $(n-1)/2$ Hamiltonian cycles thus constructed each contain n edges.

11.4 (a) Suppose that G does not contain a Hamiltonian cycle. Add edges as long as possible under the condition that G does not contain a Hamiltonian cycle. It follows that the degrees of the vertices increase and thus conditions (a), (b), and (c) are always satisfied. One can therefore suppose that G is saturated, that is, the addition of a new edge produces a new graph which contains a Hamiltonian cycle. If $[x, y]$ is not an edge of G, then the addition of this edge to G yields a Hamiltonian cycle. Thus G contains a Hamiltonian walk

$$L = [z_1 = x, z_2, \ldots, z_{n-1}, z_n = y]$$

which joins x and y.

Denote by $z_{i_1}, \ldots, z_{i_k}$ the vertices adjacent with x, where $2 = i_1 < i_2 < \cdots < i_k \leqslant n$. It follows that y is not adjacent to $z_{i_s - 1}$ for $s = 1, \ldots, k$, since otherwise G would contain a Hamiltonian cycle

$$[z_1, \ldots, z_{i_s - 1}, z_n, z_{n-1}, \ldots, z_{i_s}, z_1].$$

Thus $d(y) \leqslant n - 1 - k = n - 1 - d(x) \leqslant n - 1 - n/2 < n/2$, which contradicts the condition $d_1 \geqslant n/2$. It follows that if $\min(d_1, \ldots, d_n) \geqslant n/2$, then G contains a Hamiltonian cycle.

(c) Now suppose that condition (c) holds and G is not Hamiltonian, but that by joining each two nonadjacent vertices by an edge one does obtain a Hamiltonian cycle. Consider two nonadjacent vertices x_k and x_l such that $k < l$ and the sum $k + l$ is maximum. It can be seen that x_k is adjacent to all the vertices $x_{l+1}, \ldots, x_n$, and hence

$$d_k \geqslant n - l. \tag{1}$$

In this case one finds that x_l is adjacent to the vertices $x_{k+1}, \ldots, x_{l-1}, x_{l+1}, \ldots, x_n$, which implies that

$$d_l \geqslant n - k - 1. \tag{2}$$

Using the same reasoning as for (a), it follows that

$$d_k + d_l \leqslant n - 1. \tag{3}$$

From the last two inequalities it can be concluded that

$$d_k \leqslant n - 1 - d_l \leqslant n - 1 - (n - k - 1) = k.$$

Let $m = d_k$. It follows that $m \leqslant k$ and by the hypothesis $d_m \leqslant d_k = m$. Since $k < l$, one concludes that $d_k \leqslant d_l$, and by (3) one has

$$m = d_k < \frac{n}{2}.$$

Hypothesis (c) implies that

$$d_{n-m} \geqslant n - m = n - d_k \geqslant d_l + 1, \tag{4}$$

from which one has

$$n - d_k = n - m > l. \tag{5}$$

Otherwise it would be the case that

$$n - m \leqslant l, \quad \text{and hence} \quad d_{n-m} \leqslant d_l,$$

which contradicts (4). From (5) it follows that

$$d_k < n - l,$$

which contradicts inequality (1). This completes the proof.

(b) If (b) holds, let $d_k \leqslant k < n/2$ and $l = n - k$. If $d_l \geqslant l$, then condition (c) is satisfied. Otherwise $d_l < l$, and condition (b) and the fact that $k < n - k$ or $k < n/2$ would imply that $d_k + d_{n-k} \geqslant n$ and thus $d_{n-k} \geqslant n - d_k \geqslant n - k$. Again (c) holds, which (as has been seen) implies the existence of a Hamiltonian cycle in the graph G.

If (c) holds, it has been shown that either n is even and $G = K_{n/2, n/2}$ or G is pancyclic, that is, it has elementary cycles of every length k for $3 \leqslant k \leqslant n$.

[S. L. Hakimi, E. F. Schmeichel, *J. Combinatorial Theory*, **B17** (1974), 22–34.]

11.5 Let x, y be any two vertices of the graph G. It will be shown that G contains a Hamiltonian walk which joins x and y.

If x and y are not joined by an edge, then add the edge $[x, y]$ to G. The degree of each vertex remains greater than $n/2$, and no Hamiltonian walk with endpoints x and y uses this edge.

Insert a new vertex z on the edge $[x, y]$ to produce a new graph G_1. The graph G_1 has a Hamiltonian cycle if and only if G contains a Hamiltonian walk with endpoints x and y. The sequence of degrees of the vertices of the graph G_1 is

$$2 \leqslant d_2 \leqslant \cdots \leqslant d_{n+1},$$

where $d_2 \leqslant \cdots \leqslant d_{n+1}$ are the degrees of the vertices of G.

Solutions

By hypothesis $d_2 \geqslant (n+1)/2$, and hence, since G_1 has $n+1$ vertices, condition (c) of Problem 11.4 is satisfied, because there does not exist an index k such that $d_k \leqslant k < (n+1)/2$. In fact $d_1 = 2$ and $d_k \geqslant (n+1)/2$ for every $k \geqslant 2$. It follows that G_1 has a Hamiltonian cycle or G has a Hamiltonian walk with endpoints x and y.

11.6 Add to the graph G a new vertex y which is joined by an edge to all other vertices of G. The graph thus obtained has $2n+2$ vertices, and its degrees are at least equal to $n+1$; thus by Dirac's condition (a) of Problem 11.4, it contains a Hamiltonian cycle.

Suppressing the vertex y and the edges incident with y yields a Hamiltonian walk $[x_0, x_1, \ldots, x_{2n}]$ in the graph G.

Suppose that G does not contain a Hamiltonian cycle. It follows that if x_0 is adjacent to a vertex x_i, then x_{2n} is not adjacent to x_{i-1}, since otherwise a Hamiltonian cycle would be formed.

The vertices x_0 and x_{2n} have degree n. It is also the case that if x_0 is not adjacent to x_i, then x_{2n} is adjacent to x_{i-1}. In fact x_0 is adjacent to n vertices x_i and thus x_{2n} is nonadjacent to n vertices of the form x_{i-1}. It follows that x_{2n} is adjacent to all of the n remaining vertices.

Suppose first that x_0 is adjacent to the vertices $x_1, \ldots, x_n$ and x_{2n} is adjacent to $x_n, \ldots, x_{2n-1}$. There exists an index i with $1 \leqslant i \leqslant n$ such that x_i is not adjacent to x_n; this results from the fact that $d(x_n) = n$ and x_n is adjacent to x_0, x_{n-1}, x_{n+1}, and x_{2n}. The vertex x_i is adjacent in turn to a vertex x_j for $n+1 \leqslant j \leqslant 2n-1$, since by hypothesis $d(x_i) = n$. In this case the graph G contains a Hamiltonian cycle

$$[x_i, x_{i-1}, \ldots, x_0, x_{i+1}, \ldots, x_{j-1}, x_{2n}, \ldots, x_j, x_i]$$

(see Figure 11.2). Otherwise there would exist an index i such that $1 \leqslant i \leqslant 2n-1$ and the vertex x_{i+1} is adjacent to x_0, but x_i is not adjacent to x_0. It follows from an earlier observation that the vertex x_{i-1} is adjacent to x_{2n}, and one thus obtains an elementary cycle with $2n$ vertices in the graph G, namely

$$[x_{i-1}, \ldots, x_0, x_{i+1}, \ldots, x_{2n}, x_{i-1}].$$

Let $C = [y_1, \ldots, y_{2n}, y_1]$ be an elementary cycle with $2n$ vertices in the graph G, and y_0 the vertex of G which does not belong to the cycle C. Since G does not contain a Hamiltonian cycle, it follows that y_0 cannot be adjacent to two neigh-

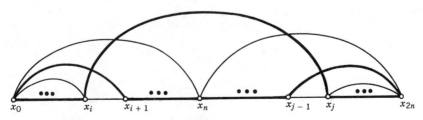

Fig. 11.2

boring vertices of C, for in this case C could be extended to a Hamiltonian cycle. In view of the fact that $d(y_0) = n$, this would imply that y_0 is adjacent to all the vertices of the cycle C whose indices differ by 2 (mod $2n$), for example, to $y_1, y_3, \ldots, y_{2n-1}$. By replacing y_{2i} with y_0 for $1 \leqslant i \leqslant n$ one obtains another elementary cycle of length $2n$. Repetition of this argument shows that y_{2i} is adjacent to $y_1, y_3, \ldots, y_{2n-1}$ for $i = 1, \ldots, n$. Finally y_1 is adjacent to $y_0, y_2, \ldots, y_{2n}$, and hence

$$d(y_1) \geqslant n+1,$$

which contradicts the hypothesis that G is regular of degree n and hence $d(y_1) = n$. (C. St. J. A. Nash-Williams.)

An extension of this result is the following: If G is an n-regular graph of order $2n$, $n \geqslant 3$, and $G \neq K_{n,n}$, or G is an n-regular graph of order $2n+1$, $n \geqslant 4$, then G is Hamiltonian-connected (every two distinct vertices of G are the endpoints of a Hamiltonian walk in G). [I. Tomescu, *J. Graph Theory*, **7** (1983), 429–436].

It can be shown analogously that there is a unique regular, non-Hamiltonian graph of degree n with $2n+2$ vertices. It consists of two disjoint copies of the complete graph K_{n+1}.

11.7 Let C be an elementary cycle of maximum length in the graph G, and suppose that C is not a Hamiltonian cycle. Denote by G_1 a connected component of the subgraph obtained from G by suppressing the vertices of the cycle C. Let $x_1, \ldots, x_s$ be vertices of C which are adjacent to vertices of G_1. No two of these are adjacent in the cycle C. If, for example, x_i and x_j are adjacent, one can replace the edge $[x_i, x_j]$ of the cycle C by a walk with endpoints x_i and x_j which passes through intermediate vertices of G_1. In this way a cycle longer than C is obtained, which contradicts the hypothesis.

Now traverse the cycle C in either sense, and let $y_1, \ldots, y_s$ be vertices which are adjacent to $x_1, \ldots, x_s$, respectively. It follows from the previous observation that $y_i \notin \{x_1, \ldots, x_s\}$ for every $1 \leqslant i \leqslant s$. It can be shown that $y_1, \ldots, y_s$ are pairwise nonadjacent. Otherwise if y_i and y_j were adjacent one could suppress the edges $[x_i, y_i]$ and $[x_j, y_j]$ of the cycle and replace them by the edge $[y_i, y_j]$ and a walk with endpoints x_i and x_j which passes through vertices of the graph G_1.

The result is an elementary cycle longer than C, whose existence contradicts the hypothesis (Figure 11.3).

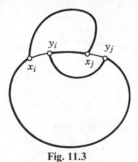

Fig. 11.3

Recall that $x_1, \ldots, x_s$ are all the vertices of C which are adjacent to vertices of G_1. It follows that if y_0 is a vertex in G_1, then the set $S = \{y_0, y_1, \ldots, y_s\}$ is an independent set of vertices. By suppressing from G the vertices $x_1, \ldots, x_s$ one obtains at least two connected components, one of which is G_1. Since the graph G is k-connected, it must contain, by definition, disconnecting sets with at least k vertices and hence $s \geqslant k$. But this implies that

$$|S| = s + 1 \geqslant k + 1,$$

which contradicts the hypothesis. Thus G is Hamiltonian. [V. Chvátal, P. Erdös, *Discrete Math.*, **2** (1972), 111–113.]

11.8 The property will be established by induction on n. If $n = 3$ then $m \geqslant 3$ and hence G is the graph K_3, which is a Hamiltonian cycle. Suppose that the property is true for all graphs with at most $n - 1$ vertices, and let G be a graph with $n \geqslant 4$ vertices and $m \geqslant \binom{n-1}{2} + 2$ edges.

It will be shown that G contains a vertex x of degree $d(x) = n - 1$ or $d(x) = n - 2$. Otherwise one would have $d(x) \leqslant n - 3$ for every vertex x, and since $\sum d(x) = 2m$, it would follow that

$$m \leqslant \frac{n(n-3)}{2},$$

which contradicts the hypothesis that $m \geqslant (n^2 - 3n + 6)/2$.

In the sequel two cases will be considered:

(a) There exists a vertex x with $d(x) = n - 2$. Suppress the vertex x and the $n - 2$ edges incident with it. The result is a subgraph G_1 with $n - 1$ vertices and m_1 edges such that

$$m_1 \geqslant \frac{n^2 - 3n + 6}{2} - (n - 2) = \binom{n-2}{2} + 2.$$

By the induction hypothesis G_1 contains a Hamiltonian cycle C_1 which passes through all of the $n - 1$ vertices of G_1, exactly once:

$$C_1 = [x_0, x_1, \ldots, x_{n-2}, x_0].$$

If there exist two adjacent vertices of the cycle, x_i and x_{i+1} (where the sum is taken modulo $n - 1$), which are also adjacent to x, then it is possible to insert x between x_i and x_{i+1} so as to obtain a Hamiltonian cycle C in the graph G.

Otherwise, for every $i = 0, \ldots, n - 2$, if x were adjacent to x_i it would follow that x is not adjacent to x_{i+1} and hence $d(x) \leqslant (n-1)/2$. It can thus be seen that $n - 2 \leqslant (n-1)/2$, which contradicts the hypothesis that $n \geqslant 4$. One can now conclude that G contains a Hamiltonian cycle.

(b) There does not exist a vertex x with $d(x) = n - 2$. It follows that there are at least two vertices y_1 and y_2 with $d(y_1) = d(y_2) = n - 1$. If not, one would have a unique vertex of degree $n - 1$ and the other $n - 1$ vertices of degree less than or equal to $n - 3$, and hence

$$\frac{(n-1)(n-2)}{2}+2 \leqslant m \leqslant \frac{(n-1)+(n-1)(n-3)}{2} = \frac{(n-1)(n-2)}{2},$$

which is a contradiction.

Now suppress the vertices y_1 and y_2 and the $2(n-1)-1=2n-3$ edges incident with them. The result is a subgraph G_2 with $n-2$ vertices and m_2 edges such that

$$m_2 \geqslant \frac{n^2-3n+6}{2} - (2n-3) = \binom{n-3}{2}.$$

If the graph G_2 contains a Hamiltonian cycle C_2 then it is evident that G also contains a Hamiltonian cycle C obtained from C_2 by inserting the vertices y_1 and y_2 in an arbitrary fashion in the cycle C_2. For otherwise G_2 would not contain a Hamiltonian cycle, and thus G_2 is obtained from K_{n-2} by suppressing at least two edges. In fact, if G_2 has vertices $x_0, \ldots, x_{n-3}$, a single pair of which (say $\{x_0, x_1\}$) are nonadjacent, then there is a Hamiltonian cycle in G_2, namely

$$[x_0, x_2, x_1, x_3, \ldots, x_{n-3}, x_0],$$

for every $n \geqslant 6$. By adding two edges u_1 and u_2 to the graph G_2 between two pairs of nonadjacent vertices one obtains a graph G_3 with $m_3 = m_2 + 2 \geqslant \binom{n-3}{2}+2$ edges, which by the induction hypothesis must then contain a Hamiltonian cycle C_3.

If C_3 does not contain the edges u_1 and u_2, then it has been shown that one can insert vertices y_1 and y_2 to obtain a Hamiltonian cycle in the graph G. Let $C_3 = [x_0, \ldots, x_i, x_{i+1}, \ldots, x_{n-3}, x_0]$, $u_1 = [x_i, x_{i+1}]$, and suppose that C_3 does not contain the edge u_2. It follows that there is a Hamiltonian cycle C for G:

$$C = [x_0, \ldots, x_i, y_1, y_2, x_{i+1}, \ldots, x_{n-3}, x_0].$$

A similar result holds when C_3 contains u_2 and does not contain u_1. Suppose that C_3 contains u_1 and u_2, and let $u_1 = [x_{i-1}, x_i]$ and $u_2 = [x_j, x_{j+1}]$ with $i \leqslant j$. In this case let

$$C = [x_0, \ldots, x_{i-1}, y_1, x_i, \ldots, x_j, y_2, x_{j+1}, \ldots, x_{n-3}, x_0].$$

Thus G is seen always to contain a Hamiltonian cycle.

If $n=4$ or $n=5$ it can be shown directly that in case (b) G contains a Hamiltonian cycle. It has already been shown that under these circumstances there are two vertices y_1 and y_2 such that $d(y_1) = d(y_2) = 3$ for $n=4$ and $d(y_1) = d(y_2) = 4$ for $n=5$. For $n=5$ it follows that $m \geqslant \binom{4}{2}+2 = 8$, and thus there also exists at least one edge connecting the vertices x_1, x_2, x_3, for example $[x_1, x_2]$. One has thus found Hamiltonian cycles $[x_1, y_1, x_2, y_2, x_1]$ and $[x_1, x_2, y_2, x_3, y_1, x_1]$ respectively (Figure 11.4), and the property is established by induction. Consider the complete graph K_{n-1}, and a vertex z different from the vertices of K_{n-1} and joined by an edge to a single vertex of K_{n-1}. The graph thereby obtained has $\binom{n-1}{2}+1$ edges and does not contain a Hamiltonian cycle, since $d(z) = 1$.

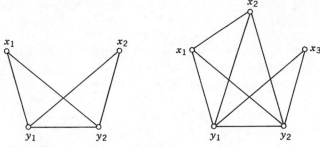

Fig. 11.4

11.9 (a) Let $L=[x_0, x_1, \ldots, x_m]$ be a longest elementary walk in the graph G. It follows from the maximality of L that all the vertices which are adjacent to x_0 belong to L. Since $d(x_0) \geqslant k$, there exists a vertex x_i which is adjacent to x_0, and such that $k \leqslant i \leqslant m$. Thus $C=[x_0, \ldots, x_i, x_0]$ is an elementary cycle of length $i+1 \geqslant k+1$.

(b) Let L be a longest elementary walk of the graph G. Suppose first that there are vertices x_i and x_j such that $i<j$, x_i is adjacent to x_m and x_j is adjacent to x_0. Similarly suppose that $j-i$ is a minimum in the set of pairs of indices with this property. Let $C=[x_0, \ldots, x_i, x_m, x_{m-1}, \ldots, x_j, x_0]$ be the elementary cycle which is formed in this way. If $j=i+1$, then the cycle C has length $m+1$ and must be a Hamiltonian cycle, since otherwise there would exist a vertex which does not belong to C, but is adjacent to a vertex of the cycle C. In this case one obtains an elementary walk longer than L, which contradicts the maximality of the walk L. Consider the case in which $j \geqslant i+2$. The vertices $x_{i+1}, \ldots, x_{j-1}$ are not adjacent to x_0 or x_m, and thus the cycle C contains, like the walk L, the vertex x_m and all vertices adjacent with x_m, and hence $k+1$ vertices. Since, by hypothesis $j \geqslant i+2$, it also contains at least $k-1$ other vertices, namely the vertices x_p with the property that x_{p+1} is adjacent to x_0 and $x_p \neq x_{j-1}$. All these vertices are pairwise distinct, and thus the cycle C contains at least $2k$ vertices.

Now consider the remaining case when x_i is the vertex of maximum index which is adjacent to x_0, x_j is the vertex of minimum index adjacent to x_m, and $i \leqslant j$. In this way one obtains elementary cycles $C_1=[x_0, \ldots, x_i, x_0]$ and $C_2=[x_j, \ldots, x_m, x_j]$. The maximality of the walk L implies that they contain, respectively, all the vertices adjacent to x_0 and x_m. Since the graph G is 2-connected, there must exist two walks L_1 and L_2 which have no vertex in common and both of which have one endpoint which belongs to C_1 and the other to C_2. Similarly one can assume that one of these walks has x_i as endpoint. If not, one could consider an elementary walk L_3 with endpoints x_i and x_j. Traverse this walk from x_i towards x_j. One will first encounter a vertex belonging to the cycle C_2 or a vertex belonging to the walk L_i ($1 \leqslant i \leqslant 2$). In the first case replace one of L_1 or L_2 by a subwalk of L_3. In the second case replace a subwalk of L_i by corresponding subwalk of L_3 which begins at x_i. Now suppose that the walk L_1 has an endpoint in x_i, and that L_2 does not have a vertex in common with L_1. If L_1 also has an endpoint in x_j, the procedure terminates. Otherwise consider an elementary walk L_3 with endpoints at x_i and x_j. Traverse L_3 from x_j towards x_i. Suppose that one first encounters a vertex of the walk L_2. In this case replace a subwalk of L_2 by the subwalk of L_3 which terminates at x_j. This produces the two walks with the desired property. On the other hand, if one first meets a vertex of the cycle C_1, then instead of L_2 consider a subwalk of L_3 contained between x_j and the first intersection of L_3 with the cycle C_1. One again obtains the two walks with the given property. Suppose there exists a walk L_i with endpoints x_i and x_j or different walks with these endpoints. In both cases, one obtains an elementary cycle which contains the vertices x_0 and x_m and all vertices adjacent with them (Figure 11.5). The cycles C_1 and C_2 which contain the vertices adjacent to x_0 and x_m respectively have at most one vertex in common (when $i=j$). It follows that the cycle which is thereby formed has a length greater than or equal to $2k+1$. [G. A. Dirac, *Proc. London Math. Soc.*, **2** (1952), 69–81.]

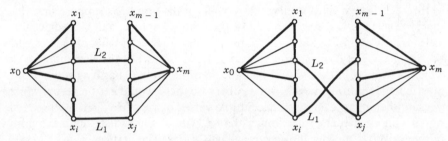

Fig. 11.5

11.10 The property will be established by induction on n. If $n \leqslant k$ then $(n-1)k/2 \geqslant \binom{n}{2}$, and thus there does not exist a graph with n vertices which has more edges than K_n. The assertion of the problem is, therefore, satisfied in this case.

Suppose now that $n > k$ and that the property is true for every graph with at most $n-1$ vertices. Let G be a graph with n vertices and $m > (n-1)k/2$ edges, and suppose that G contains a vertex x with degree less than or equal to $k/2$. Denote by G_x the subgraph obtained by suppressing the vertex x and the edges incident with x. It follows that the number of edges in G_x is greater than $(n-1)k/2 - k/2 = (n-2)k/2$. By the induction hypothesis the subgraph G_x contains an elementary cycle of length at least equal to $k+1$.

There is one remaining case: when each vertex x of the graph G has degree $d(x) \geqslant (k+1)/2$. If G is 2-connected, then property (b) of Problem 11.9 implies the existence of an elementary cycle of length at least equal to $k+1$. If G is not 2-connected, then it has at least one cut point z, and hence there exist two subgraphs G_1 and G_2 of G with vertex sets X_1 and X_2 such that $X_1 \cap X_2 = \{z\}$. The graphs G_1 and G_2 also do not have a common edge. If m_1, m_2 denote the number of edges of G_1 and G_2, respectively, then

$$m_1+m_2>\frac{k(n-1)}{2}=\frac{k}{2}(|X_1|+|X_2|-2);$$

and hence there exists an index i, $1\leqslant i\leqslant 2$, such that

$$m_i>\frac{k}{2}(|X_i|-1).$$

One can therefore apply the induction hypothesis to the subgraph G_i, which also contains fewer than n vertices. It follows that G_i and hence the graph G contains an elementary cycle of length greater than or equal to $k+1$. [P. Erdös, T. Gallai, *Acta Math. Acad. Sci. Hung.*, **10** (1959), 337–356.]

11.11 Let $C=(x_1,\ldots,x_m,x_1)$ be a longest elementary circuit of the graph G. The number of vertices of C satisfies the inequality $m>n/2$. In fact, let $Q=(z_0,\ldots,z_p)$ be a longest elementary path of G, and let $z_{i_1},\ldots,z_{i_k}$ ($i_1<\cdots<i_k$) be the vertices of G such that (z_{i_s},z_0) is an arc. It follows from the maximality of this path that $z_{i_1},\ldots,z_{i_k}$ belong to the path Q. Since $d^-(z_0)\geqslant n/2$, it follows that $i_k\geqslant k\geqslant n/2$, and hence the circuit $(z_0,z_1,\ldots,z_{i_k},z_0)$ has length greater than $n/2$, which implies that $m>n/2$.

Suppose that the circuit C is not Hamiltonian, and let $D=(y_0,\ldots,y_r)$ be a longest elementary path in the subgraph obtained from G by suppressing the vertices $x_1,\ldots,x_m$ of the circuit C. Since $d^-(y_0)\geqslant n/2$, it follows that there exist at least $n/2-r$ arcs of the form (u,y_0), where $u\notin D$, since there are at most r arcs of the form (u,y_0) with $u\in D$.

All of the arcs of the form (u,y_0) with $u\notin D$ have their initial vertex u in C. Denote these vertices by $x_{i_1},\ldots,x_{i_t}$ with $t\geqslant n/2-r$ and $i_1<\cdots<i_t$. Similarly one can find vertices $x_{j_1},\ldots,x_{j_s}$, $s\geqslant n/2-r$ such that (y_r,x_{j_p}) is an arc of the graph G for $p=1,\ldots,s$ and $j_1<\cdots<j_s$. Suppose that the vertices x_{i_p} and x_{j_q} are different. It follows that the path $(x_{i_p},\ldots,x_{j_q})$ consisting of arcs of the circuit C must have length (number of arcs) at least equal to $r+2$. Otherwise one would have a circuit C_1 of length strictly greater than that of C, which is seen to contradict the hypothesis if the path $(x_{i_p},\ldots,x_{j_q})$ is replaced by the path $(x_{i_p},y_0,\ldots,y_r,x_{j_q})$ of length equal to $r+2$.

Now consider an elementary path D_p consisting of arcs of the circuit C which originates at x_{i_p+1} (taking $x_{m+1}=x_1$) and having length equal to r for $p=1,\ldots,t$. The following two cases are possible:

(a) $x_{j_q}=x_{i_p}$ and thus $x_{j_q}\notin V(D_p)$, where $V(D_p)$ denotes the set of vertices of the path D_p. In fact $r\leqslant n-m-1\leqslant n/2-1$, and the length of the circuit C is $m>n/2$.

(b) $x_{j_q}\neq x_{i_p}$ and hence $x_{j_q}\notin V(D_p)$, since otherwise it would follow that the path $(x_{i_p},\ldots,x_{j_q})$ has length equal to $r+1$. But it has been shown that such a subpath of C must have length at least equal to $r+2$.

It is now clear that

$$x_{j_q} \notin \bigcup_{p=1}^{t} V(D_p) \quad \text{for } 1 \leqslant q \leqslant s. \tag{1}$$

On the other hand it is also the case that

$$\left| \bigcup_{p=1}^{t} V(D_p) \right| \geqslant t+r, \tag{2}$$

since the union of the sets $V(D_p)$ contains pairwise distinct vertices $x_{i_1+1}, \ldots, x_{i_t+1}$. Also there exist two vertices, say x_{i_1} and x_{i_2}, such that the path $(x_{i_1+1}, \ldots, x_{i_2})$ formed from arcs of the circuit C contains a vertex x_{j_q}. In this case it has been seen that the path D_1 of length r does not contain the vertex x_{j_q}, nor therefore the vertex x_{i_2}. It follows that the r vertices of the path D_1 which are different from $x_{i_1+1}, \ldots, x_{i_t+1}$ belong to the union of the sets $V(D_p)$ for $p=1, \ldots, t$. This observation completes the proof of (2).

Observe that (1) and (2) imply that $s \leqslant m-(t+r)$, so that

$$m \geqslant s+t+r \geqslant \frac{n}{2}-r+\frac{n}{2}-r+r = n-r. \tag{3}$$

Since the path D has $r+1$ vertices in a subgraph with $n-m$ vertices, it follows that

$$r+1 \leqslant n-m, \quad \text{or} \quad m \leqslant n-r-1. \tag{4}$$

But (3) and (4) are contradictory, which completes the proof that C is a Hamiltonian circuit.

It remains an open problem to show that under the condition of the present problem and for $n \geqslant 5$ the graph G contains at least two Hamiltonian circuits without common arcs. [C. St. J. A. Nash-Williams, *The Many Facets of Graph Theory*, Lecture Notes in Mathematics, Springer-Verlag, **110** (1969), 237–243.]

11.12 If the tournament G contains a Hamiltonian circuit, then it is strongly connected, since every two vertices of a circuit are joined by a path. Now suppose that G is strongly connected. Since for every two vertices x, y of G there is a path from x to y and from y to x, it follows that G contains a circuit.

Let $C = (y_1, \ldots, y_k, y_1)$ be an elementary circuit of G with a maximal number of vertices. Suppose that C is not a Hamiltonian circuit, and let x be a vertex which does not belong to C. Since the graph is complete, one can assume that there exists an arc (y_1, x). If there also exists an arc (x, y_2), then $(y_1, x, y_2, \ldots, y_k, y_1)$ is a circuit longer than C, which contradicts the hypothesis. It follows that the arc between x and y_2 has the form (y_2, x). In fact there exist arcs (y_i, x) for $i=1, \ldots, k$. Let A be the set of vertices x for which there is an arc (y_1, x). One can conclude that (y_i, x) is an arc for $x \in A$ and $i=1, \ldots, k$. Since G is strongly connected, there exists an arc (x, z) with $x \in A$ and $z \notin A$. It follows that $z \notin C$, and since G is complete and $z \notin A$, one can find an arc of the form (z, y_1). This yields a circuit $(x, z, y_1, \ldots, y_k, x)$ which contains more vertices than C,

Solutions

and this contradicts the hypothesis. [P. Camion, *C. R. Acad. Sci. Paris*, **249** (1959), 2151–2152.]

11.13 Let $P=[x_1, x_2, \ldots, x_n]$ and let S be a selection of k edges of P which generate j components. There are four possible cases:

(1) $[x_1, x_2] \in S$ and $[x_{n-1}, x_n] \notin S$;
(2) $[x_1, x_2] \in S$ and $[x_{n-1}, x_n] \in S$;
(3) $[x_1, x_2] \notin S$ and $[x_{n-1}, x_n] \notin S$;
(4) $[x_1, x_2] \notin S$ and $[x_{n-1}, x_n] \in S$.

It is clear that the number of edge sets S which satisfy (1) is equal to the number of solutions of the system

$$a_1 + \cdots + a_j = k,$$
$$b_1 + \cdots + b_j = n-k-1,$$

where a_i, b_i are integers and $a_i, b_i \geq 1$; this number is $\binom{n-k-2}{j-1}\binom{k-1}{j-1}$ (see Problem 1.19). In a similar manner, in case (2) one finds a system

$$a_1 + \cdots + a_j = k,$$
$$b_1 + \cdots + b_{j-1} = n-k-1$$

with $\binom{n-k-2}{j-2}\binom{k-1}{j-1}$ solutions for $a_i, b_i \geq 1$. In case (3) the corresponding system is

$$a_1 + \cdots + a_j = k,$$
$$b_1 + \cdots + b_{j+1} = n-k-1,$$

having $\binom{n-k-2}{j}\binom{k-1}{j-1}$ solutions, and for the last case the system is the same as for the first case. Hence

$$P_j(n,k) = \binom{k-1}{j-1}\left\{\binom{n-k-2}{j} + 2\binom{n-k-2}{j-1} + \binom{n-k-2}{j-2}\right\}$$
$$= \binom{k-1}{j-1}\binom{n-k}{j},$$

as required.

11.14 Let $H = [x_1, x_2, \ldots, x_n]$ be a fixed Hamiltonian walk of K_n, and denote its edges by $e_i = [x_i, x_{i+1}]$ for $1 \leq i \leq n-1$. Let A_i be the set of all Hamiltonian walks of K_n which contain the edge e_i of H for $1 \leq i \leq n-1$. Thus $H(n,k)$ is the number of Hamiltonian walks of K_n which belong to precisely k sets A_i. By using C. Jordan's sieve formula (Problem 2.3) we can show that

$$H(n,k) = \sum_{i=k}^{n-1} (-1)^{i-k} \binom{i}{k} \sum_{\substack{K \subset \{1,\ldots,n-1\} \\ |K|=i}} \left|\bigcap_{p \in K} A_p\right|. \qquad (1)$$

Let $K \subset \{1, \ldots, n-1\}$ and $|K| = i$, and suppose that the set of edges $E_k =$

$\{e_p\}_{p \in K}$ generates exactly j components $H_1, \ldots, H_j$ on H. In this case

$$\left| \bigcap_{p \in K} A_p \right| = \frac{(n-i)!}{2} 2^j \qquad (2)$$

for any $1 \leq j \leq i \leq n-2$.

In fact, let $m_1, m_2, \ldots, m_j$ be the numbers of edges in each of the components induced by E_K on H. Contract each component $H_1, \ldots, H_j$, which is a subwalk of H, to a unique vertex $y_1, \ldots, y_j$. The resulting graph has $n - (m_1 + \cdots + m_j) = n - i$ vertices. Observe that K_{n-i} has $(n-i)!/2$ Hamiltonian walks, and that any such walk may be expanded to a Hamiltonian walk of K_n by replacing every vertex y_q by the subwalk H_q of H for $1 \leq q \leq j$ in 2^j ways [and (2) follows]. Therefore from (1) one obtains

$$H(n, k) = \sum_{i=k}^{n-2} (-1)^{i-k} \binom{i}{k} \frac{(n-i)!}{2} \sum_{j=1}^{i} P_j(n, i) 2^j$$

$$+ (-1)^{n-1-k} \binom{n-1}{k},$$

where $P_j(n, i)$ is given by Problem 11.13, since $\left| \bigcap_{p \in K} A_p \right| = 1$ for $K = \{1, \ldots, n-1\}$. To obtain an expression for $DH(n, k)$ in the case of a Hamiltonian path DH of K_n^*, denote its arcs by $a_1, \ldots, a_{n-1}$ and let A_i be the set of all Hamiltonian paths of K_n^* containing arc a_i of DH for $1 \leq i \leq n-1$. The rest of the proof is similar to that for the numbers $H(n, k)$, since K_{n-i}^* has $(n-i)!$ Hamiltonian paths, and each such path may be expanded to a Hamiltonian path of K_n^* in a unique way. One also uses the identity

$$\sum_{j=1}^{i} \binom{i-1}{j-1} \binom{n-i}{j} = \binom{n-1}{i}$$

[see Problem 1.5(a)].

11.15 Since G is connected, it has a spanning tree T. It is sufficient to prove that T^3 possesses the property of the statement, which says that G^3 is Hamiltonian-connected. Now prove by induction on n that for any tree T with n vertices, T^3 is Hamiltonian-connected. For $n \leq 4$ it follows that the diameter of T is at most three, and hence T^3 is the complete graph which is Hamiltonian-connected. Suppose that all trees having at most $n-1$ vertices have as their cube a Hamiltonian-connected graph, and let T be a tree with n vertices. If x and y are two distinct vertices of T, one now considers two cases:

(a) x and y are joined in T by an edge $u = [x, y]$. Denote by T_x and T_y the two subtrees obtained from T by deleting $[x, y]$ such that T_x contains x and T_y contains y. By the induction hypothesis T_x^3 and T_y^3 are Hamiltonian-connected. Let x_1 be a vertex of T_x which is adjacent to x, and let y_1 be a vertex of T_y which is adjacent to y. If one of the trees T_x or T_y reduces to a single vertex, then let $x_1 = x$ or $y_1 = y$ respectively. The vertices x_1 and y_1 are adjacent in T^3 because $d(x_1, y_1) \leq 3$ in T. Let P_x be a Hamiltonian walk with endpoints x and

x_1 in T_x^3 (which may reduce to a single vertex), and let P_y be a Hamiltonian walk with endpoints y and y_1 in T_y^3. The walk composed of P_x followed by edge $[x_1, y_1]$ and by P_y is a Hamiltonian walk between x and y in T^3.

(b) x and y are not adjacent in T. Since T is a tree, there is a unique walk P between x and y in T. Let $v = [x, z]$ be the edge of P incident to x. By deleting v from T one obtains two subtrees: a tree T_x containing x and another tree T_z containing z. By the induction hypothesis there exists a Hamiltonian walk P_z between z and y in T_z^3. Let x_1 denote a vertex of T_x which is adjacent to x or $x_1 = x$ if T_x reduces to x, and let P_x be a Hamiltonian walk between x and x_1 in T_x^3. Since $d(x_1, z) \leqslant 2$ in T, it follows that T^3 contains the edge $[x_1, z]$. The walk composed of P_x followed by edge $[x_1, z]$ and by P_z is a Hamiltonian walk between x and y in T^3.

It follows that if G is a connected graph with at least three vertices, then the graph G^3 is Hamiltonian. [J. Karaganis, *Canad. Math. Bull.*, **11** (1969), 295–296; M. Sekanina, *Publ. Fac. Sci. Univ. Brno*, **412** (1960), 137–142.]

A. Hobbs proved that if G is a 2-connected graph, then its square G^2 is Hamiltonian-connected. [*Notices Amer. Math. Soc.*, **18** (1971), 553.]

CHAPTER 12

12.1 Suppose that the permutation p is a cycle of length k, say $p = [i_1, \ldots, i_k]$. It follows that $p(i_1) = i_2$, $p^2(i_1) = i_3, \ldots, p^{k-1}(i_1) = i_k$, and $p^k(i_1) = i_1$. Analogously one can conclude that $p^k(j) = j$ for every j, and hence k is the smallest number r such that $p^r = e$; every multiple ks of k has the property that $p^{ks} = e$.

Consider the cycles $p_1, \ldots, p_i$ in the representation of the permutation p as a product of disjoint cycles $p = p_1 \cdots p_i$. These cycles commute among themselves, and hence it follows that $p^r = p_1^r \cdots p_i^r$. Therefore if $p^r = e$ one can conclude that r is a common multiple of the lengths of the cycles of the permutation p.

12.2 Let $p(n, k)$ denote the number of permutations of n elements which have k cycles. Denote by $c(n, k)$ the coefficient of x^k in the expansion of $[x]^n$, that is,

$$[x]^n = \sum_{k=0}^{n} c(n, k) x^k. \qquad (1)$$

It is known (Problem 3.1) that

$$c(n, k) = |s(n, k)|,$$

and thus it remains to show that $c(n, k) = p(n, k)$ for every n and k.

It follows from (1) that $c(n, 1) = (n-1)!$ and $p(n, 1) = (n-1)!$, since the number of permutations which have only one cycle (circular permutations) is equal to $n!/n$, because every cycle can be written in n distinct ways by taking as the first element each of its n elements. It will be shown that $p(n, k)$ and $c(n, k)$ satisfy the same recurrence relation.

Let $X_n = \{x_1, \ldots, x_n\}$ and $X_{n+1} = X_n \cup \{x_{n+1}\}$. Every permutation of X_{n+1} with k cycles may contain the element x_{n+1} alone in a cycle, with the remaining n elements forming a permutation with $k-1$ cycles. Otherwise x_{n+1} is contained in a cycle together with other elements. The element x_{n+1} can be inserted into a cycle with p elements in exactly p distinct ways. It follows that

$$p(n+1, k) = p(n, k-1) + np(n, k). \tag{2}$$

In fact, all permutations with k cycles of the set X_{n+1} can be obtained, without repetitions, in two ways: one can add a new cycle consisting of x_{n+1} to each permutation of X_n with $k-1$ cycles, or one can insert the element x_{n+1} in one of n ways into each permutation of the set X_n with k cycles.

It follows from (1) that

$$[x]^{n+1} = [x]^n(x+n),$$

or

$$\sum_{i=0}^{n+1} c(n+1, i)x^i = \left(\sum_{i=0}^{n} c(n, i)x^i\right)(x+n).$$

By equating the coefficients of x^k in the two sides one obtains the recurrence relation

$$c(n+1, k) = c(n, k-1) + nc(n, k). \tag{3}$$

The relations (2) and (3), together with the values $c(n, 1) = p(n, 1) = (n-1)!$ and $c(n, k) = p(n, k) = 0$ for $n < k$, uniquely determine the values of $c(n, k)$ and $p(n, k)$, respectively. It follows that $p(n, k) = c(n, k) = |s(n, k)|$ for every n and k, since relations (2) and (3) are essentially identical.

12.3 It follows from Problem 12.2 that the number of permutations $p \in S_m$ which contain k cycles is equal to the coefficient of x^k in the expansion of the polynomial $[x]^m = x(x+1) \cdots (x+m-1)$. This fact can be written in the form

$$\sum_{p \in S_m} x^{c(p)} = x(x+1) \cdots (x+m-1). \tag{1}$$

By letting $x = n$ in (1) one obtains the first relation.

If x is replaced by $-x$ in (1) one sees that

$$\sum_{p \in S_m} (-1)^{c(p)} x^{c(p)} = (-1)^m x(x-1) \cdots (x-m+1). \tag{2}$$

Suppose that p is of type $1^{c_1} 2^{c_2} \cdots m^{c_m}$, where $c_1 + 2c_2 + \cdots + mc_m = m$ and $c_i \geq 0$. It will be shown that $\text{sgn}(p) = (-1)^{c_2 + c_4 + c_6 + \cdots}$, and hence the parity of p coincides with that of the number of its even cycles.

In fact, if a cycle $p = [i_1, i_2, \ldots, i_n]$ has n elements, then it can be expressed as a product of $n-1$ transpositions (which are odd permutations), say

$$p = [i_1, i_2][i_2, i_3] \cdots [i_{n-1}, i_n].$$

Solutions

If p is of type $1^{c_1} 2^{c_2} \cdots m^{c_m}$, it follows that it can be written as a product of $c_2 + 2c_3 + \cdots + (m-1)c_m$ transpositions, and hence p has the same parity as the sum $c_2 + c_4 + c_6 + \cdots$. Since $c_1 + 2c_2 + 3c_3 + \cdots + mc_m = m$, one can conclude that $c(p) + m = 2c_1 + 3c_2 + 4c_3 + 5c_4 + \cdots$, which has the same parity as $c_2 + c_4 + c_6 + \cdots$, and hence $\operatorname{sgn}(p) = (-1)^{c(p)+m}$.

By replacing $(-1)^{c(p)}$ with $(-1)^m \operatorname{sgn}(p)$ in (2) and taking $x = n$ one sees that

$$\sum_{p \in S_m} \operatorname{sgn}(p) \, n^{c(p)} = [n]_m,$$

which establishes the second identity. [M. Marcus, *American Math. Monthly*, **78**(9) (1971), 1028–1029.]

12.4 First we evaluate the number of permutations of an n-element set X containing p cycles labeled with $\alpha_1, \ldots, \alpha_p$, such that the cycle with label α_i contains x_i elements ($x_i \geq 1$ for $i = 1, \ldots, p$ and $x_1 + \cdots + x_p = n$). The number of arrangements of X in p boxes $\alpha_1, \ldots, \alpha_p$ such that box α_i contains x_i objects for $i = 1, \ldots, p$ is equal to $n!/x_1! \cdots x_p!$ (Problem 1.15). The number of circular permutations having x_i elements in a cycle is equal to $(x_i - 1)!$, and hence the number of permutations of X such that cycle α_i contains x_i elements for every $i = 1, \ldots, p$ is equal to

$$\frac{n!}{x_1! \cdots x_p!} (x_1 - 1)! \cdots (x_p - 1)! = \frac{n!}{x_1 \cdots x_p}.$$

It follows that the number of all permutations of X having p labeled cycles is obtained by summing these numbers for all representations $n = x_1 + \cdots + x_p$, where $x_i \geq 1$ for $i = 1, \ldots, p$, and where the order of the parts $x_1, \ldots, x_p$ is taken into consideration. If one erases the labels of the cycles, it turns out that there are $(1/p!) \sum n!/x_1 \cdots x_p$ permutations of X with p cycles. But, by Problem 12.2, this number is also equal to $|s(n, p)|$, which establishes the equation in the statement of this problem.

12.5 The element $n+1$ can be inserted into a cycle of length p formed from elements of the set $\{1, \ldots, n\}$ in p distinct ways. Thus $nd(n, k)$ counts the permutations $p \in S_n$ without fixed points and with k cycles which contain the element $n+1$ in a cycle of length $q \geq 3$. Also $nd(n-1, k-1)$ is the number of those permutations which contain the element $n+1$ in a cycle of length 2. In fact, the element $n+1$ can form a cycle of length 2 with each of the remaining n elements. The rest of the $n-1$ elements form a permutation of $n-1$ elements with $k-1$ cycles without fixed points. This observation implies (a).

Every permutation of $2k$ elements with k cycles and without cycles of length 1 contains only cycles of length 2 and thus is of type 2^k. It follows that

$$d(2k, k) = \frac{(2k)!}{2^k k!} = 1 \times 3 \times 5 \times \cdots \times (2k-1).$$

In order to prove (c) recall that the number of permutations $p \in S_n$ with k cycles is equal to $c(n, k) = |s(n, k)| = (-1)^{n+k} s(n, k)$, where $s(n, k)$ is the Stirling number of the first kind. (Problem 12.2). Let A_i denote the set of permutations $p \in S_n$ for which $p(i)=i$ and which have k cycles. By applying the Principle of Inclusion and Exclusion one can conclude that

$$d(n, k) = c(n, k) - |A_1 \cup A_2 \cup \cdots \cup A_n|$$

$$= c(n, k) - \sum_{i=1}^{n} |A_i| + \sum_{1 \leq i < j \leq n} |A_i \cap A_j| - \cdots.$$

But $\sum_{1 \leq i_1 < \cdots < i_j \leq n} |A_{i_1} \cap \cdots \cap A_{i_j}|$ contains $\binom{n}{j}$ terms, each of which is equal to $c(n-j, k-j)$. This is because $p \in A_{i_1} \cap \cdots \cap A_{i_j}$ implies that $p(i_1)=i_1,\ldots,$ $p(i_j)=i_j$, and because the restriction of the permutation p to the remaining elements is a permutation of $n-j$ elements with $k-j$ cycles. [P. Appell, *Arch. Math. Phys.*, **65** (1880), 171–175; J. Kaucky, *Mat. Casopis Sloven. Akad. Vied*, **21** (1971), 82–86.]

12.6 Conjugation is an equivalence relation, since it is

(1) reflexive: $s = ese^{-1}$, where e is the identity permutation;
(2) symmetric: $s = gtg^{-1}$ implies that $t = g^{-1}sg = hsh^{-1}$, where $h = g^{-1} \in S_n$;
(3) transitive: $s = gtg^{-1}$ and $t = huh^{-1}$ imply $s = ghuh^{-1}g^{-1} = (gh)u(gh)^{-1}$, where the permutation $gh \in S_n$.

In order to prove the necessity of the condition in (b), suppose that s and t are conjugate, that is, there exists $g \in S_n$ such that $s = gtg^{-1}$. Write the permutation t as a product of disjoint cycles as follows:

$$t = [t_{11}t_{12}\cdots t_{1i}][t_{21}t_{22}\cdots t_{2j}]\cdots[t_{m1}t_{m2}\cdots t_{mk}],$$

and let $g(t_{pq}) = s_{pq}$. The fact that $s = gtg^{-1}$ implies that

$$s(s_{11}) = gtg^{-1}(s_{11}) = gt(t_{11}) = g(t_{12}) = s_{12},\ldots,$$

and hence the decomposition of the permutation s as a product of cycles is given by

$$s = [s_{11}s_{12}\cdots s_{1i}][s_{21}s_{22}\cdots s_{2j}]\cdots[s_{m1}s_{m2}\cdots s_{mk}].$$

It follows that two conjugate permutations can be decomposed into the same number of cycles having respectively the same lengths.

In order to prove the sufficiency of the condition, define the permutation g by first considering the decompositions into cycles of the permutations s and t. Then let $g(t_{pq}) = s_{pq}$. The decompositions of s and t as products of disjoint cycles both contain each of the numbers $1,\ldots,n$ exactly once. It follows that $g \in S_n$ and, as in the previous calculation, one can show that $s = gtg^{-1}$. This observation completes the proof of the sufficiency of the condition.

The proof of Cauchy's formula can now be given. Let f be a permutation which has λ_s cycles of length s for $1 \leq s \leq k$. Express f as a product of cycles written in increasing order of length:

Solutions

$$f = \underbrace{[*]\cdots[*]}_{\lambda_1}\ \underbrace{[**]\cdots[**]}_{\lambda_2}\ \cdots\ \underbrace{\overbrace{[*\cdots**]}^{k}}_{\lambda_k},$$

where the stars represent the numbers $1,\ldots,n$. By suppressing parentheses one obtains a permutation $a_1 a_2 \cdots a_n$ of integers from the set $\{1,\ldots,n\}$. The number of such permutations is $n!$.

However, the same permutation f generates $\lambda_1! \cdots \lambda_k! \, 1^{\lambda_1} 2^{\lambda_2} \cdots k^{\lambda_k}$ different permutations of the set $\{1,\ldots,n\}$. In fact one can permute the λ_i cycles of length i ($1 \le i \le k$) in the representation of f as a product of cycles in $\lambda_1! \lambda_2! \cdots \lambda_k!$ distinct ways. On the other hand, each cycle of length i can be written in i distinct ways by taking as its first element each of the i elements. There are thus a total of $1^{\lambda_1} 2^{\lambda_2} \cdots k^{\lambda_k}$ representations. Starting with all the $h(\lambda_1,\ldots,\lambda_k)$ permutations which contain λ_s cycles with s elements, for $s=1,\ldots,k$ one can write each cycle of length s in s different ways. By also permuting the cycles which contain the same number of elements in all possible ways, it is easy to show that one can obtain without repetitions all the $n!$ permutations of the set $\{1,\ldots,n\}$. It follows that

$$h(\lambda_1,\ldots,\lambda_k)\lambda_1! \cdots \lambda_k! \, 1^{\lambda_1} 2^{\lambda_2} \cdots k^{\lambda_k} = n!,$$

from which Cauchy's formula follows.

An equivalence class is characterized by the number of cycles and their lengths $n_1,\ldots,n_k$, which must satisfy $n_1 + \cdots + n_k = n$. One can assume that $n_1 \ge n_2 \ge \cdots \ge n_k$, since the order of the cycles in the product is not important, inasmuch as the product of disjoint cycles is commutative. Thus the number of equivalence classes is precisely the number $P(n)$ of partitions of the integer n.

12.7 Cauchy's identity can be obtained from the relation

$$\sum_{\substack{c_1 + 2c_2 + \cdots + nc_n = n \\ c_i \ge 0}} h(c_1, c_2, \ldots, c_n) = n!,$$

by means of Problem 3.6 or by using part (c) of the preceding problem.

12.8 Calculate the number of permutations in which the number 1 is contained in a cycle of length k. The elements of this cycle can be chosen in $\binom{n-1}{k-1}$ distinct ways, since after fixing the element 1 there still remain $k-1$ elements which belong to the set $\{2,\ldots,n\}$ of cardinality $n-1$.

With these k elements one can form $(k-1)!$ distinct cycles with k elements. There are $(n-k)!$ distinct permutations of the remaining $n-k$ elements. Thus the number of permutations in which 1 occurs in a cycle with k elements is equal to

$$\binom{n-1}{k-1}(k-1)!(n-k)! = (n-1)!.$$

The desired probability is therefore $(n-1)!/n! = 1/n$ and is independent of k.

12.9 Count the permutations in which 1 and 2 belong to different cycles. If the cycle which contains 1 has length k, one can select the other $k-1$ elements from the set $\{3,\ldots,n\}$ in

$$\binom{n-2}{k-1}$$

distinct ways. There are $(k-1)!$ circular permutations of these k elements and $(n-k)!$ permutations of the $n-k$ remaining elements. The number of permutations in which 1 and 2 belong to different cycles is equal to

$$\sum_{k=1}^{n-1}\binom{n-2}{k-1}(k-1)!(n-k)!$$
$$=(n-2)!\sum_{k=1}^{n-1}(n-k)$$
$$=(n-2)!\frac{n(n-1)}{2}=\frac{n!}{2},$$

and thus the desired probability is equal to $\frac{1}{2}$.

12.10 For $1\leqslant k\leqslant n$ let $p_i(n)$ denote the number of permutations of the set $\{1,\ldots,n\}$ which contain the element i in a cycle of length k. The sum $p_1(n)+\cdots+p_n(n)$ represents the number of elements which are contained in cycles of length k in all the $n!$ permutations of the set $\{1,\ldots,n\}$. It follows that

$$C_k=\frac{1}{k}\{p_1(n)+\cdots+p_n(n)\}$$

represents the total number of cycles of length k in the $n!$ permutations.

The average number of cycles of length k in a permutation of the set $\{1,\ldots,n\}$ is equal to $(1/n!)C_k=1/k$. This follows from the fact that by Problem 12.8, $p_i(n)/n!=1/k$ for every $i=1,\ldots,n$.

One can thus conclude that the average number of cycles in a permutation of the set $\{1,\ldots,n\}$ is equal to

$$\sum_{k=1}^{n}\frac{1}{k}=1+\frac{1}{2}+\cdots+\frac{1}{n}\cong\ln n,$$

since

$$\lim_{n\to\infty}\left(1+\frac{1}{2}+\cdots+\frac{1}{n}-\ln n\right)=\gamma=0.5772\ldots\quad\text{(Euler's constant)}.$$

12.11 Since $p^2=e$, it follows that the permutation p has all its cycles of length 1 or 2. Thus the set of permutations $p\in S_n$ such that $p^2=e$ can be written in the form $A_n\cup B_n$, where A_n is the set of permutations $p\in S_n$ with $p^2=e$ which satisfy $p(n)=n$, and B_n consists of those permutations $p\in S_n$ for which $p^2=e$ and in which the element n is contained in a cycle of length 2 in p. It follows that $P_n=|A_n|+|B_n|=P_{n-1}+(n-1)P_{n-2}$. In fact, if $p(n)=n$, the other $n-1$ elements form a permutation $q \in S_{n-1}$ for which $q^2 = e$. If $p \in B_n$, then the element n can turn a cycle of length 2 with each of the other $n-1$ elements; the remaining elements form a permutation $r \in S_{n-2}$ for which $r^2 = e$.

In order to prove (b) let i denote the number of cycles of length 1 (fixed points) of the permutation p, and let j denote the number of cycles of length 2. It follows that

$$P_n = \text{Perm}(1^i 2^j; n) = n! \sum (i! j! 2^j)^{-1},$$

where the summation is taken over all representations of n in the form $i + 2j = n$ with $i, j \geq 0$. The exponential generating function of the number P_n is

$$\sum_{n \geq 0} P_n \frac{t^n}{n!} = \sum_{n \geq 0} \left(\sum_{i+2j=n} \frac{n!}{i! j! 2^j} \right) \frac{t^n}{n!} = \sum_{i,j \geq 0} \frac{t^i}{i!} \frac{(t^2/2)^j}{j!} = e^t \cdot e^{t^2/2} = \exp\left(t + \frac{t^2}{2}\right).$$

Let $P_n(m)$ be the number of permutations $p \in S_n$ for which $p^m = e$. It can be shown that

$$\sum_{n=0}^{\infty} P_n(m) \frac{t^n}{n!} = \exp\left(\sum_{k | m} \frac{t^k}{k}\right).$$

[E. Jacobstahl, *Norske Vid. Selsk. Forh.* (*Trondheim*), **21** (1949), 49–51; S. D. Chowla, I. N. Herstein, W. R. Scott, *ibid.*, **25** (1952), 29–31.]

12.12 Suppose that the permutation $p \in S_n$ has the following representation as a product of transpositions: $p = t_1 t_2 \cdots t_r$. It follows that $p t_r t_{r-1} \cdots t_1 = e$, the identity permutation in S_n.

Every permutation p can be associated with a digraph G_p with vertex set $\{1, \ldots, n\}$. The graph's connected components are elementary circuits which correspond to the cycles of the permutation p, since by definition the arc (i, j) exists if and only if $j = p(i)$. Consider the result on this graph if p is multiplied by a transposition $[a, b]$ and thus $q = p[a, b]$.

Since the transposition $[a, b]$ takes a into b and b into a, it follows that the graph G_q is obtained from G_p by replacing the arcs $(a, p(a))$ and $(b, p(b))$ with the arcs $(a, p(b))$ and $(b, p(a))$, respectively.

Suppose that a and b are located in two distinct circuits of the graph G_p. These two circuits will generate a single circuit in the graph G_q. On the other hand, if a and b belong to the same circuit in G_p, then this circuit decomposes into two circuits without common vertices in G_q, as represented in Figure 12.1.

The graph of the identity permutation $e \in S_n$ consists of n circuits of length 1. By multiplying a permutation p with a transposition t one sees that $c(pt) \leq c(p) + 1$. Thus it follows that the number r of transpositions in the representation of p as a product of transpositions satisfies the inequality

$$n = c(e) \leq c(p) + r, \quad \text{or} \quad r \geq n - c(p).$$

The number $n - c(p)$ of transpositions can be attained, since each cycle of length m can be written as a product of $m - 1$ transpositions:

$$[i_1, i_2, \ldots, i_m] = [i_1, i_2][i_2, i_3] \cdots [i_{m-1}, i_m].$$

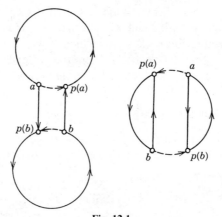

Fig. 12.1

Thus if the permutation $p \neq e$ is of type $1^{c_1} 2^{c_2} \cdots n^{c_n}$, it can be expressed as a product of $\sum_{i=1}^{n} c_i(i-1) = n - c(p)$ transpositions, since $c_1 + 2c_2 + \cdots + nc_n = n$.

12.13 In order to prove the necessity of the condition, suppose that the graph (X, T) is connected. Since this graph has n vertices and $n-1$ edges, it must be a tree. In fact, if it contained a cycle, then one could suppress an arbitrary edge of the cycle to obtain a connected graph. By repeating this process we arrive at a tree with n vertices and $n-1$ edges, which contradicts the hypothesis that (X, T) has $n-1$ edges. If (X, T) is not connected, then there exist two vertices of the graph, say a and b, which belong to different connected components. Since f is a circular permutation, it follows that there is a number $r \leq n-1$ such that $f^r(a) = b$. On the other hand the permutation f is a product of transpositions from T and transforms each vertex from X into a vertex in the same connected component as itself. Thus f^r has this property, which implies that $f^r(a) \neq b$. It follows that (X, T) is connected, that is, it is a tree.

To prove the sufficiency, suppose that (X, T) is a tree. Consider the sequence of permutations $g_1 = t_1, g_2 = g_1 t_2, g_3 = g_2 t_3, \ldots, g_{n-1} = f$.

It will be shown that the two elements i and j which define the transposition $t_q = [i, j]$ are not located in the same circuit of the graph of the permutation $t_1 t_2 \cdots t_{q-1}$ for $q = 2, \ldots, n-1$. (This graph is defined as in Problem 12.12.)

If i and j belonged to the same circuit in the graph of the permutation $t_1 t_2 \cdots t_{q-1}$, then in this sequence of transpositions there would exist transpositions $[i, i_1], [i_1, i_2], \ldots, [i_k, j](k \geq 1)$, which correspond to a walk from i to j. But in this case the graph (X, T) contains the edges $[i, i_1], \ldots, [i_k, j]$ and $t_q = [i, j]$. It therefore contains a cycle, and this contradicts the hypothesis that it is a tree.

If i and j do not belong to the same circuit in the graph of the permutation $g_{q-1} = t_1 t_2 \cdots t_{q-1}$, then the graph of $g_q = g_{q-1} t_q$ can be obtained from the graph of the permutation g_{q-1} by joining the two circuits which contain i and j,

Solutions

respectively. It follows that the number of its connected components is $c(G_{g_q}) = c(G_{g_{q-1}}) - 1$. But $c(G_{g_1}) = n - 1$, since the graph G_{g_1} consists of one circuit with two vertices and $n-2$ circuits which each contain a single vertex. Thus $c(G_{g_2}) = n-2, c(G_{g_3}) = n-3, \ldots, c(G_f) = 1$. Since G_f is connected and has a single connected component, it reduces to a circuit with n vertices, and hence f is a circular permutation.

Let $A(f)$ denote the number of representations of a circular permutation f as a product of $n-1$ transpositions. Thus the number of products of $n-1$ transpositions which generate all $(n-1)!$ circular permutations of n elements is equal to $(n-1)!A(f)$. This is because the number $A(f)$ does not depend on the circular permutation f, but is a function of n.

In view of the preceding result, it can be seen that the only products $t_1 t_2 \ldots t_{n-1}$ of $n-1$ transpositions which generate a circular permutation of n elements are those for which the graph (X, T) is a tree. It follows that the products of $n-1$ transpositions which generate all circular permutations of n elements can be obtained in the following manner: For each tree (X, T) which has its vertices labeled with the numbers $1, \ldots, n$, let $t_1 t_2 \cdots t_{n-1}$ be the product of the transpositions associated with its edges. Now permute the transpositions $t_1, t_2, \ldots, t_{n-1}$ in $(n-1)!$ distinct ways to obtain distinct products which generate circular permutations.

The number of labeled trees with n vertices is equal to n^{n-2} (Cayley's formula, Problem 6.15). Two such trees which contain at least two different edges will generate different products of transpositions. Thus $(n-1)!A(f) = (n-1)!n^{n-2}$, and hence $A(f) = n^{n-2}$, that is, the number of ways in which a circular permutation on n elements can be written as a product of $n-1$ transpositions is equal to the number of labeled trees with n vertices. [J. Dénes, *Magyar Tud. Akad. Mat. Kutató Int. Közl.*, **4** (1959), 63–71; O. Dziobek, *Sitzungsber. Berl. Math. G.*, **17** (1917), 64–67.]

12.14 In order to prove (a), observe that if $p_1 = a_1 a_2 \cdots a_n$ is a permutation in S_n and $p_2 = a_n a_{n-1} \cdots a_1$, then

$$I(p_1) + I(p_2) = \binom{n}{2}.$$

In fact, if $i < j$ then there is exactly one value of the index k for which $p_k(i) > p_k(j)$ ($1 \leq k \leq 2$). Thus the number of inversions from p_1 and from p_2 is equal to the number of pairs (i, j) with $i < j$, that is, $\binom{n}{2}$.

It follows that one can arrange the permutations in S_n in pairs so that one has $\binom{n}{2}$ inversions together. But S_n contains $n!$ permutations, and hence there are $\frac{1}{2}n!$ pairs, which implies (a).

If $I(p_1) = k$, then $I(p_2) = \binom{n}{2} - k$. The correspondence thus defined between the permutations in S_n which have k inversions and the permutations in S_n with $\binom{n}{2} - k$ inversions is injective and surjective and hence a bijection. This establishes (b).

The recurrence relation (c) will follow from (d).

In order to prove (d), start with the following equation:

$$p(n+1, k) = \left| \bigcup_{i=1}^{n+1} A_i \right| = \sum_{i=1}^{n+1} |A_i|, \qquad (1)$$

where

$$A_i = \{p \mid p \in S_{n+1}, I(p) = k, \text{ and } p(i) = n+1\}.$$

If the element $n+1$ is found at position i, then it is in inversion with all the $n+1-i$ elements which are found at positions $i+1, \ldots, n+1$. Thus if $p \in A_i$, then by suppressing the element $n+1$ from position i one obtains a permutation $p_i \in S_n$ such that $I(p_i) = k - n + i - 1$. The correspondence which is thereby defined associates with a permutation $p \in A_i$ a permutation $p_i \in S_n$. Since it is a bijection between the set A_i and the set of permutations in S_n with $k - n + i - 1$ inversions, it follows that

$$|A_i| = p(n, k - n + i - 1).$$

Part (d) now follows by using (1).

Every permutation in S_n has at most $\binom{n}{2}$ inversions. The maximum number of inversions is attained only for the permutation $n, n-1, \ldots, 1$. It follows that $p(n, \binom{n}{2}) = 1$ and $p(n, i) = 0$ for $i > \binom{n}{2}$. The only permutation with no inversions is the identity permutation $1, 2, \ldots, n$, and hence $p(n, 0) = 1$. These initial values together with recurrence relation (d) yield a row-by-row determination of the matrix of numbers $p(n, k)$.

In order to prove (c), observe that if $k < n$, then it follows from (d) that

$$p(n, k) = p(n-1, k) + p(n-1, k-1) + \cdots + p(n-1, 0),$$

$$p(n, k-1) = p(n-1, k-1) + p(n-1, k-2) + \cdots + p(n-1, 0),$$

and hence $p(n, k) = p(n-1, k) + p(n, k-1)$.

For (e), let

$$(1+x)(1+x+x^2) \cdots (1+x+\cdots+x^{n-1}) = \sum_{k \geq 0} c(n, k) x^k.$$

The term of maximum degree is $x^{1+2+\cdots+(n-1)} = x^{\binom{n}{2}}$, and thus $c(n, \binom{n}{2}) = 1$, and $c(n, k) = 0$ for $k > \binom{n}{2}$. The constant term is 1, which implies that $c(n, 0) = 1$.

In order to obtain a recurrence relation for the numbers $c(n, k)$ observe that

$$\sum_{k \geq 0} c(n+1, k) x^k = (1+x) \cdots (1+x \cdots +x^n)$$

$$= \left(\sum_{i \geq 0} c(n, i) x^i \right)(1+x+\cdots+x^n).$$

By equating the coefficient of x^k on the two sides one finds that

$$c(n+1, k) = c(n, k) + c(n, k-1) + \cdots + c(n, k-n),$$

where $c(n, i) = 0$ for $i < 0$.

Solutions

Since the numbers $c(n, k)$ and $p(n, k)$ have the same initial values and satisfy the same recurrence relation, which uniquely determine them, it follows that $c(n, k) = p(n, k)$ for every n and k. This completes the proof of equation (e).

12.15 Write a given permutation $p \in S_n$ as a product of disjoint cycles. Include cycles of length 1 (fixed points of p), and write the cycles so that their largest elements are written first. These first elements are to occur in increasing order from cycle to cycle. For example, the permutation

$$p = \begin{pmatrix} 1 & 2 & 3 & 4 & 5 & 6 & 7 \\ 3 & 4 & 5 & 2 & 1 & 7 & 6 \end{pmatrix} \in S_7$$

will be written in the form

$$p = [4, 2][5, 1, 3][7, 6].$$

In this way the last cycle will always begin with the element n. The sequence of numbers obtained in this way can be considered to be another permutation $f(p)$. In the previous example one has

$$f(p) = \begin{pmatrix} 1 & 2 & 3 & 4 & 5 & 6 & 7 \\ 4 & 2 & 5 & 1 & 3 & 7 & 6 \end{pmatrix}.$$

The permutation p can be uniquely reconstructed from the permutation $f(p) \in S_n$: The last cycle of p begins with the number n. The next to last begins with the largest number which is not contained in the last cycle, and so on.

If the permutation p contains k cycles, then the permutation $p_1 = f(p)$ will contain exactly k elements j for which $p_1(j) > p_1(i)$ for every $i < j$, and these will be the first elements in these k cycles of the permutation p.

The permutation p with k cycles can be uniquely reconstructed from the permutation $p_1 = f(p)$ which contains k elements j for which $p_1(j) > p_1(i)$ for every $i < j$. It follows that f is a bijection between the set of permutations $p \in S_n$ with k cycles and the set of permutations $p_1 \in S_n$ which have the property stated in the problem.

The number of permutations $p \in S_n$ with k cycles is equal to $|s(n, k)|$ by Problem 12.2. This observation completes the proof. [A. Rényi, *Colloquium Aarhus* (1962), 104–117.]

12.16 It is clear that $d(f, g) = 0$ implies $f(i) = g(i)$ for every $i = 1, \ldots, n$, and hence $f = g$ and $d(f, g) \geq 0$ for every $f, g \in S_n$. Furthermore $d(f, g) = d(g, f)$, and for $f, g, h \in S_n$ one has

$$|f(i) - g(i)| \leq |f(i) - h(i)| + |h(i) - g(i)|$$

for every $i = 1, \ldots, n$, which implies that

$$\max_i |f(i) - g(i)|$$
$$\leq \max_i \{|f(i) - h(i)| + |h(i) - g(i)|\}$$
$$\leq \max_i |f(i) - h(i)| + \max_i |h(i) - g(i)|,$$

or
$$d(f, g) \leq d(f, h) + d(h, g).$$

Hence $d(f, g)$ is a metric or distance on S_n. If $p \in S_n$ and $d(e_n, p) \leq 1$, where e_n is the identity permutation in S_n, then either $p(n) = n$ or $p(n) = n-1$ and $p(n-1) = n$, since otherwise one would have $d(e, p) \geq 2$.

Let q, r denote the restrictions of p to the first $n-1$ and $n-2$ elements of $\{1, \ldots, n\}$ respectively. In the first case one can conclude that $d(e_{n-1}, q) \leq 1$, and in the second case $d(e_{n-2}, r) \leq 1$. Thus $F(n, 1) = F(n-1, 1) + F(n-2, 1)$, which, together with the values $F(1, 1) = 1$ and $F(2, 1) = 2$, yields the equation $F(n, 1) = F_n$ for every $n \geq 1$.

12.17 Define the permanent of a square matrix $A = (a_{ij})_{i,j=1,\ldots,n}$ by the equation

$$\operatorname{per} A = \sum_{p \in S_n} a_{1p(1)} a_{2p(2)} \cdots a_{np(n)},$$

where p runs through the set of permutations of $\{1, \ldots, n\}$. Observe that

$$a_n = \operatorname{per} \begin{pmatrix} 1 & 1 & 1 & 0 & 0 & 0 & & & & 0 \\ 1 & 1 & 1 & 1 & 0 & 0 & & & & \\ 1 & 1 & 1 & 1 & 1 & 0 & & & & \\ 0 & 1 & 1 & 1 & 1 & 1 & & & & \\ & & \cdot & & & & \cdot & & & \\ & & & \cdot & & & & \cdot & & \\ & & & & \cdot & & & & \cdot & \\ & & & & & 0 & 1 & 1 & 1 & 1 & 1 \\ & & & & & 0 & 0 & 1 & 1 & 1 & 1 \\ 0 & & & & & 0 & 0 & 0 & 1 & 1 & 1 \end{pmatrix}.$$

In this nth-order matrix the element 1 occurs in the successive rows 3, 4, 5, 5, ..., 5, 4, 3 times, while the remaining elements are zero. In the summation which defines the permanent of this matrix, only the products which correspond to permutations p which satisfy $|p(i) - i| \leq 2$ for $i = 1, \ldots, n$ have value 1; the remaining values are zero, and this observation justifies the given equality.

From properties of permutations one can conclude that the permanent can be obtained by an expansion on a row or column analogous to the way in which a determinant is expanded, with the single difference that all the permanents in the expansion occur with a plus sign.

Expanding this permanent on the first row yields

$$a_n = a_{n-1} + b_{n-1} + c_{n-1},$$

where

$$b_n = \text{per} \begin{pmatrix} 1 & 1 & 1 & & & & & 0 \\ 1 & 1 & 1 & 1 & & & & \\ 0 & 1 & 1 & 1 & 1 & & & \\ & 1 & 1 & 1 & 1 & 1 & & \\ & & & \cdot & & & & \\ & & & & \cdot & 1 & & \\ & & & & & 1 & 1 & \\ 0 & & & & & 1 & 1 & 1 \end{pmatrix}, \quad c_n = \text{per} \begin{pmatrix} 1 & 1 & 1 & & & & & 0 \\ 1 & 1 & 1 & 1 & & & & \\ 0 & 1 & 1 & 1 & 1 & & & \\ & 0 & 1 & 1 & 1 & 1 & & \\ & & & \cdot & & & & \\ & & & & \cdot & 1 & & \\ & & & & & 1 & 1 & \\ 0 & & & & & 1 & 1 & 1 \end{pmatrix}.$$

By expanding b_n and c_n on their first columns one can see that

$$b_n = a_{n-1} + b_{n-1},$$
$$c_n = b_{n-1} + d_{n-1},$$

where

$$d_n = \text{per} \begin{pmatrix} 1 & 1 & 0 & & & & & 0 \\ 1 & 1 & 1 & 1 & & & & \\ 0 & 1 & 1 & 1 & 1 & & & \\ & 1 & 1 & 1 & 1 & 1 & & \\ & & & \cdot & & & & \\ & & & & \cdot & 1 & & \\ & & & & & 1 & 1 & \\ 0 & & & & & 1 & 1 & 1 \end{pmatrix}$$

and all the given matrices have order n. Expansion of d_n gives

$$d_n = a_{n-1} + e_{n-1},$$

where

$$e_n = \text{per} \begin{pmatrix} 1 & 1 & 1 & & & & & 0 \\ 0 & 1 & 1 & 1 & & & & \\ 0 & 1 & 1 & 1 & 1 & & & \\ & 1 & 1 & 1 & 1 & 1 & & \\ & & & \cdot & & & & \\ & & & & \cdot & 1 & & \\ & & & & & 1 & 1 & \\ 0 & & & & & 1 & 1 & 1 \end{pmatrix}.$$

Expanding e_n on the first column, one can conclude that $e_n = a_{n-1}$. These recurrence relations determine a_n, b_n, c_n, d_n, e_n for every n with initial values for $n=2$: $a_2=b_2=c_2=d_2=2$, $e_2=1$. Let

$$A = \begin{pmatrix} 1 & 1 & 1 & 0 & 0 \\ 1 & 1 & 0 & 0 & 0 \\ 0 & 1 & 0 & 1 & 0 \\ 1 & 0 & 0 & 0 & 1 \\ 1 & 0 & 0 & 0 & 0 \end{pmatrix}, \quad v_n = \begin{pmatrix} a_n \\ b_n \\ c_n \\ d_n \\ e_n \end{pmatrix}.$$

The given recurrence relations can be expressed in matrix form as follows:

$$v_n = Av_{n-1} = A^2 v_{n-2} = \cdots = A^{n-1} v_1 = A^n v_0.$$

Here

$$v_0 = \begin{pmatrix} 1 \\ 0 \\ 0 \\ 0 \\ 0 \end{pmatrix}, \quad v_1 = Av_0 = \begin{pmatrix} 1 \\ 1 \\ 0 \\ 1 \\ 1 \end{pmatrix},$$

$$v_2 = Av_1 = \begin{pmatrix} 2 \\ 2 \\ 2 \\ 2 \\ 1 \end{pmatrix},$$

which coincide with the directly obtained values for the case $n=2$. Thus a_n can be expressed as the scalar product of the vector consisting of the first row of the matrix A^n and the vector v_0, that is, $(A^n)_{11}$. [D. H. Lehmer, in: *Comb. Theory Appl., Coll. Math. Soc. J. Bolyai*, **4**, North-Holland, 1970, 755–770.]

12.18 The desired number is the permanent of the matrix

$$\begin{pmatrix} 1 & \cdots & 1 & 0 & 0 & \cdots & 0 & 0 \\ 1 & \cdots & 1 & 1 & 0 & \cdots & 0 & 0 \\ \vdots & & & & & & & \\ 1 & \cdots & 1 & 1 & 1 & \cdots & 1 & 0 \\ 1 & \cdots & 1 & 1 & 1 & \cdots & 1 & 1 \\ \vdots & & & & & & & \\ 1 & \cdots & 1 & 1 & 1 & \cdots & 1 & 1 \end{pmatrix} \begin{matrix} \\ \\ \end{matrix} \begin{matrix} n-p \\ \\ p \end{matrix}$$

$$\underbrace{\qquad}_{p} \underbrace{\qquad}_{n-p}$$

By expanding the permanent on the first row of the matrix one can obtain the recurrence relation

$$A(n, p) = pA(n-1, p).$$

Thus $A(n, p) = pA(n-1, p) = \cdots = p^{n-p}A(p, p) = p^{n-p}p!$, since $A(p, p)$ is the number $p!$ of permutations of the set $\{1, 2, \ldots, p\}$.

12.19 For an up–down permutation of the set $\{1, \ldots, n\}$ the number 1 may be found in one of the positions of rank 1, 3, 5, ..., while the number n can be found in one of the positions of rank 2, 4, 6, ...

Suppose the number 1 is found in position $2k+1$. The number of up–down permutations of the set $\{1, \ldots, n\}$ with this property is equal to $\binom{n-1}{2k}A_{2k}A_{n-1-2k}$. This follows from the fact that the numbers at positions $1, 2, \ldots, 2k$ can be chosen in $\binom{n-1}{2k}$ ways from $2, 3, \ldots, n$, and the number of up–down permutations with $2k$ positions is equal to A_{2k}. At the same time the number of up–down permutations formed with the remaining elements on the positions $2k+2, \ldots, n$ is equal to A_{n-1-2k}. If the number n occurs in the position of rank $2k$, then one can show analogously that the number of up–down permutations is equal to $\binom{n-1}{2k-1}A_{2k-1}A_{n-2k}$. From this follows the recurrence relation

$$2A_n = \binom{n-1}{0}A_0 A_{n-1} + \binom{n-1}{1}A_1 A_{n-2} + \binom{n-1}{2}A_2 A_{n-3} + \cdots$$
$$+ \binom{n-1}{n-1}A_{n-1}A_0, \quad \text{where} \quad A_0 = A_1 = 1.$$

Notice that each up–down permutation with the number 1 in position $2i+1$ and the number n in position $2j$ is counted exactly twice: once among the up–down permutations with the number 1 in position $2i+1$ and once among the up–down permutations with the number n in position $2j$. If $a_k = A_k/k!$ for $k \geq 0$, then this recurrence relation can be written

$$2na_n = a_0 a_{n-1} + a_1 a_{n-2} + \cdots + a_{n-1}a_0,$$

and hence the generating function of the numbers a_i,

$$f(x) = a_0 + a_1 x + a_2 x^2 + a_3 x^3 + \cdots,$$

satisfies the differential equation $f'(x)/\{1+f^2(x)\} = \frac{1}{2}$ with initial condition $f(0) = a_0 = 1$. The solution of this equation yields

$$\arctan f = \frac{x}{2} + C, \quad \text{or} \quad f = \tan\left(\frac{x}{2} + C\right),$$

where $C = \pi/4$. Thus

$$f(x) = \tan\left(\frac{x}{2} + \frac{\pi}{4}\right) = \frac{1 + \sin x}{\cos x} = \sec x + \tan x.$$

[D. André, *C. R. Acad. Sci. Paris*, **88** (1879), 965–967.]

There are many extensions of this classical result [L. Carlitz, *Discrete Mathematics*, **4** (1973), 273–286; L. Carlitz, R. Scoville, *Duke Mathematical J.*, **39**(4) (1972), 583–598; etc.]

12.20 It follows from the given conditions that

$$p(1)<p(3)<p(5)<p(7)<\cdots \quad \text{and} \quad p(1)<p(4)<p(6)<p(8)<\cdots.$$

Thus $p(1)<\min(p(3), p(4), p(5), p(6),\ldots)$. One can show analogously that $p(i)<\min_{j\geqslant i+2} p(j)$.

Let the number sought be denoted $g(n)$ for every $n\geqslant 1$. It can be seen that $g(1)=1$ and $g(2)=2$, since both of the permutations 1, 2 and 2, 1 satisfy the given conditions. The set of permutations of the set $\{1,\ldots,n\}$ which are 2-ordered and 3-ordered can be written in the form $A_n\cup B_n$, where A_n is the set of permutations of the set $\{1,\ldots,n\}$ which are 2-ordered and 3-ordered and for which $p(1)=1$. The set B_n consists of those permutations of the set $\{1,\ldots,n\}$ which are 2-ordered and 3-ordered and for which $p(1)=2$ and $p(2)=1$.

In fact, if the permutation p is 2-ordered and 3-ordered and $p(1)=2$, then $p(2)=1$, since otherwise the number 1 would appear in one of the positions $3, 4, \ldots, n$; this contradicts the inequality

$$p(1)<\min(p(3), p(4), \ldots, p(n)). \tag{1}$$

In the same way one can show that $p(1)\leqslant 2$, since if $p(1)\geqslant 3$ at least one of the numbers 1 or 2 appears in one of the positions $3, 4, \ldots, n$, and this again contradicts (1).

It is clear that the sets A_n and B_n are disjoint, and thus $g(n)=|A_n|+|B_n|=g(n-1)+g(n-2)$, since A_n has the same cardinality as the set of permutations of the set $\{2,\ldots,n\}$ which are 2-ordered and 3-ordered. The set B_n has the same cardinality as the set of permutations of the set $\{3,\ldots,n\}$ which are 2-ordered and 3-ordered.

Since $g(1)=F_1$ and $g(2)=F_2$ and the numbers $g(n)$ satisfy the same recurrence relation of the Fibonacci numbers, it follows that $g(n)=F_n$ for every $n\geqslant 1$. [H. B. Mann, *Econometrica*, **13** (1945), 256.]

12.21 It follows from the proof of the preceding problem that $u_i<\min_{k\geqslant 2} u_{i+k}$ and hence the sequence can contain at most two consecutive terms which are equal.

Let $F(p, q)$ denote the number of sequences $(u_1,\ldots,u_q)$ where $u_1,\ldots,u_q$ belong to a set consisting of p pairwise distinct numbers which satisfy the conditions $u_i<u_{i+2}$ $(1\leqslant i\leqslant n-2)$ and $u_i<u_{i+3}$ $(1\leqslant i\leqslant n-3)$. It follows that $f(n)=F(n, n)$.

Suppose that $u_1,\ldots,u_q \in \{1,\ldots,p\}$. If the sequence $u_1,\ldots,u_q$ begins with 1 and $u_2\geqslant 2$, then the number of such sequences is equal to $F(p-1, q-1)$, while if $u_2=1$, the number of these sequences is equal to $F(p-1, q-2)$, since in positions $3, 4, \ldots, q$ there must be numbers from the set $\{2,\ldots,p\}$ which satisfy the given conditions. If the sequence begins with $u_1=k$, then $u_1=k$ and $u_2=1$, or $u_1=k$ and $u_2=2,\ldots$, or $u_1=k$ and $u_2=k$, or $u_2\geqslant k+1$. Thus in this case the number of the sequences $(u_1,\ldots,u_q)$ is equal to $F(p-k, q-1)+kF(p-k, q-2)$. Finally one can conclude that the numbers $F(p, q)$ satisfy the recurrence relation

$$F(p, q) = F(p-1, q-1) + F(p-2, q-1) + F(p-3, q-1) + \cdots$$
$$+ F(p-1, q-2) + 2F(p-2, q-2) + 3F(p-3, q-2) + \cdots$$

for every $p, q \geq 3$.

It is also true that $F(r, 1) = r$ and $F(r, 2) = r^2$, $F(1, r) = 0$ for $r \geq 3$, $F(r, 2r) = 1$ when the unique sequence $112233\cdots rr$ is obtained, and $F(r, 2r+1) = F(r, 2r+2) = \cdots = 0$. One can also conclude that $F(2, 3) = 2$, since in this case there exist sequences 112 and 122 which satisfy the given conditions. These initial conditions yield the following table of numbers $F(i, j)$ in which the first two rows and the first two columns are determined from the initial conditions:

i \ j	1	2	3	4	5	6	7	8	9	10
1	1	1	0	0	0	0	0	0	0	0
2	2	4	2	1	0	0	0	0	0	0
3	3	9	9	8	3	1	0	0	0	0
4	4	16	24	31	22	13	4	1	0	0
5	5	25	50	85	88	75	42	19	5	1
6	6	36	126	—	—	288	—	—	—	—
7	7	49								

The number $F(p, q)$ with $p, q \geq 3$ is obtained by adding the numbers from column $q-1$ and the numbers from column $q-2$ multiplied by 1, 2, 3, 4, 5, ..., respectively, and located in row $p-1$ and above.

In particular, $f(6) = F(6, 6) = 288$.

12.22 It is clear that $A(n, n) = 1$, since there exists a unique permutation with n falls, namely $n, n-1, \ldots, 1$; and $A(n, 1) = 1$, since the unique permutation with one fall is the identity permutation. By definition, one has $A(n, 0) = 0$ and $\sum_{k=1}^{n} A(n, k) = n!$. Consider the set Q of the $A(n-1, k)$ permutations of the set $\{1, \ldots, n-1\}$ which have k falls. One can obtain $kA(n-1, k)$ permutations of the set $\{1, \ldots, n\}$ which have k falls each by inserting (in k ways) the element n between any two positions $p(i)$ and $p(i+1)$ such that $p(i) > p(i+1)$ [and also after $p(n-1)$], for every permutation $p \in Q$.

The set R of the $A(n-1, k-1)$ permutations of $\{1, \ldots, n-1\}$ which have $k-1$ falls yields $(n-k+1)A(n-1, k-1)$ permutations of the set $\{1, \ldots, n\}$ with k falls by inserting (in $n-k+1$ ways) the element n between any two positions $p(i)$ and $p(i+1)$ such that $p(i) < p(i+1)$, and also to the left of $p(1)$, for every permutation $p \in R$. In this case the number of possible ways is equal to $n-(k-1) = n-k+1$, and the number of falls increases by one. By using these two procedures one obtains all $A(n, k)$ permutations $p \in S_n$ with k falls, without repetitions, and hence (a) follows.

By using this recurrence relation one can compute in tabular form the numbers $A(n, k)$ row by row, taking into account that $A(n, k) = 0$ for $k > n$. For $n \leq 6$ the table is given below:

n \ k	1	2	3	4	5	6
1	1	0	0	0	0	0
2	1	1	0	0	0	0
3	1	4	1	0	0	0
4	1	11	11	1	0	0
5	1	26	66	26	1	0
6	1	57	302	302	57	1

In order to prove (b), let $P_{n,k}$ be the set of permutations $p \in S_n$ with k falls, and let f be the function which associates the permutation $p = p(1)p(2) \cdots p(n) \in S_n$ with the permutation $f(p) = p(n)p(n-1) \cdots p(1) \in S_n$. It is clear that p has a fall at $p(i)$ if and only if $f(p)$ does not have a fall at $p(i+1)$ for $1 \leq i \leq n-1$. By definition $f(p)$ has a fall at $p(1)$ and p has a fall at $p(n)$. Thus $f(p) \in P_{n,n+1-k}$. Since $f: P_{n,k} \to P_{n,n+1-k}$ is a bijection, it follows that $|P_{n,k}| = |P_{n,n+1-k}|$, that is, (b) holds.

In order to obtain (c) we prove that the identity

$$\sum_{k=1}^{n} A(n, k) \binom{m+k-1}{n} = m^n \qquad (1)$$

holds for fixed $n \geq 1$ and for any integer $m \geq 1$.

Consider the set of m^n words $a_1 a_2 \cdots a_n$, where the a_i are integers and $1 \leq a_i \leq m$. One can sort this sequence so as to obtain the unique sequence $a_{p(1)} \leq a_{p(2)} \leq \cdots \leq a_{p(n)}$, where $p(1)p(2) \cdots p(n)$ is a particular permutation of the set $\{1, \ldots, n\}$, under the condition that $a_{p(i)} = a_{p(i+1)}$ implies that $p(i) < p(i+1)$ for any $1 \leq i \leq n-1$. This condition is equivalent to the following: if $p(i) > p(i+1)$ then $a_{p(i)} < a_{p(i+1)}$ for any $1 \leq i \leq n-1$. It will now be shown that if the permutation $p(1) \cdots p(n)$ has k falls, then the number of words $a_1 a_2 \cdots a_n$ sorted in the abovementioned way by the permutation p is equal to $\binom{m+n-k}{n}$. Indeed, if $p \in S_n$ has k falls, then the desired number of words is equal to the number of sequences $a_1 a_2 \cdots a_n$ such that

$$1 \leq a_{p(1)} \bigcirc\!\!\!\leq a_{p(2)} \bigcirc\!\!\!\leq \cdots \bigcirc\!\!\!\leq a_{p(n)} \leq m, \qquad (2)$$

where the sign $\bigcirc\!\!\!\leq$ between $a_{p(i)}$ and $a_{p(i+1)}$ is $\leq$ if the permutation p does not have a fall at $p(i)$ and is $<$ otherwise. But the number of sequences satisfying (2) is equal to the number of sequences $b_1 b_2 \cdots b_n$ composed of positive integers which satisfy the following conditions:

$$1 \leq b_1 < b_2 < \cdots < b_{n-1} < b_n \leq m+n-k. \qquad (3)$$

Let $b_1 = a_{p(1)}$; let $b_2 = a_{p(2)}$ if p has a fall at $p(1)$ and $b_2 = a_{p(2)} + 1$ otherwise; if $b_r = a_{p(r)} + s$, let $b_{r+1} = a_{p(r+1)} + s$ if p has a fall at $p(r)$ and $b_{r+1} = a_{p(r+1)} + s + 1$

otherwise, for any $r \geq 2$. This correspondence is a bijection between the set of sequences satisfying (2) and the set of sequences satisfying (3). But the number of sequences $b_1 b_2 \cdots b_n$ which satisfy (3) is equal to the number of n-element subsets of the set $\{1, \ldots, m+n-k\}$, that is, $\binom{m+n-k}{n}$. Hence one has defined a function g which associates to every sequence $a = a_1 \cdots a_n$, where $1 \leq a_i \leq m$, a permutation $p = g(a)$ which sorts this sequence as previously described.

Let $A_k = \{a = a_1 \cdots a_n \mid g(a) \text{ has } k \text{ falls}\}$. It follows that

$$|A_k| = A(n, k) \binom{m+n-k}{n}$$

and the sets $A_1, \ldots, A_n$ are pairwise disjoint. One can write

$$m^n = \sum_{k=1}^{n} |A_k| = \sum_{k=1}^{n} A(n, k) \binom{m+n-k}{n}$$

$$= \sum_{k=1}^{n} A(n, n-k+1) \binom{m-1+(n-k+1)}{n} = \sum_{k=1}^{n} A(n, k) \binom{m+k-1}{n},$$

which is (1). Since (1) holds for every $m \geq 1$, it follows that polynomial identity (c) also holds, because the polynomials on both sides take equal values for an infinite number of values of the variable x.

(d) may be deduced from (c) as follows. For $x = 1$ one finds that $A(n, n) = A(n, 1) = 1$. For $x = 2$ it can be seen that $A(n, n-1) = A(n, 2) = 2^n - n - 1$, which is in accordance with (d). Suppose that (d) is true for $A(n, s)$, where $1 \leq s \leq k-1$. It will be shown that this relation is also true for $s = k$.

Let $x = k$ in (c) to obtain

$$k^n = A(n, n-k+1) \binom{n}{n} + A(n, n-k+2) \binom{n+1}{n} + \cdots + A(n, n) \binom{n+k-1}{n},$$

or

$$k^n = A(n, k) + A(n, k-1) \binom{n+1}{n} + \cdots + A(n, n) \binom{n+k-1}{n}.$$

The induction hypothesis yields

$$A(n, k) = \sum_{p=0}^{k-1} (-1)^p c(n, p)(k-p)^n,$$

where the coefficient $c(n, p) = \sum_{s=1}^{p} (-1)^{s+1} \binom{n+1}{p-s}\binom{n+s}{n}$ for $p \geq 1$ and $c(n, 0) = 1$. It is now necessary to show that $c(n, p) = \binom{n+1}{p}$, or equivalently that the following identity holds:

$$\sum_{s=0}^{p} (-1)^{s+1} \binom{n+1}{p-s}\binom{n+s}{n} = 0 \qquad (4)$$

for any $p \geq 1$. By using Newton's generalized binomial formula one can conclude that

$$1 = (1-x)^{n+1}(1-x)^{-(n+1)}$$

$$= \left\{1 - \binom{n+1}{1}x + \binom{n+1}{2}x^2 + \cdots + (-1)^{n+1}x^{n+1}\right\}$$

$$\times \left\{1 + \binom{n+1}{1}x + \binom{n+2}{2}x^2 + \cdots + \binom{n+s}{s}x^s + \cdots\right\}.$$

In the expansion of the product on the right-hand side the coefficient of x^p vanishes for any $p \geq 1$, so that $\sum_{s=0}^{p}(-1)^{p-s}\binom{n+s}{s}\binom{n+1}{p-s} = 0$. By multiplying both members by $(-1)^{1-p}$ and using the equality $\binom{n+s}{s} = \binom{n+s}{n}$, one obtains (4). Similarly, the Eulerian number $A(n, k)$ may be defined as the number of permutations of $\{1, \ldots, n\}$ with k rises. By definition, a permutation $p(1) \cdots p(n)$ has a rise at $p(i)$ if $p(i) < p(i+1)$; by convention, there is a rise to the left of $p(1)$. [L. Euler, *Institutiones Calculi Differentialis*, St. Petersburg, 1755, 485–487; see also L. Euler, *Opera Omnia*, 1913, Vol. 1, Section 10, 373–375.]

12.23 Let

$$\sigma = \begin{pmatrix} 1 & 2 & \cdots & n \\ n & n-1 & \cdots & 1 \end{pmatrix}.$$

Since p^2 is an even permutation, it follows that $I(\sigma) = \binom{n}{2} \equiv 0 \pmod{2}$, which implies that $n \equiv 0$ or $1 \pmod{4}$. Two cases will be considered:

(a) Let $n = 4m$. Here p has no fixed points. In fact, if there exists $a \in \{1, \ldots, 4m\}$ such that $p(a) = a$, then $\sigma(a) = p(p(a)) = a$, so $4m - a + 1 = a$, and a is not an integer. Now let $k, i \in \{1, \ldots, 4m\}$ such that $p(k) = i$. One can write $\sigma(k) = p(p(k)) = p(i)$, or $p(i) = 4m - k + 1$; $\sigma(i) = p(p(i)) = p(4m - k + 1)$, or $p(4m - k + 1) = 4m - i + 1$; $\sigma(4m - k + 1) = p(p(4m - k + 1)) = p(4m - i + 1)$, and hence $p(4m - i + 1) = k$. The numbers in the set $A = \{k, i, 4m - k + 1, 4m - i + 1\}$ are pairwise distinct. For otherwise, if $k = i$, then k would be a fixed point; if $k = 4m - k + 1$, then k would not be an integer; if $k = 4m - i + 1$, then k would be a fixed point of p; and so on.

It follows that $p(A) = A$, and the restriction p_1 of p to the set $\{1, \ldots, 4m\} \setminus A$ belongs to S_{n-4} and satisfies $p_1^2 = \sigma_1$. Here σ_1 denotes the unique permutation of the set $\{1, \ldots, 4m\} \setminus A$ having $I(\sigma_1) = \binom{n-4}{2}$ inversions.

Since $p^2 = \sigma$ implies that $p(k) \notin \{k, 4m - k + 1\}$ for every k, one can conclude that the recurrence relation satisfied by $N(4m)$ is

$$N(4m) = (4m - 2)N(4m - 4).$$

It follows that

$$N(4m) = (4m-2)(4m-6) \cdots 6N(4) = 2 \times 6 \times \cdots \times (4m-2)$$

$$= 2^m \{1 \times 3 \times 5 \times \cdots \times (2m-1)\} = \frac{(2m)!}{m!}, \quad \text{since} \quad N(4) = 2.$$

(b) If $n = 4m + 1$ then $p^2(2m+1) = \sigma(2m+1) = 2m+1$. Let $a = p(2m+1)$. It can be shown that

Solutions

$$\sigma(a)=p^2(p(2m+1))=p(p^2(2m+1))=p(2m+1)=a.$$

Since $2m+1$ is the unique fixed point of σ, it follows that $p(2m+1)=2m+1$ and p has no other fixed points. The restriction of p to the set $\{1,\ldots,4m+1\}\setminus\{2m+1\}$ has properties analogous to those of p, and hence one can conclude that

$$N(4m+1)=N(4m)=\frac{(2m)!}{m!}.$$

CHAPTER 13

13.1 Consider the axis of symmetry xx' of the rectangle which is parallel to the sides AC and BD (Figure 13.2). The desired number of configurations is equal to

$$N_1+N_2,$$

where N_1 represents the number of configurations which coincide with their images with respect to a reflection in xx'. The number N_2 represents one-half the number of configurations which are not self-corresponding under a reflec-

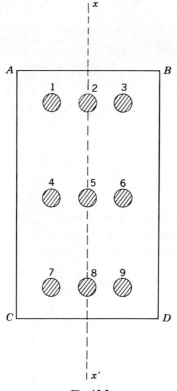

Fig. 13.2

tion in xx'. Excluding the configuration which contains no perforations, it follows that

$$N_1 + 2N_2 = 2^9 - 1 = 511.$$

In order to find the value of N_1, consider the configurations of the six perforations located on the same side of the axis xx' and on the axis xx' under a symmetry with respect to xx'. The result is that $N_1 = 2^6 - 1 = 63$. The desired number is thus

$$N_1 + N_2 = \tfrac{1}{2}(N_1 + 2N_2 + N_1) = \tfrac{1}{2}(511 + 63) = 287.$$

Suppose the side AB contains n perforations and the side AC contains m perforations, so that there are a total of mn perforations. Proceeding as before, it follows that

$$N_1 = 2^{m[(n+1)/2]} - 1$$

and $N_1 + 2N_2 = 2^{mn} - 1$. Thus the desired number of ways of performing a perforation is equal to

$$N_1 + N_2 = \tfrac{1}{2}(N_1 + 2N_2 + N_1) = \tfrac{1}{2}(2^{mn} + 2^{m[(n+1)/2]} - 2)$$
$$= 2^{mn-1} + 2^{m[(n+1)/2]-1} - 1.$$

[I. Tomescu, Problem 0:59, *Gazeta Matematica*, **84**(7) (1979).]

13.2 First we show that $x \sim y(G)$ is an equivalence relation. It satisfies:

(1) Reflexivity: $x \sim x(G)$, since $x = e(x)$, where e is the identity permutation, $e \in G$.
(2) Symmetry: $x \sim y(G)$ implies that $y \sim x(G)$, because there exists $g \in G$ such that $y = g(x)$ and hence $x = g^{-1}(y)$ with $g^{-1} \in G$.
(3) Transitivity: $x \sim y(G)$ and $y \sim z(G)$ implies $x \sim z(G)$, since there exist $g_1, g_2 \in G$ such that $y = g_1(x)$ and $z = g_2(y)$ and hence $z = g_2 g_1(x)$ with $g_2 g_1 \in G$.

In order to prove Burnside's lemma let $G_k = \{g | g \in G, g(k) = k\}$ be the set of permutations which leave fixed the element k. For every $k = 1, \ldots, n$ the set G_k is a subgroup of G, since if $f, g \in G_k$ then $fg \in G_k$, because $fg(k) = f(k) = k$. The set G_k is nonempty, since the identity permutation satisfies $e(k) = k$ for every $k = 1, \ldots, n$.

Let O_k denote the orbit of the group G which contains the element k. It will be shown that

$$|G_k| \cdot |O_k| = |G|.$$

In fact, since G_k is a subgroup of G, one can consider the set G/G_k whose classes are sets of the form $G_k a = \{ga | g \in G_k\}$, which form a partition of the set G and whose cardinality is equal to $|G/G_k| = |G|/|G_k|$. It will be shown that this number is equal to $|O_k|$ by constructing a bijection from O_k to G/G_k. Observe that

Solutions

$g(i)=h(i)=k$ with $g, h \in G$ implies that $hg^{-1}(k)=h(i)=k$, that is, $hg^{-1} \in G_k$ or $h \in G_k g$.

For every element $i \in O_k$ there exists a permutation $g_i \in G$ such that $g_i(i)=k$, since $i \sim k(G)$. The bijection $f: O_k \to G/G_k$ is defined as follows: $f(i)=G_k g_i \in G/G_k$, where the permutation g_i satisfies $g_i(i)=k$. The mapping f is well defined, since if there exist two permutations $h, g \in G$ such that $g(i)=h(i)=k$, then it has been shown that $h \in G_k g$, that is, the class of h coincides with the class of g, so $G_k h = G_k g$.

The mapping f is injective. In fact, $i, j \in O_k$ and $i \neq j$ imply the existence of two permutations $h, g \in G$ such that $g(i)=h(j)=k$ and thus $hg^{-1}(k)=h(i) \neq k = h(j)$, since h, being a permutation, is injective and $i \neq j$ implies that $h(i) \neq h(j)$.

But $hg^{-1}(k) \neq k$ means that $hg^{-1} \notin G_k$ or $h \notin G_k g$, and hence $G_k g \neq G_k h$ or $f(i) \neq f(j)$. Also, f is surjective for the class $G_k g$ if the image of the element $g^{-1}(k)=l \in O_k$, since $g(l)=k$ and hence $l \sim k(G)$. Since the mapping f is bijective, it follows that $|O_k|=|G/G_k|=|G|/|G_k|$. Now count the elements in X which are invariant under the permutations $g \in G$ in two different ways:

$$\sum_{g \in G} \lambda_1(g) = \sum_{k \in X} |G_k| = \sum_{i=1}^{q} \sum_{k \in O_i} |G_k|,$$

where $O_1, \ldots, O_q$ denote the orbits of the group G. It has been seen that if $j, k \in O_i$ then $|G_j|=|G_k|=|G|/|O_i|$, and hence

$$\sum_{g \in G} \lambda_1(g) = \sum_{i=1}^{q} |O_i| \frac{|G|}{|O_i|} = q|G|,$$

which implies that $q=(1/|G|) \sum_{g \in G} \lambda_1(g)$.

13.3 Denote by $R_1, R_2, \ldots, R_n$ the clockwise rotations of the regular polygon about its center through an angle $2\pi/n, 4\pi/n, \ldots, 2\pi$ respectively. These rotations form a group with respect to the composition of rotations; R_n is the identity element of the group. Two polygons P_1 and P_2 with k vertices are considered to be identical if there exists a rotation R_m with angle $2\pi m/n$ such that

$$P_1 = R_m(P_2).$$

By Burnside's lemma of the preceding problem, the desired number is equal to

$$\frac{1}{n} \sum_{m=1}^{n} \lambda_1(R_m), \tag{1}$$

where $\lambda_1(R_m)$ represents the number of polygons P with k vertices which are invariant under R_m, that is, $R_m(P)=P$. Observe that if one applies the rotation R_m successively to a vertex A of the regular polygon with n vertices, then the smallest index d for which $R_m^d(A)=A$ is equal to $d=n/(m, n)$. In fact it must be the case that $n \mid dm$ or $n/(m, n) \mid d \cdot \{m/(m, n)\}$. This implies that the smallest integer d which satisfies this relation is $n/(m, n)$, since the numbers $n/(m, n)$ and $m/(m, n)$ are relatively prime.

Represent the vertices of the regular polygon with n vertices by the numbers

$0, 1, \ldots, n-1$. The rotation R_m defines a permutation of these numbers, the image of the number i being the number $i+m \pmod{n}$, as is seen by considering the vertices of the regular polygon to be numbered $0, \ldots, n-1$ in the clockwise sense. Thus if the polygon P with k vertices is invariant under the rotation R_m, then this defines a permutation of the k vertices of P, which can be expressed in the following form as a product of disjoint cycles:

$$[i_1, i_2, \ldots, i_d][j_1, j_2, \ldots, j_d] \cdots [t_1, t_2, \ldots, t_d], \qquad (2)$$

where all the cycles have the same length $d = n/(m, n)$. It follows that $d \mid k$ or $n \mid k(m, n)$, and the number of cycles in (2) is equal to $k/d = k(m, n)/n$.

Consider the permutation induced by R_m on the set X of vertices of the polygon P and written as a product of disjoint cycles. It will be shown that every (m, n) consecutive numbers modulo n of the set $\{0, 1, \ldots, n-1\}$ have the property that no pair of them belong to the same cycle of the permutation induced by R_m on X. For suppose the contrary. One can then show that there exist two numbers $0 \leqslant a, b \leqslant n-1$ such that $0 < |a-b| \leqslant (m, n) - 1$ and such that they are found on the same cycle as c. It follows that $c + pm \equiv a \pmod{n}$ and $c + qm \equiv b \pmod{n}$, and hence

$$a = c + pm - rn \quad \text{and} \quad b = c + qm - sn,$$

where $r, s \geqslant 0$ are integers. Thus one has

$$|a - b| = |(p - q)m + (s - r)n| \geqslant (m, n)$$

because $a \neq b$, and this contradicts the inequality $|a - b| \leqslant (m, n) - 1$.

It follows that for every (m, n) consecutive numbers (modulo n) of the set $\{0, \ldots, n-1\}$ there are exactly $k(m, n)/n$ distinct numbers which belong to different cycles of the permutation (2) induced by R_m on X. For otherwise the number of elements in X would be smaller than $\{k(m, n)/n\}\{n/(m, n)\} = k$, which is a contradiction. In order to find the number of polygons P with k vertices which are invariant under R_m, one must find the number of permutations of form (2) where $d = n/(m, n)$ which contain $k(m, n)/n$ cycles. The permutation (2) is uniquely determined if one chooses an element in each cycle; the other elements of the cycles are obtained by repeated addition of m (modulo n) to the number chosen. Thus if $n \mid k(m, n)$, then the number $\lambda_1(R_m)$ is equal to the number of ways of choosing $k(m, n)/n$ elements from among (m, n) elements, that is, it is equal to

$$f(n, k, m) = \binom{(m, n)}{\dfrac{k(m, n)}{n}}. \qquad (3)$$

If n is not a divisor of $k(m, n)$ then $\lambda_1(R_m) = 0$. In view of (1) the desired number has the form

$$\frac{1}{n} \sum_{\substack{m=1 \\ n \mid k(m, n)}}^{n} f(n, k, m). \qquad (4)$$

Solutions

By using the notation $d = n/(m, n)$, the condition $n \mid k(m, n)$ becomes $d \mid k$, and since $d \mid n$, it follows that $d \mid (n, k)$.

Now calculate the number of terms in summation (4) which have the same value. Assume that d is fixed and m takes values between 1 and n such that $k(m, n)$ is divisible by n. If $e = m/(m, n)$ then $e \leq n/(m, n) = d$ and $(e, d) = 1$. The condition $n \mid k(m, n)$ is satisfied if $d \mid (n, k)$. Thus for fixed d, $f(n, k, m)$ takes the same value for all numbers e which are prime to and less than d. The number e uniquely determines m by $m = e(m, n) = en/d$. It follows that the number of convex polygons with k vertices which cannot be obtained from one another by a rotation is equal to

$$\frac{1}{n} \sum_{d \mid (n, k)} \varphi(d) \binom{n/d}{k/d},$$

where $\varphi(d)$ is Euler's function which gives the number of positive integers smaller than and prime to d (Problem 2.4).

13.4 Proceeding as in the preceding problem, it can be seen that the number of colorings with k colors of the vertices of the regular polygon with n vertices which cannot be obtained from one another by a rotation is equal to

$$\frac{1}{n} \sum_{m=1}^{n} \lambda_1(R_m),$$

where $\lambda_1(R_m)$ represents the number of k-colorings invariant under the rotation R_m of angle $2\pi m/n$. One can show analogously that if a k-coloring is invariant under a rotation R_m, then this rotation defines a permutation of the n vertices which can be expressed in the following form as a product of disjoint cycles:

$$[i_1^1, \ldots, i_d^1][i_1^2, \ldots, i_d^2] \cdots [i_1^p, \ldots, i_d^p], \tag{1}$$

where $d = n/(m, n)$ and $p = (m, n)$. Each cycle in (1) is formed from numbers of the set $\{0, \ldots, n-1\}$ corresponding to vertices of the regular polygon which are colored with the same color.

Also, every (m, n) consecutive numbers modulo n in the set $\{0, \ldots, n-1\}$ belong to different cycles in (1). Thus the number of k-colorings invariant under R_m is equal to the number of colorings with k colors of (m, n) consecutive numbers modulo n from $\{0, \ldots, n-1\}$, that is, to the number of functions

$$g : \{1, \ldots, (m, n)\} \to \{1, \ldots, k\}.$$

It follows that $\lambda_1(R_m) = k^{(m, n)}$. Finally the desired number of k-colorings is equal to

$$\frac{1}{n} \sum_{m=1}^{n} k^{(n, m)} = \frac{1}{n} \sum_{d \mid n} \varphi(d) k^{n/d}.$$

13.5 One first obtains a formula for g_n. If in the representation of the permutation $p \in S_n$ as a product of disjoint cycles there are d_k cycles of length k for $k = 1, \ldots, n$, then p is said to be a permutation of type $1^{d_1} 2^{d_2} \cdots n^{d_n}$, where $d_1 + 2d_2 + \cdots + nd_n = n$.

Suppose that the graph $G=(X, U)$ has vertex set $X=\{1,\ldots,n\}$. Every permutation p in S_n can be considered to be a permutation of the set of graphs G by defining $p(G)=(X, p(U))$, where $p(U)=\{[p(i), p(j)] \,|\, [i, j] \in U\}$. One first determines the number $d(p)$ of graphs G with n vertices such that $p(G)=G$, that is, the number of fixed points of p. Let G be a graph such that $p(G)=G$, and suppose that p is of type $1^{d_1}2^{d_2}\cdots n^{d_n}$. The graph G can be partitioned into subgraphs $G_1, G_2, \ldots$ such that two vertices i and j belong to the same subgraph if and only if i and j belong to the same cycle in p. Among the subgraphs G_i there are d_k which contain exactly k vertices, for $k=1,\ldots,n$.

Suppose that $\{1, 2, \ldots, k\}$ and $\{k+1, k+2, \ldots, k+l\}$ are sets of vertices for two of these subgraphs G_1 and G_2. One can assume that the permutation p contains the cycles $[1,\ldots,k]$ and $[k+1,\ldots,k+l]$. The $[k/2]$ possible edges between the vertex 1 and the vertices $2, 3, \ldots, [k/2]+1$ in the graph G_1 uniquely determine the existence of the other edges of G_1 in view of the condition $p(G)=G$.

Similarly the (k, l) possible edges between the vertex 1 and the vertices $k+i$ for $i=1,\ldots,(k, l)$ uniquely determine the existence or nonexistence of the other edges which join vertices of G_1 and vertices of G_2 in view of the condition $p(G)=G$. Thus if $p(G)=G$, one can select in an arbitrary manner the existence of a number of edges equal to

$$G_d = \sum_{k=1}^{n} (k, k)\binom{d_k}{2} + \sum_{k<l}(k, l)d_k d_l + \sum_{k \text{ odd}} \frac{1}{2} d_k(k-1) + \sum_{k \text{ even}} \frac{1}{2} k d_k$$

$$= \frac{1}{2}\left(\sum_{k,l=1}^{n}(k, l)d_k d_l - \sum_{k \text{ odd}} d_k\right).$$

Thus $d(p)=2^{G_d}$, from which (a) follows, since the number of permutations of type $1^{d_1}2^{d_2}\cdots n^{d_n}$ is equal to $n!/N_d$ and $|S_n|=n!$.

In order to prove (b) one can proceed analogously. Two vertices x, y can be nonadjacent, joined by the arc (x, y) or the arc (y, x), or joined by both the arc (x, y) and the arc (y, x). There are thus four possible cases. Also, in the expression for G_d the summation $\sum_{k \text{ even}} \frac{1}{2}k d_k$ must be replaced by $\sum_{k \text{ even}} \frac{1}{2}(k-2)d_k$. For two vertices u, v of an even cycle of length k, separated by the maximal distance $k/2$, there are only two possibilities: either u and v are nonadjacent, or u and v are joined by both arcs (u, v) and (v, u), since the existence of a single arc would contradict the invariance of the graph G under the permutation p. Thus by applying Burnside's formula one obtains the numerator of d_n, which is expressed by

$$4^{G_d - \sum_{k \text{ even}} d_k} \times 2^{\sum_{k \text{ even}} d_k} = 2^{D_d},$$

from which (b) follows.

The formula for c_n can be established analogously in view of the fact that there are three possible ways to join two vertices by arcs. Consider a tournament, and suppose the permutation p under which G is invariant contains a cycle of even length, say $[i_1, i_2, \ldots, i_{2k}]$, and that G contains the arc (i_1, i_{k+1}). Since $p(G)=G$, it follows that G also contains the arcs $(i_2, i_{k+2}), (i_3, i_{k+3}), \ldots, (i_{k+1}, i_1)$. But the existence of both arcs (i_1, i_{k+1}) and (i_{k+1}, i_1) contradicts the definition of a tournament, and hence in this case $d(p)=0$. Otherwise, if $p(G)=G$, one can choose an arbitrary direction only for

$$\sum_{k=1}^{n} \tfrac{1}{2}d_k(k-1) + \sum_{k=1}^{n} (k,k)\binom{d_k}{2} + \sum_{k<l} (k,l)d_k d_l = \frac{1}{2}\left(\sum_{k,l=1}^{n}(k,l)d_k d_l - \sum_{k=1}^{n} d_k\right) = T_d$$

arcs; the orientation of the remaining arcs is uniquely determined by the orientation of these arcs, and the numbers $d_1, d_3, d_5, \ldots$ satisfy (3).

The values of the numbers g_n, d_n, and t_n up to $n=7$ are given in the following table:

n	g_n	d_n	t_n
1	1	1	1
2	2	3	1
3	4	16	2
4	11	218	4
5	34	9,608	12
6	156	1,540,944	56
7	1 044	882,033,440	456

13.6 The relation $f_1 \sim f_2$ is an equivalence relation. It satisfies:

(1) Reflexivity: $f \sim f$ because $f = fe$, where e is the identity permutation and $e \in G$.
(2) Symmetry: $f_1 \sim f_2$ implies that $f_2 \sim f_1$, since there exists $g \in G$ such that $f_1 g = f_2$ and thus $f_2 g^{-1} = f_1$ and $g^{-1} \in G$.
(3) Transitivity: $f_1 \sim f_2$ and $f_2 \sim f_3$ imply that $f_1 \sim f_3$, since $f_1 g = f_2$ and $f_2 h = f_3$ with $g, h \in G$ implies $(f_1 g)h = f_1(gh) = f_3$, where $gh \in G$, since G is a group.

Let F denote the set of the m^n colorings $f : X \to A$. For every permutation g of X the mapping $f \to \bar{g}(f) = fg$ is an injection of F into F. In fact, $f_1 \neq f_2$ implies the existence of an object $i \in X$ such that $f_1(i) \neq f_2(i)$. Let $j = g^{-1}(i)$. It follows that $f_1 g(j) = f_1(i) \neq f_2(i) = f_2 g(j)$ and thus $f_1 g \neq f_2 g$. The mapping $\bar{g} : F \to F$ is injective, and since F is finite, this mapping must be surjective and hence a bijection. In other words $\bar{g} \in S$, where S denotes the set of permutations of the set F.

The equation $\varphi(g) = \bar{g}$ defines a mapping

$$\varphi : G \to S.$$

This mapping is injective, since if $g_1 \neq g_2$ there exists $k \in X$ such that $g_1(k) \neq g_2(k)$. It will be shown that the functions $\bar{g}_1 : F \to F$ and $\bar{g}_2 : F \to F$ are different, that is, there exists $f \in F$ such that $fg_1 \neq fg_2$, in view of the definition of $\bar{g}$.

If $m \geq 2$ there is a coloring f which uses different colors for the distinct elements $g_1(k)$ and $g_2(k)$, since F is the set of all functions $f: X \to A$.

Thus $f(g_1(k)) \neq f(g_2(k))$ or $fg_1(k) \neq fg_2(k)$, from which it follows that fg_1 and fg_2 are different and φ is an injection. If $m = 1$, then φ no longer is injective. However, in this case the number of colorings is equal to 1, so all the objects have the same color and $P(G; 1, \ldots, 1) = (1/|G|) \sum_{g \in G} 1 = 1$, which completes the proof of the property.

Let $\bar{G} = \{\bar{g} \mid g \in G\}$. Since φ is an injection, it follows that there exists a bijection $\varphi_1 : G \to \bar{G} \subset S$ defined by $\varphi_1(g) = \varphi(g)$ for every $g \in G$, and hence $|G| = |\bar{G}|$.

The set $\bar{G}$ is a subgroup of the group S of permutations of the set F of colorings, since $\bar{g}_1, \bar{g}_2 \in \bar{G}$ implies that $\overline{g_1 g_2} \in \bar{G}$. In fact $\overline{g_1 g_2}(f) = \bar{g}_1(fg_2) = f(g_2 g_1) = \overline{g_2 g_1}(f)$, and thus the product of two elements of $\bar{G}$, say $\bar{g}_1$ and $\bar{g}_2$, is an element of $\bar{G}$ which corresponds to the product $g_2 g_1 \in G$. Since S is a finite group, it follows that $\bar{G}$ is a subgroup of S.

According to the given definition, two colorings f_1 and f_2 are equivalent if there exists $g \in G$ such that $f_1 g = f_2$ or $\bar{g}(f_1) = f_2$, and thus they belong to the same orbit of the group $\bar{G}$. This implies that the number of equivalence classes is equal to the number of orbits of the group $\bar{G}$. By Burnside's theorem (Problem 13.2) this number is equal to

$$\frac{1}{|\bar{G}|} \sum_{\bar{g} \in \bar{G}} \lambda_1(\bar{g}),$$

where $\lambda_1(\bar{g})$ represents the number of fixed points of the permutation $\bar{g}$ or the number of colorings f such that $\bar{g}(f) = f$ or $fg = f$. But $fg = f$ implies that f is constant for every cycle of the permutation g, since otherwise one would have $fg \neq f$.

Thus there exist as many colorings f with the property $fg = f$ as there are functions on the set of cycles which contain $\lambda_1(g) + \cdots + \lambda_n(g)$ elements in the set of the m colors, namely, $m^{\lambda_1(g) + \cdots + \lambda_n(g)}$. In view of the fact that $|G| = |\bar{G}|$, one finds that the number of equivalence classes is equal to

$$\frac{1}{|G|} \sum_{g \in G} m^{\lambda_1(g) + \cdots + \lambda_n(g)} = P(G; m, \ldots, m).$$

13.7 Consider the set X whose six elements are the faces of a cube, which will be denoted $1, \ldots, 6$ as in Figure 13.3. The vertices of the cube are labeled a, b, c, d, e, f, g, h.

Now use Pólya's method to count the equivalence classes of the colorings, that is, the mappings $f : X \to A = \{a_1, \ldots, a_m\}$. The group G of rotations of the cube will be determined; it is a subgroup of the group of permutations of the set X.

The rotations which leave the cube invariant can be expressed as follows as a product of disjoint cycles and omitting cycles of length 1:

Solutions

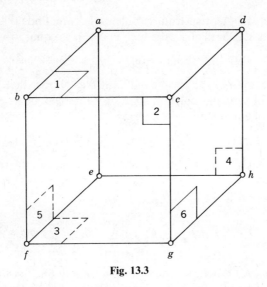

Fig. 13.3

(1) about the axis $abcd-efgh$: [2, 6, 4, 5], [2, 4][6, 5], [2, 5, 4, 6];
(2) about the axis $bcfg-adhe$: [1, 5, 3, 6], [1, 3][5, 6], [1, 6, 3, 5];
(3) about the axis $abfe-dcgh$: [1, 2, 3, 4], [1, 3][2, 4], [1, 4, 3, 2];
(4) about the axis $a-g$: [1, 4, 5][6, 3, 2], [1, 5, 4][6, 2, 3];
(5) about the axis $b-h$: [1, 5, 2][6, 4, 3], [1, 2, 5][6, 3, 4];
(6) about the axis $c-e$: [1, 2, 6][3, 4, 5], [1, 6, 2][3, 5, 4];
(7) about the axis $d-f$: [1, 6, 4][3, 5, 2], [1, 4, 6][3, 2, 5];
(8) about the axis $ab-hg$: [1, 5][3, 6][2, 4];
(9) about the axis $bc-eh$: [1, 2][3, 4][5, 6];
(10) about the axis $cd-ef$: [1, 6][3, 5][2, 4];
(11) about the axis $ad-fg$: [1, 4][2, 3][5, 6];
(12) about the axis $bf-dh$: [2, 5][6, 4][1, 3];
(13) about the axis $cg-de$: [2, 6][5, 4][1, 3].

The axis $bcfg-adhe$ means the axis determined by the centers of the squares $bcfg$ and $adhe$. The axis $ab-hg$ is the axis determined by the midpoints of the edges ab and hg, and so on. Since cycles of length 1 are not represented, the permutation [2, 6, 4, 5] is in fact [1][3][2, 6, 4, 5], the permutation [2, 4][6, 5] is [1][3][2, 4][6, 5], and so on.

This procedure, together with the identity permutation $e = [1][2][3][4][5][6]$, yields a group G of 24 permutations, written as products of cycles, and hence the cycle index polynomial of the group of rotations of the cube is

$$P(G; x_1, \ldots, x_6) = \tfrac{1}{24}(x_1^6 + 3x_1^2 x_2^2 + 6x_1^2 x_4 + 6x_2^3 + 8x_3^2).$$

By using Pólya's theorem from Problem 13.6 one finds that the number of ways of coloring the faces of a cube with m colors is equal to

$$P(G; m, \ldots, m) = \tfrac{1}{24}(m^6 + 3m^4 + 12m^3 + 8m^2).$$

13.8 In the solution to Problem 13.3 it was shown that the rotation R_m of angle $2\pi m/n$ of a regular polygon about its center can be written as a product of (n, m) cycles of equal length $d = n/(m, n)$:

$$[i_1, i_2, \ldots, i_d][j_1, j_2, \ldots, j_d] \cdots [t_1, t_2, \ldots, t_d],$$

where the numbers $i_1, \ldots, i_d, j_1, \ldots, j_d, \ldots, t_1, \ldots, t_d$ are the numbers $0, 1, \ldots, n-1$ which represent the vertices of the regular polygon. Thus the cycle index polynomial is equal to

$$\frac{1}{n} \sum_{m=1}^{n} x_{n/(n,m)}^{(n,m)} = \frac{1}{n} \sum_{d \mid n} x_d^{n/d} \sum_{\substack{m \leq n \\ (m,n) = n/d}} 1 = \frac{1}{n} \sum_{d \mid n} \varphi(d) x_d^{n/d},$$

because if $e = m/(m, n)$ then the conditions $m \leq n$ and $(m, n) = n/d$ are satisfied if and only if $e \leq d$ and $(e, d) = 1$. An application of Pólya's theorem shows that the number of colorings with k colors of the vertices of a regular polygon with n vertices which are not obtained from one another by a rotation is equal to $(1/n) \sum_{d \mid n} \varphi(d) k^{n/d}$.

In general, if G is a finite group with p elements, one can consider its representation as a group of permutations by defining for every $a \in G$

$$a(x) = xa$$

for every $x \in G$. Thus a becomes a permutation of the set G which can be decomposed into p/k cycles of length k, where k is the order of the element a in G. In fact $xa^m = x$ if and only if $a^m = 1$. Proceeding as before, one finds that the cycle index polynomial of the group G is equal to

$$\frac{1}{|G|} \sum_{d \mid p} x_d^{p/d} \varphi(G, d),$$

where $p = |G|$ and $\varphi(G, d)$ is the number of elements of order d in the group G.

Observe that in Problem 13.1 the group of rotations about the axis xx' which leave invariant the rectangle $ABCD$ is formed from permutations

$$G = \{e = [1][2][3][4][5][6][7][8][9] \text{ and } [2][5][8][1, 3][4, 6], [7, 9]\}$$

in the notation of Figure 13.1. The cycle index polynomial is thus given by

$$P(G; x_1, x_2) = \tfrac{1}{2}(x_1^9 + x_1^3 x_2^3).$$

The number of possible perforations is equal to the number of coloring schemes with two colors, excluding the case when there are no perforations on the ticket. Thus the number is equal to

$$P(G; 2, 2) - 1 = \tfrac{1}{2}(2^9 + 2^6) - 1 = 287.$$

Solutions

If there existed, for example, two types of perforations, of different diameters, then it would follow from Pólya's theorem that the number of possible perforations of a ticket is equal to

$$P(G; 3, 3) - 1 = \tfrac{1}{2}(3^9 + 3^6) - 1 = 10{,}205.$$

13.9 In the solution to Problem 13.3, it was shown that if c is a cycle of length k, then the permutation c^i has (k, i) cycles, each of length $k/(k, i)$. It follows that the contribution to the cycle index polynomial of G of the permutation f^i is equal to

$$\prod_{k=1}^{r} [x_{k/(k,i)}^{(k,i)}]^{\lambda_k(f)}.$$

The upper limit in this product may be taken to be r because r is the least common multiple of the cycle lengths of f. [J. H. Redfield, *Amer. J. Math.*, **49** (1927), 433–455.]

13.10 The automorphism group of this graph is $G = \{[1][2][3][4], [2][4][1, 3], [1][3][2, 4], [1, 3][2, 4]\}$, and hence

$$P(G; x_1, x_2, x_3, x_4) = \tfrac{1}{4}(x_1^4 + 2x_1^2 x_2 + x_2^2).$$

13.11 Denote this number by $\mathrm{MG}(m)$. It is clear that $\mathrm{MG}(m) = P(m, 1) + P(m, 2) + P(m, 3)$, since any multigraph with three unlabeled vertices and m edges corresponds to a partition of m into at most three parts. One finds that $P(m, 1) = 1$, $P(m, 2) = [m/2]$, and $P(m, 3) = P(m - 3, 1) + P(m - 3, 2) + P(m - 3, 3) = \mathrm{MG}(m - 3)$ (see Problem 5.2). Thus $\mathrm{MG}(m) = 1 + [m/2] + \mathrm{MG}(m - 3)$ for any $m \geqslant 3$, where $\mathrm{MG}(0) = 1$.

The formula for $\mathrm{MG}(m)$ now follows by induction on m, since $\mathrm{MG}(0) = \mathrm{MG}(1) = 1$, $\mathrm{MG}(2) = 2$, $\mathrm{MG}(3) = 3$, $\mathrm{MG}(4) = 4$, and $\mathrm{MG}(5) = 5$.

Suppose that the expression for $\mathrm{MG}(m)$ is valid for $m \leqslant n - 1$. The fact that it also holds for $m = n$ follows from consideration of the cases $n \equiv 0, 1, \ldots, 5 \pmod 6$.

13.12 Let $w(a) = a_1 + a_2 + \cdots + a_n$ for any Boolean vector $(a_1, \ldots, a_n)$. Then $0 \leqslant w(a) \leqslant n$, and if f is symmetric, it follows that $f(a) = f(b)$ for any $a, b \in B^n$ such that $w(a) = w(b)$. Thus it is sufficient to define any symmetric Boolean function of n variables for $v_0 = (0, \ldots, 0)$, $v_1 = (1, 0, \ldots, 0)$, $v_2 = (1, 1, 0, \ldots, 0), \ldots, v_n = (1, 1, \ldots, 1)$ to either be 0 or 1, where $w(v_i) = i$ for any $0 \leqslant i \leqslant n$. This can be done in exactly 2^{n+1} ways.

CHAPTER 14

14.1 Consider the graph of Figure 14.1 in which all edges have length equal to 1. Suppose that the property does not hold, and assume that A is colored a, B is colored b, D is colored c, and thus that F is colored a. One can conclude analogously that G is colored a and hence there are two points F and G at a distance 1 which have the same color; this contradicts the hypothesis.

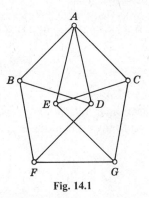

Fig. 14.1

14.2 Suppose that the two colors are red and blue. For every coloring there will be two points M and N with the same color, say red. If the midpoint P of the segment MN is red, then one has obtained three equidistant collinear points with the same color. Otherwise, suppose that P is blue. If the point N_1 which is symmetric to N with respect to M is red, then the points N_1, M, and N are red. If N_1 is blue, consider the point M_1 which is symmetric to M with respect to N. If M_1 is red, then the desired points are M, N, M_1. Otherwise M_1 is blue and thus N_1, P, and M_1 are three equidistant collinear blue points.

Thus in any coloring of the points of the plane with red and blue there will exist three equidistant collinear points of the same color, say red. Denote these points by A, B, C. Construct the equilateral triangle AFC so that B, E, D are the midpoints of its edges (Figure 14.2). If F is red, then AFC is the desired triangle. Otherwise F is blue. If D and E are blue, then the equilateral triangle DEF has blue vertices and the problem is solved. Otherwise at least one of the points D and E is red. Suppose that D is red and thus the triangle ABD has the desired property. In order to show that there exists a 2-coloring of the points of the plane in which no equilateral triangle of side 1 is monochromatic, consider the plane together with an x–y rectangular coordinate system. The lines $y = (\sqrt{3}/2)k$ with k an integer partition the plane into a network of parallel bands. Suppose that the band bounded by the lines $(\sqrt{3}/2)k$ and $(\sqrt{3}/2)(k+1)$ contains all the points of the line $(\sqrt{3}/2)k$ and none of the points of the line $(\sqrt{3}/2)(k+1)$.

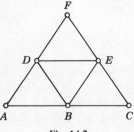

Fig. 14.2

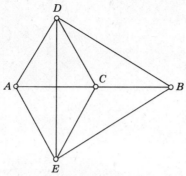

Fig. 14.3

Now color the points of the plane red and blue so that all the points of a band have the same color but each two neighboring bands have different colors. Since each band has width $\sqrt{3}/2$, which is equal to the altitude of an equilateral triangle of side 1, it follows that every equilateral triangle of side 1 has vertices in two adjacent bands and thus does not have all of its vertices the same color.

14.3 If all the points of the plane have the same color, then the property is evident. Otherwise there will exist two points A, B at a distance 2, with different colors. In fact every two points of the plane which have different colors can be joined by a polygonal line which has all of its segments of length 2, and one of these segments must have endpoints of different colors. In Figure 14.3 suppose that

$$AC = CB = AD = DC = AE = EC = 1$$

and

$$DE = BD = BE = \sqrt{3}.$$

Let A be colored red and B colored blue. One can assume that C is red, for otherwise one could make the argument below for the points which are symmetric to D and E with respect to the perpendicular bisector of AB. If D or E is red, then one obtains an equilateral monochromatic triangle of side 1. If D and E are blue, then the triangle BED has all of its vertices blue and side length equal to $\sqrt{3}$.

14.4 Consider an equilateral monochromatic (say red) triangle ABC of side a. In Figure 14.4, BDE and CHF are equilateral triangles of side b, triangle EFG is congruent to ABC, and BE and CF are perpendicular to AB.

By Problem 14.3 there exists a monochromatic equilateral triangle ABC of side $a \in \{1, \sqrt{3}\}$. If $a = 1$, then let $b = \sqrt{3}$, and if $a = \sqrt{3}$, then let $b = 1$. The triangles ABE, DBC, GFC, EFH, ACH, and DEG are all congruent to the original triangle T. Now A, B, and C are all red. If there are no monochromatic triangles congruent to T, then by considering triangles ABE, DBC, and ACH, one can see that E, D, and H must be blue. Triangle DEG forces G to be red.

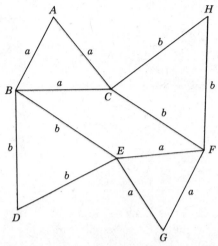

Fig. 14.4

Triangles CFG and EFH force F to be blue and red, respectively, which is a contradiction. Thus one of the six triangles must be monochromatic.

It has been conjectured that for any 2-coloring of the plane there exists a monochromatic triangle which is congruent to a given triangle T unless T is equilateral, and moreover, that any 2-coloring of the plane with no monochromatic equilateral triangle of side d in fact has monochromatic equilateral triangles of side d' for all $d' \neq d$. [P. Erdös, P. Montgomery, B. L. Rothschild, J. Spencer, E. G. Strauss, *J. Combinatorial Theory* (A), **14** (1973), 341–363.]

14.5 It will be shown that in any partition of E into two sets X and Y, at least one of the sets contains the vertices of a right triangle. Indeed, one can find points A' on BC, B' on CA, and C' on AB such that the triangle $A'B'C'$ is equilateral with edges perpendicular to the edges of ABC if one chooses

$$\frac{A'C}{BC} = \frac{B'A}{CA} = \frac{C'B}{AB} = \frac{1}{3} \qquad \text{(see Figure 14.5)}.$$

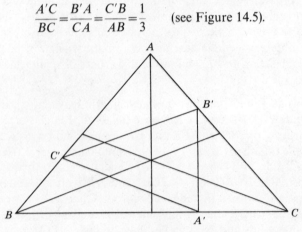

Fig. 14.5

Suppose that there exists a partition $E = X \cup Y$ such that neither X nor Y contains the vertices of a right triangle. At least two vertices from A', B', and C' belong to the same class, say $A' \in X$ and $B' \in X$. Then all points of the segment BC which are different from A' belong to Y, and hence $C' \in X$. But this again implies that all points of the segment AB which are different from C' belong to Y, and hence Y contains the vertices of a right triangle, which is a contradiction. [Problem proposed at the 24th International Mathematical Olympiad, Paris, 1983.]

14.6 Let $E(a, b) = ab(a - b)(a + b) = a^3 b - ab^3$. If one of the numbers a or b is equal to zero, it follows that $E(a, b) = 0$, which is a multiple of 10. Since $E(-a, b) = E(a, -b) = -E(a, b)$ and $E(-a, -b) = E(a, b)$, one need only consider the case in which all three numbers are positive integers.

It can easily be seen that for every a and b, $E(a, b)$ is even, since if a and b are both odd, both their sum and their difference are even.

It remains to show that among three pairwise distinct positive numbers there will be two numbers a and b such that $E(a, b)$ is a multiple of 5. If one of the three numbers is itself a multiple of 5, then the property is immediate.

Suppose on the other hand that the last digit of the three numbers belongs to the set $\{1, 2, 3, 4, 6, 7, 8, 9\}$. It will be shown that for every choice of three numbers from this set there exist two whose sum or difference is a multiple of 5.

Consider the graph with eight vertices in Figure 14.6. Two vertices i and j ($i \neq j$) are adjacent if and only if $i+j$ or $i-j$ is a multiple of 5. This graph reduces to two complete subgraphs $C_1 = \{1, 4, 6, 9\}$ and $C_2 = \{2, 3, 7, 8\}$. Thus for any choice of three vertices of this graph there will be at least two which are in the same connected component of G, and therefore have their sum or difference a multiple of 5. This completes the proof of the property.

 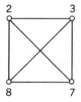

Fig. 14.6

14.7 It will be shown that every sequence of numbers $a_1, a_2, \ldots, a_{mn+1}$ contains either an increasing subsequence with at least $m+1$ terms or a decreasing subsequence with at least $n+1$ terms.

Suppose that every increasing sequence has at most m terms and every decreasing sequence has at most n terms.

For each term a_i define the numbers c_i and d_i as the numbers of terms in a longest increasing or decreasing sequence which begins with a_i. The mapping which associates with every term a_i the ordered pair (c_i, d_i) is injective. In fact, let $i < j$ and $a_i \leq a_j$. In this case one has $c_i \geq c_j + 1$ and hence $c_i \neq c_j$ or $(c_i, d_i) \neq$

(c_j, d_j). If $a_i \geq a_j$ then $d_i \geq d_j + 1$, that is, $d_i \neq d_j$, and hence $(c_i, d_i) \neq (c_j, d_j)$. But this implies that the number of elements of the given sequence is less than or equal to the number of elements in the Cartesian product $\{1, 2, \ldots, m\} \times \{1, 2, \ldots, n\}$, that is, $mn + 1 \leq mn$, which is a contradiction.

14.8 Let $t(x)$ denote the maximum length of a sequence $x = a_1 < a_2 < \cdots$ which satisfies $f(a_1) \leq f(a_2) \leq \cdots$. Since $f(1) = 1$, it follows that max $t(x) = t(1)$. It must be shown that $t(1) \geq n$. Use induction on n. For $n = 1$ the property is immediate, since $t(1) = 1$. Suppose that the property holds for every number $m \leq n - 1$, and show that $t(1) \geq n$. Under these conditions one can first show that if $t(x) \geq t(1) - k$ then $f(x) \leq 2^k$ for $k = 0, \ldots, t(1) - 1$.

Suppose on the other hand that $f(x) > 2^k$ and $t(x) \geq t(1) - k$. It follows from the fact that $1 \leq f(i) \leq i$ that $x > 2^k$ and $k < n - 1$.

Let $x = a_1 < \cdots < a_{t(x)}$ and $f(a_1) \leq \cdots \leq f(a_{t(x)})$. By the induction hypothesis there exists a sequence $1 \leq b_1 < \cdots < b_{k+1} \leq 2^k$ with $f(b_1) \leq \cdots \leq f(b_{k+1})$, since $k + 1 \leq n - 1$. One can thus conclude that

$$f(b_1) \leq f(b_2) \leq \cdots \leq f(b_{k+1}) \leq b_{k+1} \leq 2^k < f(a_1) \leq \cdots \leq f(a_{t(x)}),$$

and hence $t(1) \geq t(x) + k + 1$, which contradicts the inequality $t(x) \geq t(1) - k$.

Now observe that if $t(x) = t(y)$ and $x \neq y$, one has $f(x) \neq f(y)$. In fact, if, for example, $x < y$, then $f(x) > f(y)$, since otherwise $t(x) > t(y)$.

The inequality $t(x) \geq t(1) - k$ implies that $f(x) \leq 2^k$. It follows from this and the preceding observation that

$$|\{x \mid t(x) = t(1) - k\}| \leq 2^k \quad \text{for} \quad 0 \leq k \leq t(1) - 1.$$

Since the number of elements in the domain of f is equal to 2^{n-1}, one can conclude that

$$2^{n-1} = \sum_{k=0}^{t(1)-1} |\{x \mid t(x) = t(1) - k\}| \leq \sum_{k=0}^{t(1)-1} 2^k = 2^{t(1)} - 1,$$

and hence $t(1) \geq n$.

In order to show that 2^{n-1} is the best possible value with this property, define the integer-valued function g on the set $\{1, \ldots, 2^{n-1} - 1\}$ by

$$g(2^k + p) = 2^k - p \quad (k = 0, \ldots, n-2, \ 0 \leq p \leq 2^k - 1).$$

It follows that $g(i) \leq i$ for $i = 1, \ldots, 2^{n-1} - 1$, and the values taken on by g form $n - 1$ intervals which are strictly decreasing sequences. If $1 \leq a_1 < \cdots < a_t \leq 2^{n-1} - 1$ and $f(a_1) \leq \cdots \leq f(a_t)$, then two $f(a_i)$ and $f(a_j)$ with $i \neq j$ cannot both belong to the same interval, and hence $t \leq n - 1$. [E. Harzheim, *Publ. Math. Debrecen*, **14** (1967), 45–51.]

14.9 If the convex covering of the nine points (the smallest convex polygon which contains the nine points in its interior or on the sides) has at least five vertices, then the property is immediate.

Now analyze the remaining cases in which the convex covering is a quadrilateral or a triangle.

Solutions

(a) Suppose that the convex covering is the quadrilateral $MNPQ$ (Figure 14.7). If the other five points form a convex pentagon, the proof is finished. Otherwise there exist four points which do not form a convex quadrilateral. Let D be the interior point of the triangle ABC. Among the angles ADB, BDC, CDA, there exists one which contains two vertices of the quadrilateral, for example, M and N. One has thus obtained a convex pentagon $MNBDA$.

(b) If the convex covering of the nine points is the triangle ABC, then one must analyze two subcases:

    (1) The convex covering of the six remaining points is a quadrilateral $MNPQ$. Let U, V be the two other points in the interior of the quadrilateral. If the line UV intersects two adjacent edges of $MNPQ$, then a convex pentagon is produced (Figure 14.8).

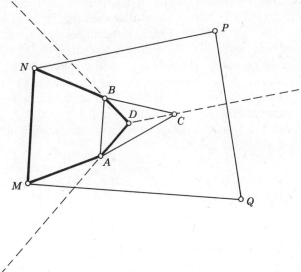

Fig. 14.7

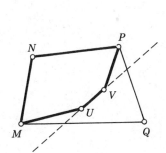
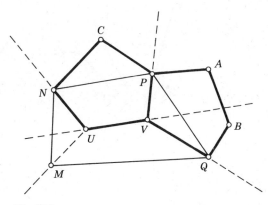

Fig. 14.8

Suppose that UV intersects the opposite edges MN and PQ. The half lines UM, UN, VQ, VP divide the exterior of the quadrilateral $MNPQ$ into four regions. If the region bounded by the lines UN, NP, VP or the region bounded by the lines UM, MQ, VQ contains at least one of the points A, B, C, then a convex pentagon is obtained (Figure 14.8). Otherwise one of the regions bounded by the half lines UN, UM or VP, and VQ contains two of the points A, B, C. The result is a convex pentagon with these points as vertices. The line AB no longer cuts the segments VP, VQ, since the convex covering of the nine points is the triangle ABC.

(2) Suppose that the convex covering of the six points in the interior of the triangle ABC is another triangle MNP, and let U, V be two of the remaining points in the interior of the triangle MNP (Figure 14.9). Now assume that the line UV intersects the segments MN and MP. If one of the angles MUN and MVP contains at least two of the points A, B, C, then one obtains a convex pentagon as in Figure 14.9, since, by hypothesis, the convex covering of the nine points is the triangle ABC. Otherwise, at least one of the points A, B, C is found in the region bounded by the segment UV and the half lines UN and VP, which again yields a convex pentagon. It is indicated with heavy lines in Figure 14.9.

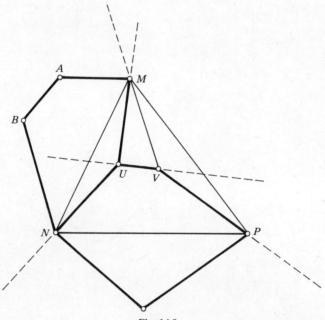

Fig. 14.9

It has been conjectured that for every choice of $2^{m-2}+1$ points in the plane such that no three are collinear there are m which are the vertices of a convex polygon. The conjecture has been verified through $m=5$. It has also been shown that there is a choice of 2^{m-2} points in the plane with no three collinear such that no m points form a convex polygon.

14.10 Suppose that there exists a coloring of the edges of the graph G which does not induce a monochromatic triangle and which uses different colors for the edges of the cycle C_n. Thus there exist two adjacent edges of C_n which have different colors, say $[A_1, A_2]$ red and $[A_2, A_3]$ blue.

Consider an arbitrary vertex B_j of the cycle C_m. One can assume, for example, that the edge $[B_j, A_2]$ is red. Since there is no monochromatic triangle, it follows that the edge $[A_1, B_j]$ is blue.

Let B_{j+1} be a vertex joined by an edge to B_j in the cycle C_m (taking $B_{m+1} = B_1$). If the edge $[B_{j+1}, A_2]$ is red, then the edge $[B_{j+1}, A_1]$ is blue, and hence, no matter whether the edge $[B_j, B_{j+1}]$ is colored red or blue, one of the triangles $B_j B_{j+1} A_1$ or $B_j B_{j+1} A_2$ is monochromatic (see Figure 14.10), which contradicts the hypothesis.

Thus if the edge $[B_j, A_2]$ is colored red, then the edge $[B_{j+1}, A_2]$ is colored blue. Analogously if the edge $[B_j, A_2]$ is colored blue, then one finds that the edge $[B_{j+1}, A_2]$ must be colored red by replacing A_1 with A_3 in the preceding argument. Traverse the cycle C_m, starting from the vertex B_j with the edge $[B_j, A_2]$ colored red and passing through the neighboring vertices. One will again meet B_j, which implies that $[B_j, A_2]$ must be colored blue, since m is odd. This yields a contradiction which establishes that all the edges of the cycle C_n have the same color. It can be shown analogously, since n is also odd, that all the edges of the cycle C_m must also be colored with the same color.

It remains to show that the two colors of the cycles C_n and C_m are identical. Suppose that there exists a coloring without monochromatic triangles with the property that all the edges of the cycle C_n are blue and all the edges of the cycle C_m are red. Suppose, without loss of generality, that the edge $[A_1, B_1]$ is red (otherwise the colors red and blue can be interchanged). (See Figure 14.11.)

It follows from the nonexistence of a monochromatic triangle that $[A_1, B_2]$ is blue, $[A_2, B_2]$ is red, $[A_2, B_3]$ is blue, and hence $[A_1, B_3]$ is red. Now replace

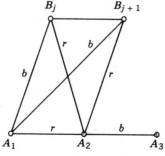

Fig. 14.10

B_1 by B_3 and repeat the argument. It turns out that for every two adjacent vertices B_j and B_{j+1} of the cycle C_m, the colors of the edges $[A_1, B_j]$ and $[A_1, B_{j+1}]$ are different. But this conclusion leads to a contradiction, since the cycle C_m is odd. It follows that the $m+n$ edges of the cycles C_m and C_n are colored with the same color.

(For $m=n=5$ this problem was proposed at the 21st International Mathematical Olympiad in London in 1979.)

14.11 The property will be established by induction on $p+q$. For $p=1$ or $q=1$ it is immediate. Suppose now that $p, q > 1$. Let x be a vertex of the complete graph with n_0 vertices which is colored with two colors. Denote by $d_r(x)$ the number of red edges which have an endpoint at x, and by $d_b(x)$ the number of blue edges which have an endpoint at x.

Since

$$d_r(x) + d_b(x) = n_0 - 1 = \binom{p+q}{p} - 1$$

$$= \binom{p+q-1}{p-1} + \binom{p+q-1}{p} - 1,$$

it follows that either

$$d_r(x) \geq \binom{p+q-1}{p-1} \quad \text{or} \quad d_b(x) \geq \binom{p+q-1}{p}.$$

Suppose, for example, that the first inequality holds, and let G be the complete subgraph induced by the vertices which are joined by red edges to x. (The second inequality can be treated analogously.) Since G has at least $\binom{p+q-1}{p-1}$ vertices and its edges are colored red and blue, it follows from the induction hypothesis that G contains either a complete red subgraph with p vertices or a complete blue subgraph with $q+1$ vertices. In the second case, the proof is finished. But in the first case G contains a complete red subgraph with vertex set H, where $|H|=p$. However, in this case $H \cup \{x\}$ is a complete red subgraph with $p+1$ vertices, and the proof is finished. It is clear that $R(p,q) = R(q,p)$ and $R(p,2) = p$. The only known nontrivial values of the numbers $R(p,q)$ with

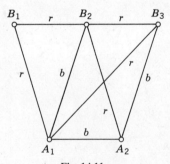

Fig. 14.11

$p, q \geqslant 3$ are given in the following table, where $\frac{a}{b}$ signifies the fact that only a lower bound a and an upper bound b are known:

$p \backslash q$	3	4	5	6	7	8	9
3	6	9	14	18	23	$\frac{28}{29}$	36
4	9	18	$\frac{25}{28}$	$\frac{34}{36}$			
5	14	$\frac{25}{28}$	$\frac{38}{55}$	$\frac{38}{94}$			
6	18	$\frac{34}{36}$	$\frac{38}{94}$	$\frac{102}{178}$			
7	23						
8	$\frac{28}{29}$						
9	36						

14.12 Consider the complete graph K_5 with five vertices as being represented by a regular pentagon together with all its diagonals. Color the sides of the pentagon red and the diagonals blue. There are no monochromatic triangles, and hence $R(3, 3) \geqslant 6$.

In order to prove the opposite inequality, it remains to show that every coloring of the edges of K_6 with red and blue will yield at least one monochromatic triangle.

Let x be a vertex of K_6. There are five edges which originate at x, and hence at least three of these have the same color, say red. Thus there exist three vertices a_1, a_2, a_3 which are joined to x by red edges. If one of the edges determined by a_1, a_2, a_3, say $[a_1, a_2]$, is red, then there is a red triangle x, a_1, a_2. Otherwise the triangle with vertices a_1, a_2, a_3 has all of its edges blue.

Thus every coloring of K_6 with two colors contains a monochromatic triangle.

14.13 The upper bound follows from Problem 14.11. In fact, it has been seen that

$$R(k, k) \leqslant \binom{2k-2}{k-1}$$

for $p=q=k-1$.

Observe that
$$\binom{2k-2}{k-1} = \frac{2(2k-3)}{k-1}\binom{2k-4}{k-2} < 4\binom{2k-4}{k-2}. \qquad (1)$$

For $k=2$ one has
$$\binom{2k-2}{k-1} = 2 = 2^{2k-3}.$$

Suppose that
$$\binom{2k-2}{k-1} \leq 2^{2k-3} \qquad (2)$$

for every $2 \leq k \leq p$. It then follows from (1) that
$$\binom{2p}{p} < 4\binom{2p-2}{p-1} \leq 4 \times 2^{2p-3} = 2^{2p-1},$$

and thus (2) is true for every $k \geq 2$. In order to obtain the lower bound one can suppose that the index $k \geq 4$. In fact for $k=2$, $R(2,2) = 2 = 2^{k/2}$, and for $k=3$ one finds that $R(3,3) = 6 > 2\sqrt{2}$ by Problem 14.12.

Let $n = \{2^{k/2}\}$, that is, the smallest integer which is greater than or equal to $2^{k/2}$. Let X be the set of $\binom{n}{2}$ edges of the complete graph K_n, and let $E = \{E_1, \ldots, E_m\}$ be the family of subsets of X defined as follows: E_i is the family of edges of a complete subgraph with k vertices in K_n. Since there are $\binom{n}{k}$ such complete subgraphs, it follows that $m = \binom{n}{k}$. By a similar argument one can show that $r = |E_i| = \binom{k}{2}$ for every $1 \leq i \leq m$, since K_k has $\binom{k}{2}$ edges.

By the given definition, it is possible to color the elements of X with two colors so that no subset in the family E has all elements of the same color if and only if the complete graph K_n can be colored with two colors so that there is no complete monochromatic subgraph with k vertices. Thus $n < R(k,k)$ if one can color the elements of X with two colors so that no subset in the family E is monochromatic. By Problem 4.24 this condition holds for $m \leq 2^{r-1}$, or
$$\binom{n}{k} \leq 2^{\binom{k}{2}-1}. \qquad (3)$$

The validity of inequality (3) follows from the fact that $\binom{n}{k} < (n-1)^k/2^{k-1}$ for every $k \geq 3$ and $n - 1 < 2^{k/2}$. One can therefore conclude that
$$\binom{n}{k} < \frac{2^{k^2/2}}{2^{k-1}} = 2^{k^2/2 - k + 1} \leq 2^{\binom{k}{2}-1},$$

since $\binom{k}{2} - 1 - (k^2/2 - k + 1) = k/2 - 2 \geq 0$ if $k \geq 4$.

Finally, it follows from (3) that
$$R(k,k) > n \geq 2^{k/2}.$$

14.14 Problem 14.11 shows that

Solutions

$$R(a_1, a_2) = R_2(a_1, a_2) \leqslant \binom{a_1+a_2-2}{a_1-1},$$

which is the first step in the proof by induction on k.

Let $k \geqslant 3$, and suppose that the numbers $R_{k-1}(b_1, \ldots, b_{k-1})$ exist for all values $b_1, \ldots, b_{k-1} \geqslant 1$. It will be shown that

$$R_k(a_1, \ldots, a_k) \leqslant R_{k-1}(a_1, \ldots, a_{k-2}, R_2(a_{k-1}, a_k)).$$

In order to do this, let G be the complete graph with $R_{k-1}(a_1, \ldots, a_{k-2}, R_2(a_{k-1}, a_k))$ vertices. Color its edges arbitrarily with the colors $c_1, \ldots, c_k$. For the time being recolor the edges which have colors c_{k-1} and c_k with a new color d.

By the induction hypothesis there are two possible cases. Either there exists an index i, $1 \leqslant i \leqslant k-2$, such that the graph G contains a complete subgraph with a_i vertices and with all edges colored c_i (in which case the proof is finished), or G contains a complete subgraph G_1 with $R_2(a_{k-1}, a_k)$ vertices and with all edges colored d. In the graph G_1 replace the color d with color c_{k-1} or c_k, which originally existed for the edges of G. From the definition of the Ramsey number $R_2(a_{k-1}, a_k)$ it follows that G_1 (and hence G) contains either a complete subgraph with a_{k-1} vertices and with all edges colored c_{k-1}, or a complete subgraph with a_k vertices and with all edges colored c_k.

14.15 Observe that

$$[k!\,e] = \left[\sum_{j=0}^{\infty} \frac{k!}{j!}\right] = \sum_{j=0}^{k} \frac{k!}{j!}.$$

The proof will use induction on k.

For $k=2$ it follows from Problem 14.12 that $R_2(3, 3) = 6$.

Let x be an arbitrary vertex of a complete graph G with $[k!\,e]+1$ vertices whose edges are colored with the colors $c_1, \ldots, c_k$. The vertex x is thus an endpoint of $[k!\,e]$ edges in G. Since

$$[k!\,e] = \sum_{j=0}^{k} \frac{k!}{j!} = 1 + k \sum_{j=0}^{k-1} \frac{(k-1)!}{j!} = 1 + k[(k-1)!\,e],$$

it follows that among the $[k!\,e]$ edges with an endpoint at x there will exist at least $1 + [(k-1)!\,e]$ edges with the same color, say c_1. Let X be the set of vertices which are joined to x by an edge of color c_1. If X contains two vertices joined by an edge of color c_1, then these two vertices, together with x, form a monochromatic triangle of color c_1, and the property is proved.

Otherwise X induces a complete subgraph G_1 with $1 + [(k-1)!\,e]$ vertices of G, whose edges are colored with $k-1$ colors. By the induction hypothesis G_1 (and hence G) contains a monochromatic triangle whose sides all have one of the colors $c_2, \ldots, c_k$; this completes the proof of the inequality.

Observe that $R_3(3, 3, 3) \leqslant [3!\,e] + 1 = 17$. In order to prove the opposite inequality one must find a coloring with three colors of the edges of the complete graph K_{16}, which does not contain a monochromatic triangle.

Thus let $M = \{1, 2, 3, 4, 5\}$ be the set of vertices of a pentagon P, and let X be the set of vertices of the graph K_{16}. Denote by Y the family of the 16 subsets of M of even cardinality (empty set, ten subsets with two elements, and five subsets with four elements). Let $A \triangle B = (A \setminus B) \cup (B \setminus A)$ denote the symmetric difference of the sets A and B. It is clear that if $A, B \in Y$ and $A \neq B$, then $A \triangle B$ is a subset of M with two or four elements. Let f be any bijection from X to Y.

The edges of K_{16} will be colored as follows: If $a, b \in X$ and $a \neq b$, color the edge $[a, b]$ according to the following rule: With color c_1 if $f(a) \triangle f(b)$ is a side of P; with color c_2 if $f(a) \triangle f(b)$ is a diagonal of P; with color c_3 if $|f(a) \triangle f(b)| = 4$. It must be shown that in this case there are no monochromatic triangles.

Suppose that there is a triangle with sides colored c_1 and having vertices a, b, c. It can be seen that $f(a) \triangle f(b)$, $f(a) \triangle f(c)$, and $f(b) \triangle f(c)$ are sides of the polygon P. But $\{f(a) \triangle f(b)\} \triangle \{f(b) \triangle f(c)\} = f(a) \triangle f(c)$, and thus $f(a) \triangle f(c)$ cannot be a side of the polygon P. In fact, if the sides $f(a) \triangle f(b)$ and $f(b) \triangle f(c)$ have a common vertex, then their symmetric difference is a diagonal of the pentagon P and hence the side $[a, c]$ is colored c_2. If the sides have no common vertex, then their symmetric difference has cardinality 4 and thus the side $[a, c]$ is colored c_3, which is a contradiction. An analogous argument can be used if it is supposed that the monochromatic triangle is colored c_2 or c_3.

14.16 Denote by n_k the maximum number of vertices of a complete graph which admits a coloring of its edges with k colors and without a monochromatic triangle. It will be shown that $n_k \geq 2n_{k-1}$.

Consider two copies of a complete graph with n_{k-1} vertices with edges colored $c_1, \ldots, c_{k-1}$ and without a monochromatic triangle, and color with a new color c_k all the edges which join pairs of vertices which belong to the two copies of the complete graph with n_{k-1} vertices. One thus obtains a complete graph with $2n_{k-1}$ vertices with its edges colored with k colors and which does not contain a monochromatic triangle. This justifies the inequality $n_k \geq 2n_{k-1}$. Since $n_1 = 2$, it follows that $R_k(3) \geq n_k + 1 \geq 2^k + 1$. [R. E. Greenwood, A. M. Gleason, *Canadian J. Math.*, **7** (1955), 9–20.]

14.17 Since $n \geq k!\, e$, by Problem 14.15 it follows that in any coloring of the complete graph K_{n+1} with k colors at least one monochromatic triangle will appear. Let A_i denote the set of natural numbers in class i for $i = 1, 2, \ldots, k$. Denote the vertices of K_{n+1} by the numbers $1, 2, \ldots, n+1$, and color the edge $[i, j]$ with color p if $|i - j| \in A_p$. If a monochromatic triangle has vertices i, j, k, then $x = |i - j|$, $y = |j - k|$, and $z = |i - k|$ belong to the same class. Suppose, for example, that $i < j < k$. Then $x + y = z$, which completes the proof. [I. Schur, *Jahresb. Deutschen Math.-Ver.*, **25** (1916), 114–117.]

The Schur function is defined as follows: $S(k) = \max \{r \mid \{1, 2, \ldots, r\}$ can be partitioned into k subsets (possibly empty) with the property that none of them contain numbers x, y, z such that $x + y = z\}$. The result of this problem implies that $S(k) \leq [k!\, e] - 1$. The only known values of the Schur function are the following: $S(1) = 1$, $S(2) = 4$, $S(3) = 13$, $S(4) = 44$. The last value was found in 1961 by using a computer.

Solutions

14.18 Let $n_0(k)=[k!\,e]$. If the nonempty subsets of the set $\{1, 2, \ldots, n\}$ are colored with k colors, then color the edge $[i,j]$ (where $1 \leqslant i < j \leqslant n+1$) of the complete graph K_{n+1} with vertices $1, 2, \ldots, n+1$ with the same color as the subset $\{i, \ldots, j-1\}$. According to Problem 14.15, K_{n+1} contains a monochromatic triangle.

Suppose that the vertices of this triangle are p, q, r where $1 \leqslant p < q < r \leqslant n+1$. It follows that $X = \{p, \ldots, q-1\}$, $Y = \{q, \ldots, r-1\}$, and $X \cup Y = \{p, \ldots, r-1\}$ have the same color.

14.19 The property will be established by induction on r. Let $r=2$, denote the vertices of K_∞ by $x_1, x_2, \ldots, x_n, \ldots$, and let $r(x_i)$ denote the number of red edges incident with x_i; the two colors will be called red and blue.

Two cases will be examined:

(a) There exist a countable infinity of vertices $x_{i_1}, x_{i_2}, \ldots$, which satisfy $r(x_{i_j}) < \infty$. Define inductively the vertices $y_1, y_2, \ldots$, of the complete infinite monochromatic subgraph as follows: Let $y_1 = x_{i_1}$; y_2 is the first of the vertices x_{i_j} which is not joined to y_1 by a red edge, and therefore y_2 is joined to y_1 by a blue edge. There will exist such a vertex y_2 because there exist only a finite number of red edges which are incident with y_1. The vertex y_3 is the first vertex among the vertices with indices greater than the index of y_2 which are joined by a blue edge to y_1 and y_2, and so on. The process can be continued indefinitely, since the set of vertices x_{i_j} with $j \geqslant 1$ is infinite and each such vertex is incident with a finite number of red edges. It follows that the complete infinite subgraph generated by the set of vertices $\{y_1, y_2, \ldots\}$ has all of its edges blue.

(b) Suppose that (a) does not hold, and assume that the graph K_∞ does not contain a complete infinite subgraph with all of its edges blue. Let $X_1 = \{x \mid r(x) = \infty\}$. It follows that X_1 is an infinite set. Choose $y_1 \in X_1$, and denote by $X'_1 \subset X_1$ the subset of vertices of X_1 which are joined to y_1 by a red edge. Since (a) does not hold, it follows that the vertices of K_∞ which do not belong to X_1 are finite in number, and thus X'_1 is an infinite set. Denote by $X_2 \subset X'_1$ the subset of vertices of X'_1 which are incident with an infinite number of red edges with both endpoints in X'_1. If X_2 is finite, it follows that $X'_1 \setminus X_2$ is an infinite set of vertices which are incident with a finite number of red edges which have both endpoints in X'_1. The problem is thus reduced to case (a) for the complete infinite subgraph with vertex set X'_1. There is a complete infinite subgraph of K_∞ with all edges blue which contradicts the hypothesis. Thus X_2 is infinite and one can choose $y_2 \in X_2$.

It has been shown that the hypothesis that $X'_1 \setminus X_2$ is an infinite set leads to a contradiction, and thus $X'_1 \setminus X_2$ must be finite. Since y_2 is incident with an infinite number of red edges with endpoints in X'_1, and since $X'_1 \setminus X_2$ is finite, it follows that the subset $X'_2 \subset X_2$ of vertices in X_2 which are joined to y_2 by a red edge is infinite. Let $X_3 \subset X'_2$ be the subset of vertices of X'_2 which are incident with an infinite number of red edges with both endpoints in X'_2. As before, one finds that X_3 is an infinite set, so that one may choose $y_3 \in X_3$, and so on.

By induction, a complete infinite graph generated by the set of vertices

$\{y_1, y_2, \ldots\}$ is produced. It has all of its edges red. This establishes the result for $r=2$.

Suppose that the property holds for all colorings of K_∞ with at most $r-1$ colors, and consider a coloring of K_∞ with r colors: $c_1, c_2, \ldots, c_r$, where $r \geqslant 3$. Recolor the edges which are colored c_{r-1} or c_r with a new color c_{r+1}. By the induction hypothesis for $r-1$ colors, there exists a complete infinite monochromatic subgraph. If the color of the edges of this graph is one of the colors $c_1, \ldots, c_{r-2}$, then the proof is finished. Otherwise there exists a complete infinite subgraph with all edges colored c_{r+1}. By recoloring the edges of this graph which originally were colored c_{r-1} and c_r with these colors and applying the property for two colors, one obtains a complete infinite subgraph with all edges having the same color (c_{r-1} or c_r). The property is thus demonstrated for every r.

14.20 Denote the vertices of the complete graph K_∞ by numbers from the set $\{1, 2, \ldots\}$. The edge $[i, j]$ is colored red if $i < j$ and $a_i < a_j$. It is colored yellow if $i < j$ and $a_i > a_j$, and blue if $i < j$ and $a_i = a_j$. By using the preceding problem one can show the existence of a complete infinite monochromatic subgraph. If its color is red, then there will be an infinite strictly increasing subsequence; for yellow one finds an infinite strictly decreasing subsequence, and for blue an infinite constant subsequence.

14.21 First we show that for every infinite set A of points in the plane there is an infinite subset A_1 of collinear points or an infinite subset A_2 of points such that no three points are collinear. Consider all the lines determined by pairs of points from A. If one of them contains an infinite number of points of A, then the property is demonstrated. Otherwise each line determined by two points of A contains a finite number of points of A.

In this case carry out the following construction: Let x_1 and x_2 be two points of A. Denote by B_1 the set obtained from A by eliminating all the points of the line $x_1 x_2$, including x_1 and x_2. It follows that B_1 is an infinite set. Let $x_3 \in B_1$. Denote by B_2 the set obtained from B_1 by eliminating all points on the lines $x_3 x_1$ and $x_3 x_2$ which belong to A. It follows that B_2 contains an infinite number of points. If a set of points $\{x_1, \ldots, x_r\}$ has been obtained with the property that no three are collinear, and if B_{r-1} is the infinite set of points which belong to the set A and are not found on any of the lines determined by pairs of points from $\{x_1, \ldots, x_r\}$, then let $x_{r+1} \in B_{r-1}$. Denote by B_r the infinite subset of points of B_{r-1} which are not found on any line $x_1 x_{r+1}, \ldots, x_r x_{r+1}$, and so on. It has thus been shown by induction that this construction can be continued indefinitely, and thus A contains an infinite subset $\{x_1, x_2, \ldots\}$ of points with the property that no three are collinear.

Now let A be an infinite set of points in space. If there exists a line determined by a pair of points in A which contains an infinite set A_1 of points of A, then A_1 is a set of the type discussed in case (1). Otherwise, if there exist three points in A which determine a plane containing an infinite number of points of A, then by applying the previous result one obtains a set A_2 as discussed in case (2).

Solutions

Otherwise, every pair of points of A determines a line which contains only a finite number of points of A, and every three noncollinear points of A determine a plane which contains a finite number of points of A.

Now make the following construction: Let $x_1, x_2 \in A$. Eliminate from the set A all the points of the line $x_1 x_2$ to obtain an infinite set B_1. Let $x_3 \in B_1$. Eliminate from B_1 all points of the plane $x_1 x_2 x_3$ and obtain an infinite set B_2. Let $x_4 \in B_2$. It follows that x_4 does not belong to the plane $x_1 x_2 x_3$, nor therefore to the lines $x_1 x_2, x_1 x_3$, and $x_2 x_3$. Eliminate from B_2 all points of the planes $x_1 x_2 x_4$, $x_1 x_3 x_4$, and $x_2 x_3 x_4$ to obtain an infinite subset B_2 of A.

If one has found points $x_1, \ldots, x_r$ with the property that no four are coplanar and an infinite subset B_{r-1} of A with the property that it does not contain any point of the $\binom{r}{3}$ planes determined by $x_1, \ldots, x_r$, then let $x_{r+1} \in B_{r-1}$. The infinite set B_r is obtained from B_{r-1} by eliminating all points of the planes determined by x_{r+1} and the $\binom{r}{2}$ lines determined by pairs of points from the set $\{x_1, \ldots, x_r\}$.

It has been shown by induction that this construction can be continued indefinitely, and hence A contains an infinite subset $A_3 = \{x_1, x_2, \ldots\}$ of points with the property that no four are coplanar.

14.22 Suppose that the property is false. Let the two classes be A and B, and suppose that $5 \in A$. It follows that the numbers 1 and 9 do not both belong to A. In view of the symmetry, without loss of generality it is sufficient to consider only the following two cases:

(a) $1 \in A$ and $9 \in B$. Since 1 and 5 are in class A, it follows that $3 \in B$; $3, 9 \in B$ implies that $6 \in A$; $5, 6 \in A$ implies that $4 \in B$; $3, 4 \in B$ implies that $2 \in A$; $5, 6 \in A$ implies that $7 \in B$; and $7, 9 \in B$ implies that $8 \in A$. Thus $\{2, 5, 8\} \subset A$, and hence the class A contains an arithmetic progression with three terms.

(b) $1 \in B$ and $9 \in B$. There are two subcases:

(1) $7 \in A$. In this case $5, 7 \in A$ implies that $6 \in B$ and $3 \in B$. Thus $\{3, 6, 9\} \subset B$, and B contains an arithmetic progression with three terms.

(2) $7 \in B$. From the fact that $7, 9 \in B$ it follows that $8 \in A$; $1, 7 \in B$ implies $4 \in A$; $4, 5 \in A$ implies $3 \in B$; and $1, 3 \in B$ implies $2 \in A$. One has again found an arithmetic progression $\{2, 5, 8\} \subset A$.

The property is no longer true if one considers the set $\{1, \ldots, 8\}$. This result is a particular case of a theorem due to van der Waerden which states that for every two positive integers k, t, there exists a natural number $W(k, t)$ which is the smallest integer with the following property: If the set $\{1, 2, \ldots, W(k, t)\}$ is partitioned into k classes, there will always exist a class of the partition which contains an arithmetic progression with $t+1$ terms. [B. L. van der Waerden, *Nieuw. Archief voor Wiskunde*, **15** (1927), 212–216.]

The following values of the van der Waerden numbers are known: $W(1, t) = t+1$, $W(k, 1) = k+1$, $W(2, 2) = 9$, $W(2, 3) = 35$, $W(3, 2) = 27$, $W(4, 2) = 76$,

$W(2,4) = 178$. [M. D. Beeler, E. P. O'Neil, *Discrete Mathematics*, **28**(2) (1979), 135–146.]

14.23 Let $M = M_1 \cup M_2$ be a partition of M into two classes, and let $P = \{0, 1, \ldots, 8\}$. Define a decomposition $P = P_1 \cup P_2$ as follows: $k \in P_i$ if and only if $2^k \in M_i$ for every $0 \leq k \leq 8$ ($1 \leq i \leq 2$). It follows that $P_1 \cap P_2 = \varnothing$. If $P_1 = \varnothing$, then P_2 will contain three numbers a, b, and c in arithmetic progression. If $P_1 \neq \varnothing$ and $P_2 \neq \varnothing$, then $P_1 \cup P_2$ is a partition of P into two classes, and by Problem 14.22 at least one class, say P_1, will contain an arithmetic progression with three terms a, b, and c. It follows that 2^a, 2^b, and 2^c is a geometric progression and 2^a, 2^b, $2^c \in M_1$.

14.24 The answer is yes. To see this define a sequence of sets A_n, $n = 1, 2, \ldots$, as follows:

$$A_1 = \{1, 2\} \quad \text{and} \quad A_{n+1} = A_n \cup (A_n + 3^n) \quad \text{for } n \geq 1,$$

where $A + b$ denotes the set $\{a + b \mid a \in A\}$.

Let a_n be the greatest element of A_n. These numbers satisfy the following recurrence formula: $a_{n+1} = a_n + 3^n$, and hence by induction $a_n = \frac{1}{2}(3^n + 1)$, since $a_1 = 2$. One can show, also by induction, that none of these sets contains an arithmetic triple. This is obviously true for A_1. Assume that A_n contains no arithmetic triple and that A_{n+1} does contain such a triple: $x, y, z \in A_{n+1}$ such that $z - y = y - x > 0$. By the induction hypothesis, this triple cannot be contained in A_n or in $A_n + 3^n$, and, since $A_n + 3^n$ is located on the real axis to the right of A_n, it follows that $x \in A_n$ and $z \in A_n + 3^n$. Thus

$$x \in [1, a_n], \quad z \in [3^n + 1, 3^n + a_n],$$

whence

$$y = \tfrac{1}{2}(x + z) \in [\tfrac{1}{2}(3^n + 2), \tfrac{1}{2}(3^n + 2a_n)] = [\tfrac{1}{2}(3^n + 2), \tfrac{1}{2}(2 \times 3^n + 1)].$$

Thus this interval is disjoint both from A_n and $A_n + 3^n$, since $\frac{1}{2}(3^n + 2) > \frac{1}{2}(3^n + 1)$ and $\frac{1}{2}(2 \times 3^n + 1) < 3^n + 1$. It follows that $y \notin A_{n+1}$, which is a contradiction. One can conclude from the construction of the A_n's that A_{n+1} has twice as many elements as A_n (the sets A_n and $A_n + 3^n$ being disjoint). Hence the cardinality of A_n is 2^n.

For $n = 11$ this reduces to a problem proposed at the 24th International Mathematical Olympiad (Paris, 1983): Is it possible to choose 1983 distinct positive integers, all less than or equal to 10^5, and no three of which are consecutive terms of an arithmetic progression?

14.25 First we establish the following inequality:

$$R(3, t) \leq R(3, t-1) + t. \tag{1}$$

By definition there exists a graph G with $R(3, t) - 1$ vertices which does not contain a triangle (subgraph K_3) or an independent set with t vertices. Let x be a vertex of G. Since G does not contain a triangle, every two vertices which

Solutions

are adjacent to x are themselves nonadjacent. Let $d(x)=d$. It follows that $d \leqslant t-1$, since G does not contain an independent set with t vertices. Let G_0 denote the subgraph obtained from G by suppressing the vertex x and the d vertices adjacent to x. It follows that G_0 contains $R(3, t)-d-2=p_0$ vertices. Since $d \leqslant t-1$, one has

$$p_0 \geqslant R(3, t)-t-1.$$

The graph G_0 does not contain a triangle, since G has this property. Similarly, G_0 cannot contain an independent set with $t-1$ vertices, since if it did, the same set together with the vertex x would yield a set of t pairwise nonadjacent vertices in G. Thus $R(3, t-1) > p_0$, which implies (1). Now apply induction on $t \geqslant 2$. For $t=2$ one has $R(3, t)=3$ while $(t^2+3)/2 > 3$, and thus the property holds. Suppose that it is valid for $t \leqslant n-1$, $n \geqslant 3$, and let $t=n$. Take n to be odd, and let $n=2k+1$. It follows from (1) that

$$R(3, 2k+1) \leqslant R(3, 2k)+2k+1,$$

and hence, by the induction hypothesis,

$$R(3, 2k+1) \leqslant \frac{4k^2+3}{2}+2k+1,$$

or

$$R(3, 2k+1) \leqslant 2k^2+2k+2 = \frac{(2k+1)^2+3}{2}.$$

Thus the property is established for n odd.

Now let $n=2k$ be even. By again applying (1) and the induction hypothesis it can be shown that

$$R(3, 2k) \leqslant R(3, 2k-1)+2k \leqslant \frac{(2k-1)^2+3}{2}+2k = 2k^2+2.$$

The proof is concluded by showing that the last inequality for $R(3, 2k)$ is strict, that is, that in fact $R(3, 2k) \leqslant 2k^2+1 < (4k^2+3)/2$. Suppose that there exists a value of the index $k \geqslant 2$ for which $R(3, 2k)=2k^2+2$. Thus there exists a graph H with $2k^2+1$ vertices which contains neither a triangle nor an independent set with $2k$ vertices.

If there existed a vertex y with $d(y) \geqslant 2k$, then no two vertices adjacent to y would be selfadjacent. It follows that H contains an independent set with $2k$ vertices, which contradicts the hypothesis. Hence for every vertex x of H one has $d(x) \leqslant 2k-1$. Since H has an odd number of vertices and the sum of the degrees of its vertices is even, it follows that not all the vertices of H can have the even degree $2k-1$. Hence H contains a vertex z such that $d(z)=d \leqslant 2k-2$. Consider the graph H_0 obtained from H by suppressing the vertex z and all the vertices adjacent to z. It can be seen from the induction hypothesis that H_0 has q_0 vertices, where

$$q_0 = 2k^2 - d \geqslant \frac{(2k-1)^2 + 3}{2} \geqslant R(3, 2k-1).$$

Thus H_0 contains either a triangle or an independent set with $2k-1$ vertices. If H_0 contains a triangle, then H has the same property. If H_0 contains an independent set with $2k-1$ vertices, then this set together with the vertex z yields an independent set with $2k$ vertices in the graph H. This leads to a contradiction and completes the proof of the theorem.

M. Ajtai, J. Komlós, and E. Szemerédi proved that $R(3, t) < 100 t^2 / \ln t$ [*J. Combinatorial Theory* **A29**(3) (1980), 354–360], and C. Nara and S. Tachibana recently showed that $R(3,t) \leqslant \binom{t}{2} - 5$ for every $t \geqslant 13$ [*Discrete Math.*, **45** (1983), 323–326].

14.26 Suppose that the partite sets of G are A and B and hence $|A| = |B| = 2p+1$. Assume that the two colors are a and b, and let $u \in A$. It follows that u is adjacent to at least $p+1$ vertices of B by edges having the same color, say a. Let $U \subset B$ denote the set of vertices of B which are adjacent to u by edges having color a, so that $|U| \geqslant p+1$. Let C_a be the connected component composed only of edges with color a and containing the vertex $u \in A$. Suppose that $|C_a \cap A| = x \geqslant 1$. It follows that every vertex of A that does not belong to C_a is adjacent to all vertices of U by edges of color b only. Denote the number of these vertices which are found in $A \setminus C_a$ by $y \geqslant 0$. It follows that the x vertices of $A \cap C_a$ together with the vertices of U are included in the connected component C_a having all edges of color a, and the y vertices of A which do not belong to C_a together with the vertices of U are contained in a connected component of G having all edges of color b. Let $|U| = r \geqslant p+1$. One must show that $\max(x+r, y+r) \geqslant 2p+2$. Suppose that $x+r \leqslant 2p+1$ and $y+r \leqslant 2p+1$. In this case $x+y+2r \leqslant 4p+2$, but one can write $x+y+2r = 2p+1+2r \geqslant 2p+1+2(p+1) = 4p+3$, which is a contradiction. This completes the proof.

In order to see that the bound $2p+2$ cannot be improved, consider the partitions $A = A_1 \cup A_2$ and $B = B_1 \cup B_2$, where $|A_1| = |B_1| = p$, $|A_2| = |B_2| = p+1$. Now color with a all the edges between A_1 and B_1 and between A_2 and B_2, and with b all the remaining edges of $K_{2p+1, 2p+1}$.

14.27 (a) It is necessary to show that if G is not connected, then its complement $\bar{G}$ is connected. Let x and y be two vertices of G. If x and y are not adjacent in $\bar{G}$, they are adjacent in G, and therefore x and y belong to the same component C_1 of G. Since G is not connected, there exists a vertex $z \notin C_1$, which implies that z is not adjacent to x and to y in G. It follows that $[x, z, y]$ is a walk of length 2 in $\bar{G}$ between x and y, and that $\bar{G}$ is connected.

(b) Suppose that the edges of K_n are colored with three colors a, b, c. It will first be shown that there exists a monochromatic connected spanning subgraph of K_n with at least $[(n+1)/2]$ vertices. If one of the three colors is not used, it follows from (a) that this property holds. Suppose that K_n contains edges having colors a, b, and c, and let R denote a connected component of the spanning subgraph of K_n composed of all edges with the color a. If $|R| = n$, the property is

Solutions

verified. Otherwise, let x be a vertex of K_n such that $x \notin R$. It follows that all edges joining x with the vertices of R have color b or c. Hence one can assume that there exist at least $\frac{1}{2}|R|$ vertices of R which are adjacent to x by edges with color b. Let V be the set of these vertices of R, and let W be the connected component of the spanning subgraph of K_n composed of all edges with color b, which contains x. If y is a vertex of K_n such that $y \notin R$ and $y \notin W$, it follows that all edges between y and the vertices of V have the color c. Let Q be the connected component of the spanning subgraph of K_n consisting of all edges with the color c, which contains y. If there do not exist vertices y having the above-mentioned property, then $Q = \varnothing$. The sets R, W, Q together contain all the vertices of K_n. Indeed, every vertex $z \notin R$ is connected with any vertex of V by an edge with color b or c, and hence $z \in W$ or $z \in Q$.

If $Q = \varnothing$ then $|R| + |W| \geq n$, and therefore $\max(|R|, |W|) \geq n/2$, which implies that $\max(|R|, |W|) \geq [(n+1)/2]$. Otherwise, since $V \subset R$, $V \subset W$, and $V \subset Q$, one can write

$$|R| + |W| + |Q| \geq n + 2|V| \geq n + 2\frac{|R|}{2} = n + |R|,$$

or $|W| + |Q| \geq n$, which implies that $\max(|W|, |Q|) \geq [(n+1)/2]$. Thus it has been shown that $f_3(n) \geq [(n+1)/2]$. If $n \not\equiv 2 \pmod 4$, the opposite inequality also holds. Let X denote the vertex set of K_n, and consider an equipartition

$$X = X_1 \cup X_2 \cup X_3 \cup X_4$$

such that $-1 \leq |X_i| - |X_j| \leq 1$ for every $i, j = 1, \ldots, 4$. Color the edges of K_n in the following way: all edges between X_1, X_2 and between X_3, X_4 with the color a; all edges between X_1, X_4 and between X_2, X_3 with the color b; and all edges between X_1, X_3 and between X_2, X_4 with the color c. All edges having both ends in a set X_i where $1 \leq i \leq 4$ will be colored arbitrarily with the colors a, b, or c. If $n \not\equiv 2 \pmod 4$ it is clear that the maximum number of vertices of a monochromatic connected spanning subgraph of K_n is equal to $[(n+1)/2]$, and hence in this case it follows that $f_3(n) = [(n+1)/2]$. If $n \equiv 2 \pmod 4$, then for the above-defined coloring of the edge set of K_n the maximum number of vertices of a monochromatic connected spanning subgraph is equal to $n/2 + 1$, and therefore in this case $f_3(n) \leq n/2 + 1$. In order to prove the opposite inequality consider an arbitrary coloring with the colors a, b, and c of the edges of K_n for $n = 4p + 2$ ($p \geq 1$). It has been shown that there exists a monochromatic connected spanning subgraph of K_n with at least $[(n+1)/2] = 2p + 1$ vertices. Let H be such a subgraph, and suppose that the edges of H have the color c. If A denotes the connected component composed of edges with the color c, which contains the vertices of H, and if $|A| \geq 2p + 2$, it follows that $f_3(n) \geq n/2 + 1$. Hence in this case the equality $f_3(n) = n/2 + 1$ holds for $n \equiv 2 \pmod 4$. Otherwise, one has $|A| = 2p + 1$, and if B denotes the set of the remaining vertices, it follows that $|B| = 2p + 1$. No edge having one extremity in A and the other in B is colored with the color c, and hence the bipartite complete graph whose partite sets are A and B has edges with the colors a or b only. By Problem 14.26 there exists a monochromatic connected spanning subgraph of K_n with $2p+2=n/2+1$ vertices. This observation completes the proof. [L. Gerencsér, A. Gyárfás, *Annales Univ. Sci. Budapest. R. Eötvös, Sect. Math.*, **X** (1967), 167–170; B. Andrásfai, *Ibid.*, **XIII** (1970), 103–107 (1971).]

14.28 We prove that $R(K_{1,m}, K_{1,n}) \leq m+n$ by considering a coloring of the edges of K_{m+n} using red and green. At each vertex there are $m+n-1$ incident edges, so that if fewer than m are red, then at least n are green. If m is odd, then $m-1$ is even and hence there exists a regular graph of degree $m-1$ on $m+n-1$ vertices (see Problem 8.8). Moreover, the complementary graph is regular of degree $n-1$. One can therefore 2-color K_{m+n-1} so that there is no red $K_{1,m}$ and no green $K_{1,n}$. Thus in this case $R(K_{1,m}, K_{1,n}) = m+n$. If m and n are both even, then there is no regular graph of degree $m-1$ on $m+n-1$ vertices, in view of the fact that there would be an odd number of vertices of odd degree. It follows that $R(K_{1,m}, K_{1,n}) \leq m+n-1$. But there does exist a regular graph of degree $m-1$ on $m+n-2$ vertices, so that $R(K_{1,m}, K_{1,n}) = m+n-1$. [F. Harary, *Graph Theory and Applications*, Proceedings of the Conference at Western Michigan University, May 10–13, 1972, Lecture Notes in Math., Springer-Verlag, 1972, 125–138.]

14.29 Let $k=(n-1)/(m-1)$. Form a 2-coloring of K_{m+n-2} by taking $k+1$ copies of K_{m-1} all having only red edges, and interconnecting them by blue edges. No red T_m has been formed, since T_m has m vertices. Also, no blue $K_{1,n}$ has been formed, since the largest blue degree in K_{m+n-2} is $k(m-1)=n-1$. This shows that $R(T_m, K_{1,n}) \geq m+n-1$. Next we show by induction on m that

$$R(T_m, K_{1,n}) \leq m+n-1 \quad \text{for} \quad m \geq 2. \tag{1}$$

For $m=2$, (1) is immediate, since if K_{n+1} has no red T_2, it follows that it has only blue edges, and hence any vertex of K_{n+1} is a center of a blue star $K_{1,n}$. Assume (1) to be true for all $m' \leq m-1$, and form the tree T_{m-1} by removing an endpoint x of T_m and the edge $[x, y]$ incident to x in T_m. In a 2-colored K_{m+n-1} one can assume by induction that there is either a blue $K_{1,n}$ or a red T_{m-1}. Suppose that the latter choice obtains. Since there are $m+n-1-(m-1)=n$ vertices v_i of K_{m+n-1} which are not vertices of the red T_{m-1}, and K_{m+n-1} contains no blue $K_{1,n}$, it follows that some edge from y to some v_i must be red. But this forms a red T_m in K_{m+n-1}. Thus, by induction (1) holds and the proof is complete.

The corresponding results for the case in which $m-1$ does not divide $n-1$ are much more complicated, and a complete solution has not been obtained. However, in this case (1) still holds and, in fact, for almost all trees T_m, $R(T_m, K_{1,n}) = m+n-2$ for n sufficiently large. [S. A. Burr, *Graphs and Combinatorics*, Lecture Notes in Mathematics, 406, Springer-Verlag, Berlin, 1974, 52–75.]

14.30 Form a 2-coloring of $K_{(m-1)n}$ by taking $m-1$ copies of K_n all having only green edges, and interconnect them by red edges. This coloring contains no

Solutions

red K_m and no green $K_{1,n}$, and hence $R(K_m, K_{1,n}) \geq (m-1)n+1$. The inequality $R(K_m, K_{1,n}) \leq (m-1)n+1$ can be obtained for $m \geq 2$ and $n \geq 1$ as a corollary of Turán's theorem (Problem 9.9). Indeed, in a graph G with $p=(m-1)n+1$ vertices and minimum degree $(m-2)n+1$ (the complement has maximum degree $n-1$), the number q of edges is bounded below by $q \geq (mn-n+1)(mn-2n+1)/2$. The number of edges in Turán's graph with $(m-1)n+1$ vertices and $m-1 \geq 1$ parts is equal to

$$M(mn-n+1, m) = \frac{m-2}{m-1} \cdot \frac{(mn-n+1)^2 - 1}{2}$$

$$= \frac{n(m-2)(mn-n+2)}{2}.$$

The inequality

$$\frac{(mn-n+1)(mn-2n+1)}{2} > \frac{n(m-2)(mn-n+2)}{2}$$

is equivalent to $n+1 > 0$, which holds for $n \geq 1$. Hence $q > M(mn-n+1, m)$ and by Turán's theorem G must contain K_m. For $m=1$ the result is obvious.

This formula was generalized to $R(K_m, T_{n+1})$, where T_{n+1} is any tree with $n+1$ vertices, by V. Chvátal [*J. Graph Theory* **1**(1) (1977), 93].

14.31 It will be shown that $r(m) = m^2 - m - 1$. Let $\{1, \ldots, m^2 - m - 2\} = A \cup B$ be a partition into two classes, where

$$A = \{1, 2, \ldots, m-2, (m-1)^2, (m-1)^2 + 1, \ldots, m^2 - m - 2\}$$

and

$$B = \{m-1, m, \ldots, (m-1)^2 - 1\}.$$

Each of A and B is m-sum-free, that is, the equation

$$x_1 + \cdots + x_{m-1} = x_m$$

has a solution which is neither in A nor in B.

If $x_1, \ldots, x_{m-1}$ belong to $\{1, \ldots, m-2\}$, then

$$m-1 \leq x_1 + \cdots + x_{m-1} \leq (m-1)(m-2) < m(m-2) = (m-1)^2 - 1,$$

and hence $x_1 + \cdots + x_{m-1} \in B$. If the sum $x_1 + \cdots + x_{m-1}$ also contains terms from the set $\{(m-1)^2, \ldots, m^2 - m - 2\}$, then the smallest of its values is $1 + \cdots + 1 + (m-1)^2 = m^2 - m - 1 \notin A$.

Similarly if $x_1, \ldots, x_{m-1}$ are terms from B, one can conclude that $x_1 + \cdots + x_{m-1} \geq (m-1)^2$, or $x_1 + \cdots + x_{m-1} \notin B$. It follows that $r(m) \geq m^2 - m - 1$.

It will now be shown that the opposite inequality also holds. In other words, every partition of the set $\{1, \ldots, m^2 - m - 1\}$ contains a class which is not m-sum-free.

Suppose that this property does not hold. Let $1 \in A$. It follows that $m-1 \in B$ and $(m-1)^2 \in A$. Now consider two cases: (a) $m \in A$ and (b) $m \in B$.

(a) If $m \in A$ one can see that $m^2 - 2m + 1 = m + \cdots + m + 1$, that is, A is not m-sum-free.

(b) If $m \in B$, then since $(m-1) + m + \cdots + m = m^2 - m - 1$ it follows that $m^2 - m - 1 \in A$. In this case one can write

$$1 + \cdots + 1 + m^2 - 2m + 1 = m^2 - m - 1;$$

thus A is not m-sum-free, and this contradicts the hypothesis.

[A. Beutelspacher, W. Brestovansky, *Combinatorial Theory*, Proceedings, Lecture Notes Math., Springer-Verlag, Berlin, **969** (1982), 30–38.]

Bibliography

M. Aigner, *Combinatorial Theory*, Springer-Verlag, Berlin–Heidelberg, 1979.

C. Berge, *Principes de Combinatoire*, Dunod, Paris, 1968.

C. Berge, *Graphes et Hypergraphes*, Dunod, Paris, 1970.

M. Behzad, G. Chartrand, L. Lesniak-Foster, *Graphs and Digraphs*, Prindle, Weber and Schmidt, Boston, 1979.

B. Bollobas, *Graph Theory. An Introductory Course*, Springer-Verlag, New York–Heidelberg, 1979.

L. Comtet, *Analyse Combinatoire, I, II*, Presses Universitaires de France, Paris, 1970.

L. R. Ford, Jr., D. R. Fulkerson, *Flows in Networks*, Princeton University Press, 1962.

A. M. Gleason, R. E. Greenwood, L. M. Kelly, *The W.L. Putnam Mathematical Competition. Problems and Solutions: 1938–1964*, The Mathematical Association of America, 1980.

J. E. Graver, M. E. Watkins, *Combinatorics with Emphasis on the Theory of Graphs*, Springer-Verlag, New York–Heidelberg–Berlin, 1977.

M. Hall, Jr., *Combinatorial Theory*, Blaisdell, Waltham, Massachusetts–Toronto–London, 1967.

F. Harary, *Graph Theory*, Addison-Wesley, Reading, Massachusetts, 1969.

D. E. Knuth, *The Art of Computer Programming, Vol. 1: Fundamental Algorithms*, 2nd ed., Addison-Wesley, Reading, Massachusetts, 1973.

L. Lovász, *Combinatorial Problems and Exercises*, Akademiai Kiado, Budapest, 1979.

J. W. Moon, *Counting Labelled Trees*, Canadian Mathematical Monographs, 1, Canadian Mathematical Congress, 1970.

J. W. Moon, *Topics on Tournaments*, Holt, Rinehart and Winston, New York, 1968.

H. Ryser, *Combinatorial Mathematics*, Wiley, New York, 1963.

I. Tomescu, *Introduction to Combinatorics*, Collet's, London–Wellingborough, 1975.

N. Vilenkin, *Combinatorial Mathematics for Recreation*, Mir, Moscow, 1972.

R. J. Wilson, *Introduction to Graph Theory*, Longman, London, 1975.